Springer Theses

Recognizing Outstanding Ph.D. Research

Aims and Scope

The series "Springer Theses" brings together a selection of the very best Ph.D. theses from around the world and across the physical sciences. Nominated and endorsed by two recognized specialists, each published volume has been selected for its scientific excellence and the high impact of its contents for the pertinent field of research. For greater accessibility to non-specialists, the published versions include an extended introduction, as well as a foreword by the student's supervisor explaining the special relevance of the work for the field. As a whole, the series will provide a valuable resource both for newcomers to the research fields described, and for other scientists seeking detailed background information on special questions. Finally, it provides an accredited documentation of the valuable contributions made by today's younger generation of scientists.

Theses are accepted into the series by invited nomination only and must fulfill all of the following criteria

- They must be written in good English.
- The topic should fall within the confines of Chemistry, Physics, Earth Sciences, Engineering and related interdisciplinary fields such as Materials, Nanoscience, Chemical Engineering, Complex Systems and Biophysics.
- The work reported in the thesis must represent a significant scientific advance.
- If the thesis includes previously published material, permission to reproduce this must be gained from the respective copyright holder.
- They must have been examined and passed during the 12 months prior to nomination.
- Each thesis should include a foreword by the supervisor outlining the significance of its content.
- The theses should have a clearly defined structure including an introduction accessible to scientists not expert in that particular field.

More information about this series at http://www.springer.com/series/8790

Sumanta Chakraborty

Classical and Quantum Aspects of Gravity in Relation to the Emergent Paradigm

Doctoral Thesis accepted by
Jawaharlal Nehru University, Pune, India

Springer

Author
Dr. Sumanta Chakraborty
Department of Theoretical Physics
Indian Association for the Cultivation of Science
Jadavpur, Kolkata, West Bengal
India

Supervisor
Prof. Thanu Padmanabhan
Inter-University Centre for Astronomy and Astrophysics (IUCAA)
Pune
India

ISSN 2190-5053 ISSN 2190-5061 (electronic)
Springer Theses
ISBN 978-3-319-63732-7 ISBN 978-3-319-63733-4 (eBook)
DOI 10.1007/978-3-319-63733-4

Library of Congress Control Number: 2017946667

Printed on acid-free paper

This Springer imprint is published by Springer Nature
The registered company is Springer International Publishing AG
The registered company address is: Gewerbestrasse 11, 6330 Cham, Switzerland

To my grandparents—Prakash, Jharna, Samir and Binapani—*for their love, inspiration and knowledge.*
To my parents—Subenoy and Archana—*for everything.*
To my cousins—Sukriti and Poulomi—*for their love and support.*

Supervisor's Foreword

Combining the principles of General Relativity and Quantum Theory—the two key developments in theoretical physics in twentieth century—is a major challenge which has attracted the attention of the community for more than half a century. These decades of exploration have essentially led to a series of resounding failures, very often after a period of initial hope and hype with regard to each of the approaches. (In fact, at any given time during last 50 years or so, there was a set of people who were 'true believers' in some particular approach they were pursuing, with the hope that the solution was just around the corner!). The current situation actually has a historic parallel. At least twice in the history of physics, very talented people have got themselves stuck in a "local minimum" of a completely wrong idea. The first was in the case of the calorific theory for heat; the second was in the theory of aether with matter arising as vortices and excitations of the aether. I believe the currently popular paradigms of quantum gravity belong to yet another case of very smart physicists pursuing very wrong ideas.

So what could be a possible alternative route which can lift us above the "local minimum"? A major paradigm shift is required, which I can best illustrate with an analogy. We know that the elasticity (or fluid mechanics) can be described as a self-consistent physical theory in terms of a set of dynamical equations. If one takes these equations as fundamental and applies the principles of quantum theory to them, one will certainly discover interesting new phenomena, like, for example, phonons and their interactions. However, such an approach will not lead us to the quantum structure of matter. Similarly, there is sufficient evidence to suggest that the dynamical equations describing gravity (in a large class of models including—but not limited to—Einstein's theory) have the same conceptual status as the equations of elasticity or fluid mechanics. Reinterpreting gravitational field equations using the principles of quantum theory is analogous to attempts in which one is applying the quantum principles to the equations of, say, elasticity. While some interesting new results will emerge, we will not discover the quantum nature of spacetime by such an approach.

What is required is to re-examine closely the structure of gravitational dynamics and try to learn what it tells us about the microscopic structure of spacetime. Such a top-down approach (in length scales) is precisely what Boltzmann used, to figure out that matter is made of discrete degrees of freedom, *because* it exhibits thermal phenomenon. The fact that null surfaces in a spacetime appear to be hot to certain observers holds the key to the understanding of the microstructure of spacetime. Only a formalism which (i) recognizes this fact up front and (ii) can adopt itself to an *observer dependent* thermodynamics of spacetime will ever have a chance of succeeding in combining the principles of gravity and quantum theory. Most of the popular models of quantum gravity, which came and went (or on their way-out!) in the last five decades, did not do justice to this fact.

I and my collaborators have been exploring, since 2002, several aspects of this alternative paradigm in which the thermodynamics of null surfaces plays a crucial role. Sumanta Chakraborty's thesis dealt with several aspects of this paradigm in a nice and comprehensive manner. The first part of the thesis provides the necessary broad introduction and the remaining chapters of the thesis deal directly with several questions which are vital to the emergent paradigm of gravity.

In particular, Sumanta has made a significant amount of progress in extending the emergent paradigm to a class of models called the Lanczos-Lovelock models of gravity. These are theories of gravity in which the field equations remain second order in time but the entropy of the horizon is *not* proportional to its area. In the literature, one often finds very imaginative pieces of research which—unfortunately—are strongly linked to the idea that the entropy of a horizon is proportional to its area. The Lanczos-Lovelock models of gravity take you out of the comfort-zone in which entropy is such a nice geometrical quantity, viz. the horizon area. So they act as a viable test-bed for the ideas of emergent gravity. If any of the ideas or the results which one come up with, holds only in Einstein's General Relativity but not in the Lanczos-Lovelock models of gravity, I would treat it with caution. It is therefore remarkable—and gratifying—that virtually every key idea within the emergent paradigm developed by our group could be extended to Lanczos-Lovelock models of gravity and Sumanta's thesis contains several examples of this extension. In a broader context, this fact tells you that the thermodynamic connection transcends Einstein's theory and should be viewed as a more fundamental feature of spacetime, not too closely linked to the specific field equations describing the gravitational dynamics.

I enjoyed working with Sumanta through his Ph.D. phase and I wish him all the best in his future career. I also congratulate Springer for having this programme, of bringing out good theses as books, which both provides an encouragement to the students as well as creating topical reference works for the community.

Prof. Thanu Padmanabhan
Thesis Advisor
Inter-University Centre for Astronomy
and Astrophysics (IUCAA), Pune, India

Preface

General Relativity (GR) is a very successful theory and is the best formalism we have to describe the geometrical properties of the spacetime. It has passed all the experimental and observational scrutinies so far, ranging from local tests like perihelion precession and bending of light to precision tests using pulsars.

In spite of these outstanding successes there still remain some unresolved issues, suggesting that general relativity is not complete. The most important reason is the presence of singularities in many physical situations leading to a loss of predictability. Another reason has to do with the fact that the horizons in general relativity possess thermodynamic properties like temperature and entropy. Within the framework of general relativity, there is no natural explanation for this "thermodynamic" interpretation and it provides motivation to take a fresh look at the theory. A third reason arises from the fact that all the other known interactions (electromagnetic, weak and strong) are described by quantum theories, while gravity alone is still described by a classical theory. This laid the foundation of the belief that "quantum theory of gravity" awaits discovery. The attempts to obtain a perturbative quantum general relativity, taking a cue from the quantization of the other forces, has not succeeded. This has the unavoidable conclusion: we need to modify our understanding of quantum field theory or the understanding of general relativity or both.

In this thesis, we try to understand the thermodynamic nature of general relativity better by taking a closer look at the structure of general relativity and its higher curvature cousins, collectively called Lanczos-Lovelock gravity. If one can derive a result in the context of Lanczos-Lovelock gravity, the result for general relativity is encompassed by it as well. We shall analyze the geometrical structure of Lanczos-Lovelock gravity (which has general relativity as a special case) leading to the inescapable connection between gravity and thermodynamics. We will also have occasion to talk about Virasoro algebra associated with an arbitrary null surface and associated entropy in this context.

As a complementary approach towards a quantum theory of gravity, we study some aspects of quantum field theory in curved spacetime. The specific issues addressed in this context include: (a) What can we say about classical singularities from the viewpoint of quantum theory? This specifically requires one to probe quantum fields inside the black hole horizon. (b) Is the retrieval of information from an evaporating black hole possible? (We will show that distortions to the thermal spectra of a particular kind, referred to as non-vacuum distortions can be used to fully reconstruct a subspace of initial data.) (c) Can Rindler effect be present for geodesic observers? We illustrate for a specific (1+1) black hole spacetime, there are geodesic observers who are confined to a flat region of the spacetime and hence will experience Rindler effect.

Finally, in order to capture some quantum gravity effects, we have introduced a zero point length to the spacetime and have discussed its geometrical consequences. In particular, we have shown that at the Planck scale the spacetime becomes essentially two-dimensional.

Kolkata, India

Sumanta Chakraborty

Publications

The thesis is based on the following papers:

1. **S. Chakraborty** and T. Padmanabhan, Geometrical Variables with Direct Thermodynamic Significance in Lanczos-Lovelock Gravity. *Phys. Rev. D***90** (October, 2014) 084021, arXiv:1408.4791.
2. **S. Chakraborty** and T. Padmanabhan, Evolution of Spacetime Arises due to the Departure from Holographic Equipartition in all Lanczos-Lovelock Theories of Gravity. *Phys. Rev. D***90** (December, 2014) 124017, arXiv:1408.4679.
3. **S. Chakraborty**, S. Singh and T. Padmanabhan, A Quantum Peek Inside the Black Hole Event Horizon. *JHEP***06** (June, 2015) 192, arXiv:1503.01774.
4. **S. Chakraborty**, Lanczos-Lovelock Gravity from a Thermodynamic Perspective. *JHEP***08** (August, 2015) 029, arXiv:1505.07272.
5. **S. Chakraborty** and T. Padmanabhan, Thermodynamical Interpretation of Geometrical Variables Associated with Null Surfaces. *Phys. Rev. D***92** (November, 2015) 104011, arXiv:1508.04060.
6. K. Lochan, **S. Chakraborty**, and T. Padmanabhan, Information retrieval from black holes. *Phys. Rev. D***94** (August, 2016) 044056, arXiv:1604.04987 [gr-qc].
7. K. Lochan, **S. Chakraborty**, and T. Padmanabhan, Dynamic realization of the Unruh effect for a geodesic observer. Submitted for publication, arXiv:1603.01964 [gr-qc].
8. T. Padmanabhan, **S. Chakraborty**, and D. Kothawala, Spacetime with zero point length is two-dimensional at the Planck scale. *Gen. Rel. Grav.***48** no. 5, (April, 2016) 55, arXiv:1507.05669 [gr-qc].
9. **S. Chakraborty**, S. Bhattacharya and T. Padmanabhan, Entropy of a generic null surface from its associated Virasoro algebra. *Phys. Lett. B***763**, (2016) 347, arXiv:1605.06988 [gr-qc].

Acknowledgements

My grandfather always said, "finding a proper friend, philosopher and guide is the most difficult thing in one's life". I am fortunate enough to find at least two of them in my life. My thesis supervisor, Prof. Padmanabhan is definitely one of them. I believe it is not only difficult but impossible to find a better supervisor than him, despite being extremely busy, he heard my problems with patience and with his amazing insight gave advises which in most of the cases resolved them. He gave me complete freedom on my work and I am grateful for that. Bengali's are well-known for being food critique, of which I am no exception, still the dishes prepared by his wife Vasanthi were awesome and his daughter Hamsa always acted as a constant source of inspiration to me. Thanks to them as well for making my stay in IUCAA enjoyable.

The second person, who completely changed my perception towards physics and acted as an auxiliary guide to me is Prof. Soumitra SenGupta. I fell in love with physics mainly due to him. His excellent explanatory skills and dedication for physics helped me in many ways in the early days of my career.

Prof. Naresh Dadhich, Prof. Sanjeev Dhurandhar and Prof. Sukanta Bose have also helped me in many ways, suggesting exciting problems, providing wonderful explanations are a few among them. I have learned a lot from the numerous discussions we had.

I thank my collaborators Bibhas da, Dawood, Kinjalk, Krishna, Sourav da and Suprit (in alphabetical order) for spending a large amount of their time helping me to understand various aspects of physics. Suprit became an elder brother to me and always will remain so. I have learned quantum field theory in curved spacetime from him, his creative thinkings are always one step ahead. In the course of time Kinjalk became one of my best "Quantum" friends and I know only a handful of people who have as clear idea about physics as Kinjalk's. Krishna's dedication and perfection towards physics is unparalleled, I hope to absorb some of his qualities into me. Bibhas da, Dawood and Sourav da have been my mentor throughout, I have learnt a lot from them. I am really grateful that I met these people in my life.

I thank my batchmates: Javed, Kabir, Labani, Nikhil and Shabbir (in alphabetical order) for their help, support and various discussions from which I have learnt a lot. Best wishes for their research career ahead.

I fondly remember two of my best friends Soumojit Roy and Chandan Hati, for innumerable physics discussions we had among ourselves during our daily journey to Calcutta University.

I thank the other students and post docs, rest of the academic staff and non-academic staff for the pleasure and support they have provided me during the past three years. I specially acknowledge the facilities at IUCAA library, the best place in IUCAA itself. The calm and quiet environment provided by the library helped me a lot during long calculations.

As usual I acknowledge my family members at the end. My parents Subenoy and Archana were ready to sacrifice anything for me, the thesis becomes complete with love and support from them. My grandparents through their life made me understand the value of life. Sadly, three of them would not be able to see this thesis but I know they will always be my side. My uncles Jayanta, Subho, Dumpi and aunt Sonali, thanks to you for being supportive and to my cousins Sukriti, Poulomi, Sukanya, Indira, Anupam, Jayeeta and Brishti, well, for being what they are. Finally to Bandana and Eva, my aunts, their concerns were no less than my mother.

My research leading up to this thesis is supported by the Shyama Prasad Mukherjee fellowship of the Council of Scientific and Industrial Research, Government of India.

Contents

Part IV Classical Gravity, Quantum Matter

Notations and Conventions

The notations and conventions used in this thesis are as follows:

- We use the metric signature $(-, +, +, +)$.
- The fundamental constants $16\pi G$, $\hbar$ and c have been set to unity (Sometimes, when we switch to $G = 1$ units, it will be mentioned specifically).
- The Latin indices, $a, b, \ldots$, run over all spacetime indices, and are hence summed over four values (or D values depending on dimension of the spacetime). Greek indices, $\alpha, \beta, \ldots$, are used when we specialize to indices corresponding to a codimension-1 surface, i.e a $3-$ surface (or $D-1$ surface), and are summed over three values (or $D-1$ values). Upper case Latin symbols, $A, B, \ldots$, are used for indices corresponding to two-dimensional hypersurfaces (or, $D-2$ surface), leading to sums going over two values (or, $D-2$ values).

Part I
Introduction and Motivation

Chapter 1
It Is All About Gravity

Even after hundred years, general relativity is still referred to as one of the very successful and beautiful theory human race has ever witnessed [1, 2]. The hundred years of appreciation for general relativity is not only due to its theoretical beauty, but also because it has passed all the experimental tests so far. The aesthetic appeal of general relativity is due to its bold predictions—*existence of black holes*, *prediction of gravitational waves*, *bending of light*, *modification of the precession angle of orbits*, *expansion of the universe*, which are the most appealing among many others. These predictions scan a vast spectrum of length scales, from a few hundreds of megapersecs (cosmological scale) to a few minutes of light travel time (solar system scale). The excitement about general relativity has reached new heights after the first direct detection of gravitational waves.

Despite these outstanding successes of general relativity there are still unresolved issues. The correctness of any theory crucially depends on its predictive power and a singularity free description. There exists at least two situations where general relativity is known to break down and loses its predictive power. These two situations where the theory breaks down, correspond to black hole singularity and the big bang singularity, respectively.

The break down of general relativity at the singularity suggests that at such small length scales and hence high energy one needs to replace the theory with a better one. Further, general relativity is a classical theory, which needs to be merged with quantum mechanics in order to yield a consistent theory at such small scales and high energy. Hence the general hope is that if one succeeds in bringing general relativity and quantum mechanics together, the singularity problem would be resolved, which has not come true yet. Quantum theory, on the other hand in the past hundred years has grown and flourished, bringing all the other forces, namely, strong, weak and electromagnetic, under one roof. But it has not succeeded in bringing the other "pillar of modern physics", general relativity on its canvas. As the standard treatment fails, there have been numerous other approaches to "quantize gravity", but still we are missing the quantum origin of gravity.

S. Chakraborty, *Classical and Quantum Aspects of Gravity in Relation to the Emergent Paradigm*, Springer Theses, DOI 10.1007/978-3-319-63733-4_1

The reason for a "quantum theory of gravity" to remain elusive till date could be twofold. First, classical general relativity might have its own issues. Or, alternatively, it is also possible that quantum theory itself needs modification before it can be merged with gravity. In this thesis we will be mainly concerned about the first approach and will have a few occasions to comment on the second in passing.

In this introductory chapter, we shall discuss various existing schemes of "quantizing" gravity, which will motivate us to look for some new avenues and interpreting gravity from a new perspective. We have divided this chapter into three major sections and each into several subsections. In the first section we will present some basic aspects of both classical general relativity and quantum field theory. In Sect. 1.1.1 we will present a broad overview of general relativity along with its possible shortcomings. Then we turn to quantum field theory by providing a brief introduction and its problems in Sect. 1.1.2. Armed with these limitations we will highlight some features that sets gravity apart from the other three interactions in Sect. 1.1.3 and possible difficulties one might face while gluing quantum theory and gravity together, in Sect. 1.1.4.

In the second part of the introduction we will present an intermediate situation, in which one considers quantum matter in a classical gravitational background. Since one does not have a full quantum theory of gravity, it is legitimate to ask, whether we can say anything about the quantum theory even before obtaining it? That is, can we detect features in the interaction of classical gravity with quantum matter, which, to the leading order, can provide some hints about the complete theory of quantum gravity? We will discuss this situation and shall provide a broad overview in Sect. 1.2.1. Semi-classical gravity also encounters a strange tussle—possible violation of unitarity and hence information loss. We will comment on these issues, specifically on information recovery, in Sect. 1.2.2.

In the third part of this chapter we will start with the question "why gravity should be quantized?" in Sect. 1.3.1. Then we will go on and discuss several candidates for a quantum theory of gravity in Sect. 1.3.2. After getting the broad picture we will discusses the view that gravity may be an emergent phenomenon and not a fundamental interaction. This view has been termed as *the emergent gravity paradigm*. Pressing further we delineate one possible effect of quantum gravity—introduction of a zero point length to the spacetime and its possible consequences in Sect. 1.3.4. This will conclude the discussion on quantum theories of gravity.

1.1 Approach to Classical Gravity and Quantum Field Theory

1.1.1 Classical Gravity and Its Limitations

General relativity in its standard form has the metric, g_{ab}, as the dynamical variable. The most natural scalar that one can build out of the metric corresponds to the Ricci scalar R and thus one goes forward to build the Einstein-Hilbert action (including $16\pi G$ for the moment):

$$A_{\rm EH} = \frac{1}{16\pi G}\int d^4x\,\sqrt{-g}R \tag{1.1}$$

The above action has been written with $c = 1$, as per our conventions. The field equations for gravity, which in this case corresponds to equations for the metric g_{ab}, is obtained by varying the total action, the Einstein-Hilbert action plus the matter action, with respect to the metric. The equations so obtained are

$$G_{ab} = 8\pi G T_{ab} \tag{1.2}$$

which are the celebrated Einstein's equations. The Einstein's equations (in particular, the solutions derived from it) have withstood all experimental and observational scrutinies over the past century [3]. Thus, the theory of general relativity is considered to be well-established.

In spite of its success with both observations and experiments, general relativity, as already emphasized, cannot be regarded as a complete theory of nature, since it predicts existence of singularities (for a review, see [4]). Broadly speaking, a singularity is a spacetime event where the curvature invariants diverge [5]. (Note that pathological behavior of metric could originate due to the use of pathological coordinates and does not guarantee existence of singularity. Hence one should always consider invariants to decide whether a spacetime event is singular or not.) Singularity also corresponds to an event at which the predictability of the theory breaks down. That is, Einstein's equations can no longer be used in order to predict what is going to happen in the vicinity of that event. There are also spacetimes in which the curvature invariants do not diverge, but still can have observers whose future is uncertain, even after techniques like changing the coordinates or extending the spacetime have been carried out. For a spacetime with singularities or without predictive future evolution, one can ask reasonable questions for which the theory can not provide adequate answers. For example, if one ask [6]: "As a massive star at the end of its life collapses and forms a black hole, how will the physical phenomena appear with respect to a hypothetical observer on the surface of the star at arbitrarily late times as measured by the observer's clock?" There is no satisfactory answer to this question in the premises of general relativity, since the observer reaches the singularity in a finite proper time and after that there is no way classical general relativity can predict what happens to the observer. Thus, general relativity reaches its limit at the singularities. One expects that this breakdown of predictability is not a feature of nature but occurs just because we have an incomplete theory (see, for example Chap. 44 in [7]). Therefore, the presence of singularities provides a strong hint that classical general relativity is not a complete theory but should give way to a more complete theory.

1.1.2 Quantum Field Theory and Its Issues

The standard formulation of classical general relativity has been discussed in the above section, signaling its breakdown at high enough energies, requiring modifications. On the other hand, the last decades have also witnessed remarkable progress in the search of a unified theory for the forces of nature. The electrodynamics and weak interaction were merged together into "electro-weak" theory, which lends its way to a "marriage" with strong force, leading to grand unified theories. All these progress result from quantum field theory, which broadly speaking describes particles as excitations of an underlying field.

The grand unified theories are build up on the premise of quantum field theory, which is by no means complete. The admirers of general relativity mainly emphasize its aesthetic and conceptual beauty, while the appreciation for quantum field theory originates mainly from the accuracy with which it has matched experimental data. Even though many physicists, appreciate the accuracy and concepts involved with quantum field theory, but there are also examples of physicists being uncomfortable with quantum field theory as well. Einstein, for example was dissatisfied with the conceptual structure of quantum mechanics, while Dirac was a critic of the renormalization programme in quantum field theory. It is often mentioned that the theoretical beauty offered by general relativity is missing in quantum field theory which is mainly due to some procedures used routinely in it, e.g., the renormalization procedure. Following these limitations/conceptual problems, one might argue that not only general relativity, but quantum field theory as well needs modifications (see e.g., [6], Chap. 4 in [8]).

1.1.3 Gravity Is Peculiar

Gravity always has the feature of standing alone, compared to its three cousins, weak, strong and electromagnetic forces. This has to do with certain peculiar features of gravity, that are not shared by the other three interactions. Some of these peculiarities are listed below (see also [6]):

- *Gravity is universal*—all kinds of energies produce a gravitational field. A simple corollary being that all matter particles produce gravitational fields, since all known matter particles possess energies. Further, every material object must be affected by the gravitational fields, which follows directly from the equivalence principle, a cornerstone for any theories of gravity.
- *Gravity affects the spacetime structure*—In the context of other interactions, the spacetime is just an arena where these forces play out their roles. But inclusion of gravity makes this arena to join the troop of actors, i.e., makes it dynamical. Concepts of distance between objects or time between events are affected by gravity which can warp spacetime structure to generate regions of spacetime, whose boundaries can act as a one-way membrane for some class of observers,

such that no information from that region can reach a particular observer. For example, no information from the region inside a black hole horizon can reach an observer staying outside the black hole. Surprisingly, this can happen even in flat spacetime. If an observer accelerates with a constant acceleration and moves with the same acceleration indefinitely, there will be a region of the spacetime causally inaccessible to him/her. (The observer is known as Rindler observer and the surface which acts as a one-way membrane is known as Rindler horizon.) However note that this result is observer-dependent. In the case of black hole spacetime, we do have radially in-falling observers who has access to the full spacetime. Also in the case of accelerated observers in Minkowski spacetime, the Minkowski geodesic observers will have access to the full spacetime. Thus existence of one-way membrane is very much observer-dependent, a crucial fact we will use later.

- *At large scales gravity dominates*—Both the weak and strong forces are of very short range and their effects are only felt at the microscopic scales having no effect on macroscopic physics. On the other hand, electromagnetic force has a long range. But due to charge neutrality, large bodies essentially contain almost equal amounts of positive and negative charges. Only gravity has a single kind of charge; hence it always attracts and cannot be shielded. This is why, even though it is weak compared to all other forces, at large scales it is the dominating one and hence dynamics at large scales in the universe is determined by gravity.

1.1.4 Gravity and Quantum Theory: Chaos Out of Order

It turns out that there are features present in both the classical general relativity and quantum field theory that could help each other, when one tries to combine the principles of gravity with quantum theory. As we have already emphasized, even though singularities are of concern in general relativity, there is a general consensus that when quantum theory and gravity are combined, the singularities would be removed, perhaps through introduction of a fundamental minimal length scale (may be the Planck length). Indications along similar directions has already been obtained in the case of loop quantum gravity [9–11]. Introduction of such a fundamental minimal length would help quantum field theory as well, by introducing a natural ultraviolet cutoff to the theory. This will regularize the divergent integrals appearing routinely in quantum field theory calculations.

Even with these encouraging hopes, one should realize that general relativity and quantum field theory are different at a fundamental level and as a consequence combining them will not be an easy task to perform [6]. As a first step towards combining quantum field theory with general relativity, one must upgrade Lorentz covariance symmetry in quantum field theory to general covariance. In absence of Lorentz covariance there will be no preferred slicing of spacetime and hence the concept of vacuum states, particles etc. will become ambiguous. This suggests that such quantum field theory concepts have to be modified or replaced with new generally covariant concepts. At the same time quantum field theory should also be modified to take into account the fact that, for black hole spacetimes in general relativity, there exists a class of observers who do not have access to the information beyond a horizon. Another important thing to address corresponds to the issue of the zero point energy of the fields. In general relativity, all energy gravitates, hence the subtraction of the zero point energy is not at all justified in the presence of gravity. We will discuss this more in Sect. 1.2.1. On the gravity side, one expects to lose the notion of a smooth spacetime, since quantum field theory brings with it the inherent vacuum fluctuations. For quantum gravity, these vacuum fluctuations would lead to a spacetime which itself should be fluctuating. Thus, the classical picture of a continuous spacetime geometry has to break down, to be replaced by a new picture of fluctuating quantum spacetime.

1.2 Quantum Matter in Classical Gravity

1.2.1 Quantum Fields in Curved Spacetime—What Can We Learn About Quantum Gravity?

The success of general relativity provides convincing evidence that gravitational phenomena can be understood by considering spacetime to be curved. The matter fields present in the spacetime tells spacetime how to curve and spacetime dictates those matter fields how to move. Since at present we are not even close to a complete quantum theory of gravity, it is important to study quantum fields in curved spacetime in search for new effects of gravitation, which can shed some light on various quantum gravity features. At this level, one treats gravity classically while the methods/techniques of flat spacetime quantum field theory are carried over to curved spacetime as much as possible. Interestingly, this simple extension turns out to yield rich consequences.

Among various results, the above extension of quantum field theory to curved spacetime leads to the physically important process of particle creation in cosmological and black hole spacetimes. The process of particle creation can happen in early phase of nearly exponential expansion of the universe and hence the vacuum fluctuations of the quantum field can generate primordial fluctuations acting as seeds to generate large scale structures in the universe. Further, in black hole spacetimes, creation of particles leads to gradual evaporation of the black holes. This is an important observation in the search for microscopic origin of black hole entropy, which has the potential of leading to new physics connecting both gravitation and quantum theory.

The next layer of thought has to do with the fact that the right hand side of Einstein's field equations contain energy-momentum tensor for the matter fields (see 1.2), which have to be replaced by operators. However we have not quantized gravity, so there is no way one can replace the left hand side of Einstein's equations by operators.

The standard way out of this problem is to consider expectation value of the energy momentum tensor at the right hand side of the gravitational field equations,

$$G_{ab} = 8\pi G\langle T_{ab}\rangle \tag{1.3}$$

where the expectation value is taken with respect to some state $|\Psi\rangle$ of the quantum field. But T_{ab} being a quadratic operator in the field, its expectation value even in Minkowski vacuum state is divergent. In Minkowski spacetime one ignores this infinite expression by arguing that only energy differences are observable, so that one can subtract them out (a sophisticated way of subtracting corresponds to normal ordering). However when gravity is present this approach is not satisfactory, since all energies gravitate. One cannot throw away the zero point energy—rescaling of energy is now not allowed. Further, in curved spacetime ignoring the zero point energy is not the only source of divergence, there exist additional divergent terms, which needs to be carefully subtracted. Thus Eq. (1.3) should be rewritten as,

$$G_{ab} = 8\pi\langle T_{ab}\rangle_{\text{ren}} \tag{1.4}$$

where $\langle T_{ab}\rangle_{\text{ren}}$ stands for the expectation value of T_{ab} in a state $|\Psi\rangle$ with all the infinities subtracted out. It would be interesting to construct scalar quantities out of $\langle T_{ab}\rangle_{\text{ren}}$ and study how they behave compared to their classical counterparts. If it turns out that in situations where general relativity predicts singularity the quantum effects dominate over the classical background, then it can be taken as a good hint that when quantum effects are properly incorporated the issue of singularity would be resolved. Hence without quantizing gravity itself, one can extract fair amount of information about the nature of quantum gravity. Later, in Chap. 9 , we would again have opportunities to discuss these effects.

1.2.2 Tussle of Titans: Black Hole Evaporation Versus Loss of Information

Evaporation of black holes, for many decades, has caused conceptual discomfort for the otherwise very successful quantum theory. The most basic and fundamental feature of the standard quantum theory, namely unitary evolution, is seemingly threatened if one tries extrapolating the results obtained at the semi-classical level [12]. Insights of Bekenstein [13, 14], suggested that the black holes must have an entropy proportional to the area of their event horizons, for the second law of thermodynamics to work. Hawking [12] proved that quantum effects could lead to the evaporation of a black hole which involves a radiation of positive energy, flux of particles with (nearly) thermal spectrum which an asymptotic observer can detect and a flux of negative energy flowing into the black hole decreasing its mass. So the mass lost by the black hole appears in the form of energy of the thermal radiation.

Although this effect completes the thermodynamic description of black holes, yet such a process, together with other properties of black holes, appears to violate the standard unitary quantum mechanics [15].

The particles in the outgoing flux, received by the asymptotic observer, remain entangled with the particles in the in-going flux. The resulting Hawking radiation is thermal precisely because we trace over the modes which entered the horizon. By such a process the black hole shrinks, losing the mass in the form of Hawking radiation. However, once the black hole completely evaporates by this process, there is an apparent paradox. At the end, there is nothing left for the outgoing particles to remain entangled with; yet they are in a mixed state since at no stage of the evaporation their entanglement with the interior modes was explicitly broken. This process, wherein a pure state evolves into a mixed state, is contrary to the standard unitary quantum evolution.

There is also a related issue of the information content of the matter which had undergone the collapse to form the black hole or even matter which falls into the black hole after it is formed. No-hair conjectures [7] suggest that no other information apart from the mass, charge and angular momentum of the matter that enters the event horizon can be available to the outside observers at any stage. Therefore, all other information about matter crossing the horizon would end up in the singularity and get destroyed. Thus, all the information about the initial state of matter which is falling into the horizon (other than those captured by mass, charge or angular momentum) is not coded in the Hawking radiation and is not available to the asymptotic future observers. This part of information, which ended up in singularity, is lost forever. Such a situation seems to require non-unitary evolution [16].

An initial resolution of the paradox stemmed from the idea that we might be making an error in trusting the semi-classical Hawking process all the way to the complete evaporation of the black hole. In principle semi-classical description should work fine when the black hole is large enough. But as the black hole becomes smaller and smaller, the curvature at horizon begins to rise and at very high curvatures the quantum nature of gravity must become important, and the semi-classical approximation must break down. Therefore, quantum gravity—rather than the semi-classical physics—should govern the final moments of the black hole evaporation. One then expects an overriding correction to the semi-classical description $\mathcal{O}(l_{\mathrm{P}}^2)$ which makes it non-thermal, and only becomes dominant when the black hole becomes of the Planck size. It must be noted that there exist many other sources of distortions to the thermal Hawking radiation [17], apart form the quantum gravity induced corrections. These non-thermal corrections can, in principle, store some information. However, it can be shown [15] that, since all such correction terms are sub-dominant in nature, none of these can help in making the theory unitary. Only corrections of $\mathcal{O}(1)$ can provide a possibility for unitary description, and we could identify no distortions of that kind. Thus, in its new avatar the paradox seems more robust as far as restoring unitarity to the quantum evolution is concerned.

In the literature, there are many different proposals to handle this issue. There are suggestions advocating radical modification to the unitary quantum theory itself, to accommodate non-unitary processes [18]. Such modifications to the unitary quantum theory have also been argued for, using some other conceptual considerations [19, 20]. However, these non-unitary quantum evolution models can also be applied to many other physical scenarios [21] where the predictions will be at variance from the standard unitary theory, constraining the models. There are also suggestions that the black hole evaporation must halt at the Planck level and leave behind a Planck size remnant at the end of the process. However, irrespective of the size, mass and other classical features of the black hole formed initially, the end product always has to be a Planck size remnant. This remnant should house all the information, which the out-going modes lack, in order to completely specify the state. Thus the complete description of a remnant and the Hawking radiation should be a pure state. This looks like a viable option. Still, it is not clear how a Planck size remnant could accommodate the vast landscape of varying initial configurations which could have formed the initial black hole. Other interesting suggestions include pinching of the spacetime [22] which could, in principle, restore faith in the essential tenets of both classical gravity and the quantum theory. However, the implication of such pinching effects for other types of horizons (e.g. Rindler, de Sitter) remains to be understood satisfactorily.

So we can summarize the crux of the information paradox as follows. When the black hole evaporates completely without leaving any remnant behind, one is justified in assuming that the entire information content of the collapsing body gets either destroyed or must be encoded in the resulting radiation. However, remnant radiation in this process is (dominantly) thermal, which is thermodynamically prohibited to contain much of the information and also incapable of making the theory unitary. Therefore, most of the information content of the matter which made the black hole in the first place is not available to the future asymptotic observers.

In this thesis, we argue that this version of the paradox—concerning the information content of the initial data—stems from a hybrid quantum/classical analysis of a process which is fully quantum mechanical in nature. That is, it arises from an artificial division between a quantum test field and the classical matter which collapses to form a black hole. When an event horizon is formed, the quantum field residing in its vacuum state at the beginning of collapse, gradually gets populated, erasing the black hole through a negative energy flux into the horizon with a corresponding positive energy flux appearing at infinity as thermal radiation [23, 24]. However, the matter which forms a black hole in the first place, is also fundamentally quantum mechanical in nature and should follow a quantum evolution. This, we believe, holds the key to the resolution of the paradox and has been elaborated in Chap. 10. We expect the classical description to be true, at lowest order, leading to formation of an event horizon. However, the fact that the collapsing material was inherently quantum mechanical in nature (e.g. a coherent state of the field which is collapsing) should not be completely ignored in studying this process. The matter which forms the black hole, if treated quantum mechanically, will populate its modes at future asymptotia non-thermally, in a manner which depends on its initial state. In this thesis, we demonstrate the presence of this effect at the semi-classical level. The result indicates that the no-hair theorems will be superseded at the full quantum gravity level.

1.3 The Final Frontier: A Quantum Theory of Gravity

Quantum gravity, or to be precise, the quantum theory of gravity, involves a series of problems that have remained unsolved for many years. Most of these problems boil down to the fact that, unlike any other interaction, gravity affects the global arena, viz., spacetime in which everything takes place. Quantization of other fields except gravity leaves this global arena unchanged. But as gravity is brought onto the scene, the arena itself becomes dynamical. Besides suffering from the quantum fluctuations of the other interactions, it further introduces fluctuations of its own. Even though general relativity changed our perception about spacetime, the events were still sharply defined. Bringing quantum fluctuations into picture changed this as well. In such a situation exact location of events lose their meaning and become fuzzy; exact locations are being substituted by probabilities of finding an object in a given region of space at some given instant of time.

1.3.1 The Need to Quantize Gravity

All these difficulties in the way to quantize gravity raised the question—is it necessary that gravity be quantized? Before answering this question, let us recall that the other three interactions, strong, weak and electromagnetic, are described by quantum theories. On the other hand, we have also discussed that general relativity is riddled with singularities, signaling the incompleteness of the theory and general hope is that quantization may remedy this inadequacy. But, there is no way to tell whether this is the only way to remedy the shortcomings of general relativity. Since it is a difficult problem, and people have tried to cure these issue without quantizing gravity as well. Thus one cannot ignore the possibility that gravity alone among all the interactions need not be quantized and that the singularities may be removed by some non-quantum modification/generalization of general relativity.

People have also come up with various arguments as to why one needs to quantize gravity. One such argument in favour of quantization of gravitational field involves coupling of gravitation to quantum systems and the fact that since the quantum system intrinsically is in a superposed state, that would require that the gravitational field also to be in a superposed state. Hence it follows that gravitation should be quantized [25]. There exist various other arguments as well in favour of quantization of gravity, based on various thought experiments. However [26] and [27] conclude that none of these arguments can prove the logical necessity for quantization of gravity and hence the question can only be settled by experiment.

1.3.2 Approaches to Quantize Gravity

Despite of the difficulties, there have been numerous attempts to achieve a quantum theory of gravity. In this section, we shall list some of the major attempts towards a theory of quantum gravity that are still surviving.

The most natural choice while constructing a quantum theory of gravity is to apply the perturbative quantum field theory methods that have been hugely successful with the other three interactions. Unfortunately, this program has failed—gravity is non-renormalizable. The above consequence originates through the following two results—(a) general relativity when coupled to matter fields is a non-renormalizable theory [28–31] and (b) even pure general relativity as a perturbative quantum field theory is non-renormalizable [32]. Hence people have more or less abandoned the program to develop standard general relativity as a perturbative quantum field theory.

Among the many candidate theories of quantum gravity, string theory is perhaps the best contender for the title of "quantum gravity" and hence is the most famous and popular one. By and large the origin of string theory goes back to quantum field theory, while later it was found to include gravity as well and hence providing a theory of quantum gravity. The basic assumption in string theory is that all the fundamental particles are actually various vibrating modes of one-dimensional strings. Using these strings as a building block string theory has claimed to be the sought-after "Theory Of Everything" unifying all the forces. With its power comes the weakness of the theory as well—requiring features like supersymmetry and additional dimensions which have not been observationally verified till now. In fact, it seems that supersymmetry is not realized in the nature (at least not at low energy scales as predicted by string theory) from recent Large Hadron Collider (LHC) experiments. (For a review, see [33].)

Loop quantum gravity is the other major candidate theory of quantum gravity. It is mainly interested in canonically quantizing gravity albeit in a non-perturbative manner. Last two decades have witnessed the formulation and major breakthroughs of this theory. Some of the major results under its territory are—the discovery that area and volume of a region of spacetime has to be quantized [34] and that quantum effects may lead to resolution of singularities in some cases (in particular the big bang singularity) [9–11].

Besides these two major approaches to quantum gravity, there are also many other candidates. Some such candidate theories are: Supergravity [35], Euclidean quantum gravity [36], Causal sets [37], Causal dynamic triangulations [38] etc., which we will not pause to comment on.

1.3.3 Gravity May Not Be a Fundamental Interaction

1.3.3.1 Black Hole Thermodynamics—The Inescapable Connection

In the early seventies, a curious connection was discovered between the dynamics of gravity and thermodynamics. This connection between gravity and thermodynamics first came to light with the work of Bekenstein and draw serious attention of the community subsequently [13, 39, 40]. In these works Bekenstein ascribed to a black hole an entropy which is proportional to the surface area of its horizon. This connection was soon exploited in order to formulate the four laws of black hole mechanics [41]. At that time, it was not at all clear whether the black hole area should really be considered as an entropy. This being the prime reason for people to consider area to be analogous to entropy but not actually the entropy. (This is why these were called four laws of black hole mechanics instead of the four laws of black hole thermodynamics.) The fact that classically black holes are not supposed to have temperature since they do not radiate anything also bolstered this claim. But soon it was discovered, by Hawking, that black hole horizons do possess temperature and radiate when quantum effects (not quantum gravity effects) are taken into consideration [12, 42]. This provides ample evidence that thermodynamics is really at work here and that the black hole area should really be interpreted as its entropy.

In the past four decades since then, this intriguing connection between gravity and thermodynamics has steadily become stronger and stronger (see e.g., [43, 44]). These thermodynamic features associated with horizons do not have a natural explanation in the standard framework of general relativity. There have been numerous approaches to explain this intriguing connection, by deriving the entropy-area relation from first principles. In the framework of string theory, this has been achieved along with the correct proportionality factor in the case of certain extremal black holes [43, 45]. On the other hand loop quantum gravity has also managed to produce the area-entropy relation but without determining the proportionality factor [43, 46]. In recent times, deriving the entropy-area law is taken to be a definitive test for any quantum gravity model.

1.3.3.2 What Emerges Is Emergent—The Emergent Gravity Paradigm

There is another interpretation for the curious connection between gravitational dynamics and horizon thermodynamics. According to this interpretation the thermodynamics of gravitational theories is arising due to the fact that gravity itself is not a fundamental interaction but is emergent, just like elasticity is a phenomenon emergent from the interactions between molecules in a solid. This particular interpretation of gravitational dynamics originates back to Sakharov ([47], reproduced in [48]) and was resurrected in 1995 by Ted Jacobson, who used the concept of local Rindler observers and the physical process version of the first law of black hole thermodynamics to derive the field equations of general relativity. At the very end of his paper he compared gravitational field equations with the equations that govern sound waves in a gas and thus speculating that gravitational field equations may not be fundamental but emergent from some underlying fundamental dynamics. This suggests a dramatic revolution in perceiving general relativity, that it may not be correct to canonically quantize the Einstein's equations as is done in almost all the quantization schemes. The situation is analogous to quantizing sound waves in a material medium, which of course is not correct, even if they describe a phenomenon that is ultimately quantum mechanical in nature. Subsequently, Padmanabhan and his collaborators worked on these ideas (see [49] for a review) and shaped it in more elegant way. This particular framework where gravity is not considered as a fundamental interaction has been labeled "the emergent paradigm" and till date a large amount of work has been done in this area. Below we summarize some of the claims of this research endeavour:

1. The gravitational field equations reduce to thermodynamic identities on horizons for a wide class of gravity theories that are more general than Einstein gravity [50–55];
2. Gravitational field equations can also be obtained from thermodynamical extremum principles [56, 57];
3. It is possible to obtain the density of microscopic degrees of freedom through equipartition arguments [58, 59];
4. The action functional itself for gravitation in a class of theories can acquire thermodynamic interpretation [50, 60–62];
5. Einstein's equations when projected on a null surface reduce to Navier-Stokes equations of fluid dynamics in any spacetime by generalizing previous results for black hole spacetime [63–66];
6. The euclidean path integral associated with the gravitational action in the Lanczos-Lovelock models when interpreted as a partition function has provided expressions for free energy, energy and entropy [50, 67, 68];
7. In [69], a pair of conjugate variables $f^{ab} = \sqrt{-g}g^{ab}$ and $N^c_{ab} = -\Gamma^c_{ab} + (1/2)(\Gamma^d_{bd}\delta^c_a + \Gamma^d_{cd}\delta^a_b)$ have been introduced in terms of which the gravitational action can be interpreted as a momentum space action. Also variation of these variables has very natural thermodynamic interpretation. δf^{ab} is related to variation of entropy and δN^c_{ab} is related to variation of temperature when evaluated on an arbitrary null surface.
8. This paradigm also offers a possible solution to the cosmological constant problem. In this paradigm, (a) the field equations are invariant under addition of a constant to the matter Lagrangian, (b) the cosmological constant appears as an integration constant and finally (c) its value can be determined by postulating a dimensionless number (known as CosMIn) to have a value 4π [70, 71]. This dimensionless number counts the number of modes that cross the Hubble volume at the end of inflation and re-enters the Hubble volume at the beginning of late-time acceleration phase [71].
9. More recently, it has been shown [72] that the total Noether charge in a 3-volume $\mathcal{R}$ related to the time evolution vector field can be interpreted as the heat content of the boundary $\partial\mathcal{R}$ of the volume and the time evolution of the spacetime itself can be described in an elegant manner as being driven by the departure from holographic equipartition measured by $(N_{\rm bulk} - N_{\rm sur})$. Here, the number of bulk degrees of freedom $N_{\rm bulk}$ is related to the Komar energy density while the number of surface degrees of freedom $N_{\rm sur}$ is related to the geometrical area of the boundary surface.

All the recent derivations have been performed in the context of general relativity only. However, at high energies it is believed that the gravitational action should have higher curvature terms. If we further impose the condition that field equations should be second order in the dynamical variable then it uniquely picks out the Lanczos-Lovelock Lagrangian (see Chap. 2 for details). Thus it is important to ask whether all these results holding for general relativity also holds for Lanczos-Lovelock gravity as well, which will enhance the domain of applicability to a very large class of gravitational theories. We shall examine the properties of certain variables we introduce in the framework of Lanczos-Lovelock gravity (see Chap. 3) in the context of the emergent paradigm in Chap. 4. We shall also look at the generalizations of various aspects of the emergent gravity paradigm, derived for general relativity to Lanczos-Lovelock theories of gravity—Holographic equipartition and spacetime evolution in Chap. 5; and Noether current, gravitational momentum and its thermodynamic interpretation in Chap. 6. Finally, we present in Chap. 7 a possible unification of various thermodynamic results through gravitational momentum and null surfaces, with Chap. 8 exploring the connection between area and entropy.

1.3.4 Trademark of Quantum Gravity—Existence of Zero Point Length

The marriage of quantum mechanics and special relativity—uncertainty principle and the finite speed of light, when put together, lead to a minimum length for relativistic quantum mechanics: *it is not possible to localize a particle with a better accuracy than its Compton wavelength.* This can be understood easily from the following consideration, for relativistic particles the position-momentum uncertainty relation can be written in terms of its energy E, as $\Delta x \Delta E \gtrsim 1$. Thus, for a position uncertainty of the particle which is smaller compared to its Compton wavelength (i.e., $\Delta x \lesssim E^{-1}$), we get $E^{-1}\Delta E \gtrsim 1$, which is larger than the rest mass of the particle, leading to pair creation and annihilation requiring multi-particle description [73].

The next step is to introduce gravity which results in a dynamical spacetime. It is affected by, and also affects, the objects and particles that co-exist within it. Due to mutual interactions between matter and gravity, any quantum uncertainty in the position of a particle implies an uncertainty in its momentum and hence, leads to an uncertainty in the geometry as well. This in turn introduces an additional uncertainty in the position of the particle. From this simple reasoning, it follows that the expression for the uncertainty in position of a particle will harbor two pieces—(a) one coming from the standard Heisenberg uncertainty principle and (b) the second one can be attributed to the response of spacetime due to the presence of this quantum uncertainty.

The first question that one might ask is about the scale of these fluctuations. These can be obtained by a simple thought experiment, illustrating basic features involved. Let us say we want to resolve a spherical region of extent l and for that purpose it is necessary to have a photon having wavelength smaller than l. Hence its energy will be greater than $1/l$ with associated energy density ρ being greater than $1/l^4$. From Einstein's equations, on the other hand, the gravitational potential generated by the energy density of this photon corresponds to $g \gtrsim \ell_p^2/l^2$, where ℓ_p stands for the Planck length. Thus the length that is being measured will have an uncertainty due to gravity as, $\sqrt{gl^2} \gtrsim \ell_p$ [73]. Hence, independent of any particular way of measuring the position, the distance between two events will always have a minimum uncertainty $\mathcal{O}(\ell_p^2)$. The same conclusion, can also be obtained from various other thought experiments and theoretical analyses of quantum gravity models.

The existence of a fundamental length $\mathcal{O}(\ell_p)$ beyond which the very concepts of space and time lose their meaning has a conceptual similarity with the speed barrier experienced by an observer in special relativity. In the case of special relativity one can accelerate forever but will never be able to reach the speed of light. In the same spirit, given a coordinate frame, one can reduce the coordinate distance between two events as much as one wants but still the proper distance between them will not decrease beyond Planck length ℓ_p. This means that any notion based on a metric structure, e.g. distance, causality, etc., lose their meaning at the Planck's scale. The remarkable fact about the above result is that many different analyses based on different ways of looking at the problem of quantum gravity yield the same prediction regarding the nature of spacetime at the small length scale (or at high energy). This result seems to be a model-independent feature of quantum gravity, being a direct consequence of the physical laws that affect the spacetime structure. To introduce zero point length we will invoke an effective metric which naturally leads to a spacetime interval with a cutoff length scale $\mathcal{O}(\ell_p)$. In this thesis we will explore possible consequences of this effective metric in Chap. 12.

References

1. L. D. Landau, E. M. Lifshitz, Course of theoretical physics series in *The Classical Theory of Fields*, Vol. 2, 4th edn. (Butterworth-Heinemann, 1980)
2. S. Chandrasekhar, The general theory of relativity-why it is probably the most beautiful of all existing theories. J. Astrophy. Astron. **5**, 3–11 (1984)
3. C.M. Will, The Confrontation between general relativity and experiment. Living Rev. Rel. **9**, 3 (2006). arXiv:gr-qc/0510072 [gr-qc]
4. A.D. Rendall, The nature of spacetime singularities, in *100 Years Of Relativity : space-time structure: Einstein and beyond* PP. 76–92 (2005). arXiv:gr-qc/0503112 [gr-qc]
5. R.M. Wald, *General Relativity*, 1st edn. (The University of Chicago Press, 1984)
6. T. Padmanabhan, Conceptual issues in combining general relativity and quantum theory, in *The Universe*, (Springer, 2000), pp. 239–251
7. C.W. Misner, K.S. Thorne, J.A. Wheeler, *Gravitation*, 3rd edn. (W. H. Freeman and Company, 1973)
8. S. Hawking, R. Penrose, *The Nature Of Space And Time*, (Princeton University Press, 2010)
9. A. Ashtekar, A. Barrau, Loop quantum cosmology: from pre-inflationary dynamics to observations. arXiv:1504.07559 [gr-qc]
10. C. Rovelli, F. Vidotto, Evidence for maximal acceleration and singularity resolution in covariant loop quantum gravity. Phys. Rev. Lett. **111**, 091303 (2013). arXiv:1307.3228 [gr-qc]
11. R. Gambini, J. Pullin, Loop quantization of the Schwarzschild black hole. Phys.Rev.Lett. **110**(21) 211301, (2013). arXiv:1302.5265 [gr-qc]
12. S. Hawking, Particle creation by black holes. Commun. Math. Phys. **43**, 199–220 (1975)
13. J. Bekenstein, Black holes and the second law. Lett. Nuovo Cimento Soc. Ital. Fis. **4**, 737–740 (1972)
14. W.G. Unruh, R.M. Wald, What happens when an accelerating observer detects a Rindler particle. Phys. Rev. D **29**, 1047–1056 (1984)
15. S.D. Mathur, The Information paradox: a pedagogical introduction. Class. Quant. Grav. **26**, 224001 (2009). arXiv:0909.1038 [hep-th]
16. R.B. Mann, T.G. Steele, Thermodynamics and quantum aspects of black holes in (1+1)-dimensions. Class. Quant. Grav. **9**, 475–492 (1992)

17. M. Visser, Thermality of the Hawking flux. JHEP **07**, 009 (2015). arXiv:1409.7754 [gr-qc]
18. S.K. Modak, L. Ortiz, I. Pena, D. Sudarsky, Non-Paradoxical loss of information in black hole evaporation in a quantum collapse model. Phys. Rev. D **91**(12), 124009 (2015). arXiv:1408.3062 [gr-qc]
19. S. L. Adler, A.C. Millard, Generalized quantum dynamics as prequantum mechanics. Nucl. Phys. B **473**(199–244) (1996). arXiv:hep-th/9508076 [hep-th]
20. A. Bassi, G.C. Ghirardi, Dynamical reduction models. Phys. Rept. **379**, 257 (2003). arXiv:quant-ph/0302164 [quant-ph]
21. A. Bassi, K. Lochan, S. Satin, T.P. Singh, H. Ulbricht, Models of wave-function collapse, underlying theories, and experimental tests. Rev. Mod. Phys. **85**, 471–527 (2013). arXiv:1204.4325 [quant-ph]
22. S.D. Mathur, Tunneling into fuzzball states. Gen. Rel. Grav. **42**, 113–118 (2010). arXiv:0805.3716 [hep-th]
23. S. Chakraborty, S. Singh, T. Padmanabhan, A quantum peek inside the black hole event horizon. JHEP **1506**, 192 (2015). arXiv:1503.01774 [gr-qc]
24. S. Singh, S. Chakraborty, Black hole kinematics: The âinâ- vacuum energy density and flux for different observers. Phys.Rev. D **90**(2), 024011 (2014). arXiv:1404.0684 [gr-qc]
25. C.M. DeWitt, D. Rickles, *The Role Of Gravitation In Physics: Report From The 1957 Chapel Hill Conference*, vol. 5. epubli, 2011
26. M. Albers, C. Kiefer, M. Reginatto, Measurement analysis and quantum gravity. Phys. Rev. D **78**, 064051 (2008). arXiv:0802.1978 [gr-qc]
27. S. Carlip, Is quantum gravity necessary? Class. Quant. Grav. **25**, 154010 (2008). arXiv:0803.3456 [gr-qc]
28. G. 't Hooft, An algorithm for the poles at dimension four in the dimensional regularization procedure. Nucl.Phys. B **62**(444–460) (1973)
29. G. 't Hooft, M. Veltman, One loop divergencies in the theory of gravitation. Ann. Poincare Phys.Theor. A **20**(69–94) (1974)
30. S. Deser, P. van Nieuwenhuizen, One loop divergences of quantized Einstein-Maxwell fields. Phys. Rev. D **10**, 401 (1974)
31. S. Deser, P. van Nieuwenhuizen, Nonrenormalizability of the quantized Dirac-Einstein system. Phys. Rev. D **10**, 411 (1974)
32. M.H. Goroff, A. Sagnotti, The ultraviolet behavior of Einstein gravity. Nucl. Phys. B **266**(3), 709–736 (1986)
33. S. de Haro, D. Dieks, E. Verlinde et al., Forty years of string theory reflecting on the foundations. Found. Phys. **43**(1), 1–7 (2013)
34. C. Rovelli, L. Smolin, Discreteness of area and volume in quantum gravity. Nucl. Phys. B **442**(3), 593–619 (1995)
35. D.Z. Freedman, A. Van Proeyen, *Supergravity* (Cambridge University Press, 2012)
36. G. Esposito, A. Y. Kamenshchik, G. Pollifrone, *Euclidean Quantum Gravity On Manifolds With Boundary*, vol. **85**. (Springer Science & Business Media, 1997)
37. R. D. Sorkin, *Causal sets: Discrete gravity* in Lectures on quantum gravity, 305–327. Springer, 2005
38. J. Ambjorn, A. Goerlich, J. Jurkiewicz, R. Loll, Quantum gravity via causal dynamical triangulations. arXiv:1302.2173 [hep-th]
39. J.D. Bekenstein, Black holes and entropy. Phys. Rev. D **7**, 2333–2346 (1973)
40. J.D. Bekenstein, Generalized second law of thermodynamics in black hole physics. Phys. Rev. D **9**, 3292–3300 (1974)
41. J.M. Bardeen, B. Carter, S. Hawking, The Four laws of black hole mechanics. Commun. Math. Phys. **31**, 161–170 (1973)
42. S. Hawking, Black Holes and Thermodynamics. Phys. Rev. D **13**, 191–197 (1976)
43. R.M. Wald, The thermodynamics of black holes. Living Rev. Rel. **4**, 6 (2001). arXiv:gr-qc/9912119 [gr-qc]
44. T. Padmanabhan, Gravity and the thermodynamics of horizons. Phys.Rept. **406**(49–125) (2005). arXiv:gr-qc/0311036 [gr-qc]

45. G.T. Horowitz, Quantum states of black holes. arXiv:gr-qc/9704072 [gr-qc]
46. C. Rovelli, Loop quantum gravity: the first twenty five years. Class. Quant. Grav. **28**, 153002 (2011). arXiv:1012.4707 [gr-qc]
47. A.D. Sakharov, Vacuum quantum fluctuations in curved space and the theory of gravitation. Gen. Relativ. Gravit. **32**(2), 365–367 (2000)
48. A.D. Sakharov, Vacuum quantum fluctuations in curved space and the theory of gravitation. Sov. Phys.-Dokl. **12**, 1040–1041 (1968)
49. T. Padmanabhan, Thermodynamical aspects of gravity: new insights. Rept. Prog. Phys. **73**, 046901 (2010). arXiv:0911.5004 [gr-qc]
50. T. Padmanabhan, Classical and quantum thermodynamics of horizons in spherically symmetric space-times. Class.Quant.Grav. **19**(5387–5408) (2002). arXiv:gr-qc/0204019 [gr-qc]
51. R.-G. Cai, S.P. Kim, First law of thermodynamics and Friedmann equations of Friedmann-Robertson-Walker universe. JHEP **0502**, 050 (2005). arXiv:hep-th/0501055 [hep-th]
52. A. Paranjape, S. Sarkar, T. Padmanabhan, Thermodynamic route to field equations in Lancos-Lovelock gravity. Phys. Rev. D **74**, 104015 (2006). arXiv:hep-th/0607240 [hep-th]
53. M. Akbar and R.-G. Cai, Friedmann equations of FRW universe in scalar-tensor gravity, f(R) gravity and first law of thermodynamics. Phys. Lett. B **635**(7–10) (2006). arXiv:hep-th/0602156 [hep-th]
54. T. Padmanabhan, Dark energy: mystery of the millennium. AIP Conf. Proc. **861**(179–196) (2006). arXiv:astro-ph/0603114 [astro-ph]. [,179(2006)]
55. D. Kothawala, T. Padmanabhan, Thermodynamic structure of Lanczos-Lovelock field equations from near-horizon symmetries. Phys. Rev. D **79**, 104020 (2009). arXiv:0904.0215 [gr-qc]
56. T. Padmanabhan, A. Paranjape, Entropy of null surfaces and dynamics of spacetime. Phys. Rev. D **75**, 064004 (2007). arXiv:gr-qc/0701003 [gr-qc]
57. T. Padmanabhan, Dark energy and gravity. Gen.Rel.Grav. **40**(529–564) (2008). arXiv:0705.2533 [gr-qc]
58. T. Padmanabhan, Equipartition of energy in the horizon degrees of freedom and the emergence of gravity. Mod. Phys. Lett. A **25**(1129–1136) (2010). arXiv:0912.3165 [gr-qc]
59. T. Padmanabhan, Surface density of spacetime degrees of freedom from equipartition law in theories of gravity. Phys. Rev. D **81**, 124040 (2010). arXiv:1003.5665 [gr-qc]
60. T. Padmanabhan, Holographic gravity and the surface term in the Einstein-Hilbert action. Braz.J.Phys. **35**(362–372) (2005). arXiv:gr-qc/0412068 [gr-qc]
61. A. Mukhopadhyay, T. Padmanabhan, Holography of gravitational action functionals. Phys. Rev. D **74**, 124023 (2006). arXiv:hep-th/0608120
62. S. Kolekar, T. Padmanabhan, Holography in action. Phys. Rev. D **82**, 024036 (2010). arXiv:1005.0619 [gr-qc]
63. T. Damour, Surface effects in black hole physics in *Proceedings of the Second Marcel Grossmann Meeting on General Relativity* (1982)
64. K.S. Thorne, R.H. Price D.A. MacDonald, Black Holes: The Membrane Paradigm. (Yale University Press, 1986)
65. T. Padmanabhan, Entropy density of spacetime and the Navier-Stokes fluid dynamics of null surfaces. Phys. Rev. D **83**, 044048 (2011). arXiv:1012.0119 [gr-qc]
66. S. Kolekar, T. Padmanabhan, Action principle for the fluid-gravity correspondence and emergent gravity. Phys. Rev. D **85**, 024004 (2012). arXiv:1109.5353 [gr-qc]
67. S. Kolekar, D. Kothawala, T. Padmanabhan, Two aspects of black hole entropy in Lanczos-Lovelock models of gravity. Phys. Rev. D **85**, 064031 (2012). arXiv:1111.0973 [gr-qc]
68. G. Gibbons, S. Hawking, Action integrals and partition functions in quantum gravity. Phys. Rev. D **15**, 2752–2756 (1977)
69. K. Parattu, B.R. Majhi, T. Padmanabhan, Structure of the gravitational action and its relation with horizon thermodynamics and emergent gravity paradigm. Phys. Rev. D **87**124011, (Jun, 2013). arXiv:gr-qc/1303.1535 [gr-qc], doi:10.1103/PhysRevD.87.124011
70. T. Padmanabhan, H. Padmanabhan, CosMIn: the solution to the cosmological constant problem. Int. J. Mod. Phys. D **22**, 1342001 (2013). arXiv:1302.3226 [astro-ph.CO]

71. T. Padmanabhan, H. Padmanabhan, Cosmological Constant from the Emergent Gravity Perspective. Int. J. Mod. Phys. D **23**(6), 1430011 (2014). arXiv:1404.2284 [gr-qc]
72. T. Padmanabhan, General relativity from a thermodynamic perspective. Gen. Rel. Grav. **46**, 1673 (2014). arXiv:1312.3253 [gr-qc]
73. L.J. Garay, Quantum gravity and minimum length. Int. J. Mod. Phys. A **10**(145–166), (1995). arXiv:gr-qc/9403008 [gr-qc]

Part II
Geometrical Aspects of Gravitational Action

Chapter 2
Setting the Stage: Review of Previous Results

2.1 Introduction

It is well-known that general relativity has some peculiar features, introducing difficulties (these are not encountered in other field theories, e.g., non-Abelian gauge theories) when we try to derive the Einstein's equations from the Einstein-Hilbert action principle (see, e.g. Chap. 6 of [1]). Based on the usual theoretical prejudice [2], it is often conjectured that the field equations which are second order in derivatives of the coordinates, are obtained from an action which is quadratic in the first derivatives of the dynamical variables. However, the situation is entirely different in the case of gravity. Due to equivalence principle, the generally covariant Lagrangian for gravity is forced to have at least second order derivatives of the metric. The usual and the simplest choice, R, has a special structure (viz. linearity in second derivatives), which allows one to obtain the field equations which still involves only second derivatives of the metric, if (and only if) we fix metric *and* its derivatives on the boundary. This arises since one can separate out the second derivatives in the Lagrangian leading to a total divergence, which becomes a surface term in the action. Variation of the surface term does not contribute to the field equations however affects the boundary value problem and hence we need to set both the variations of the metric *and* its derivatives to be zero on the boundary.

The main conceptual difficulty with this program—which makes the gravitational action different from those in other field theories—is that we need to fix both the dynamical variable and its derivative at the boundary in order to arrive at the field equations. This procedure encounters a few difficulties at the conceptual level. Let the spacetime region we are considering is between two spacelike surfaces and also assume that all quantities of interest fall off at spatial infinites. If we now fix both the metric and its derivative on the earlier spacelike hypersurface, the Einstein's equations should tell us the corresponding values on the latter hypersurface. Thus it is clear that we do not really have the freedom to fix arbitrary values for the metric and its derivatives on both the boundaries. Another option is to add a boundary term to the gravitational action, such that the variation of this term exactly cancels the terms

S. Chakraborty, *Classical and Quantum Aspects of Gravity in Relation to the Emergent Paradigm*, Springer Theses, DOI 10.1007/978-3-319-63733-4_2

with variation of the normal derivative. An well-known example of such a boundary term is the Gibbons-Hawking-York counter-term [3, 4], even though it is not a unique choice [5]. We should emphasize that addition of a boundary term, though leads to a generally covariant gravitational action but it turns out to be foliation dependent.

So far we have been concerned about general relativity alone. However it is well known that unlike the kinematics (which can be described in an elegant manner using the principles of equivalence and principle of general covariance), there is no natural principle to determine the field equations for gravity. Hence it seems reasonable to study all possible field equations, which are second order in the independent variable, the metric. The criteria—field equations should be second order differential equations in the dynamical variable—automatically fixes the gravitational Lagrangian to be Lanczos-Lovelock Lagrangian (this was first pointed out by Lanczos and Lovelock [6–10]). The field equations for Lanczos-Lovelock gravity also follow from an action in higher dimensions, which is polynomial in the curvature tensor, reducing to Einstein-Hilbert Lagrangian—*uniquely*—when $D = 4$. In higher dimensions the Lanczos-Lovelock Lagrangian has additional higher curvature terms and hence differ from the Einstein-Hilbert Lagrangian. Due to existence of higher curvature terms, the Lanczos-Lovelock Lagrangian also exhibits a rich structure in comparison to the Einstein-Hilbert Lagrangian.

In this chapter, we will mainly review earlier results and shall collect all the essential relations that will be used extensively in the subsequent chapters in relation to both Einstein-Hilbert and Lanczos-Lovelock gravity.

2.2 A Fresh Look at the Einstein-Hilbert Action

Use of g_{ab} and its conjugate momentum as the dynamical variables, leads to the conclusion—the surface variation *involves only the variation of the canonical momenta*. This is a non-trivial result and arises *only because* the surface and bulk terms of the Einstein-Hilbert Lagrangian are intertwined with each other. This, in turn, implies that one might be able to treat the Einstein-Hilbert action as a momentum space action. This suggests that it may be better to deal with gravitational physics in the space of the canonical momenta corresponding to the metric rather than in the space of the metric itself. It was first demonstrated in [11], that this program will also work with a new dynamical variable, a tensor density, $f^{ab} = \sqrt{-g}g^{ab}$. The variable $f^{ab} = \sqrt{-g}g^{ab}$ has already been used in earlier literatures [12–15], but have not attracted much attention in recent years. In this section we will content ourselves with introduction of these new canonically conjugate variables in general relativity and their various geometric properties following [11]. In particular we will present Einstein-Hilbert action as a momentum space action in terms of these variables. We will also discuss Noether current and gravitational momentum for the Einstein-Hilbert action [16] and its relation to this set of canonically conjugate variables.

2.2.1 On the Structure of the Einstein-Hilbert Action

In this section, we shall rapidly review various existing ideas and relations present in the literature associated with the action principle for general relativity, which corresponds to the Einstein-Hilbert action and is given by

$$A_{\mathrm{EH}} \equiv \int_{\mathcal{V}} d^4x \sqrt{-g}R. \tag{2.1}$$

where R corresponds to the Ricci scalar and $\mathcal{V}$ stands for the four-dimensional spacetime volume of interest. It is useful to define a quantity Q^{cd}_{ab} such that,

$$Q^{ab}_{cd} = \frac{1}{2}\delta^{ab}_{cd}; \qquad \delta^{ab}_{cd} = \delta^a_c\delta^b_d - \delta^a_d\delta^b_c \tag{2.2}$$

where δ^{ab}_{cd} stands for the completely antisymmetric determinant tensor. Using this tensor Q^{ab}_{cd}, it is possible to write the Einstein-Hilbert action (see, Eq. (2.1)) in an equivalent form as:

$$A_{\mathrm{EH}} \equiv \int_{\mathcal{V}} d^4x\sqrt{-g}\; Q_a^{\ bcd} R^a_{\ bcd} \tag{2.3}$$

$$= \int_{\mathcal{V}} d^4x\sqrt{-g}\; \left(L_{\mathrm{quad}} + L_{\mathrm{sur}}\right) \tag{2.4}$$

where we have defined the bulk Lagrangian (since this is quadratic in the connection, it is also referred to as the quadratic Lagrangian) and the surface Lagrangian respectively as

$$L_{\mathrm{quad}} \equiv 2Q_a^{\ bcd}\Gamma^a_{dk}\Gamma^k_{bc}; \qquad L_{\mathrm{sur}} \equiv \frac{2}{\sqrt{-g}}\partial_c\left[\sqrt{-g}\,Q_a\Gamma^a_{bd}\right]. \tag{2.5}$$

The advantage of writing the Einstein-Hilbert action in this particular form is that it can be readily generalized to gravitational theories in more than 4 dimensions (for more details see Sect. 2.3). One of the striking feature of the above action stems from the fact that the variation of the bulk term alone can furnish the field equations (as explicitly demonstrated in Chap. 6 of [1]). Once the field equations and hence its solutions are obtained, which includes black hole solutions as well, one can evaluate the surface term on the horizon, which surprisingly reproduces the entropy of the horizon in the Euclidean sector. (In general it gives τTS where τ is the range of integration; we get S when $\tau = \beta$, the inverse temperature.) The fact that the surface term, which is not supposed to carry any information about the field equations, yields the entropy when integrated over a horizon is a direct hint that gravitational action principle contains information about horizon thermodynamics. An algebraic origin to this peculiar behavior lies in the fact that the bulk term L_{quad} and the surface term L_{sur} are not independent, but related to each other by the relation

$$\sqrt{-g}L_{\text{sur}} = -\left[\partial_c\left(g_{ab}\frac{\partial\left(\sqrt{-g}L_{\text{quad}}\right)}{\partial(\partial_c g_{ab})}\right)\right]. \tag{2.6}$$

Since it relates a quantity on the surface with a quantity in the bulk, this relation has been termed "holographic" [17–19].

2.2.2 *In Search of Alternative Variables in General Relativity*

A natural question is whether g_{ab} is unique in providing us with a neat and clean variational principle for general relativity or are there other variables that does the same. It turns out that the Einstein-Hilbert action can be written in terms of a new variable $f^{ab} = \sqrt{-g}g^{ab}$ and we obtain the following structure [11]

$$\sqrt{-g}R = \sqrt{-g}L_{\text{quad}} - \partial_c\left[f^{ab}\frac{\partial(\sqrt{-g}L_{\text{quad}})}{\partial(\partial_c f^{ab})}\right] = \sqrt{-g}L_{\text{quad}} - \partial_c\left[f^{ab}N^c_{ab}\right] \tag{2.7}$$

where we have introduced the object N^a_{bc} to be the momentum conjugate to f^{ab}, since

$$N^a_{bc} \equiv \frac{\partial(\sqrt{-g}L_{\text{quad}})}{\partial(\partial_a f^{bc})} \tag{2.8}$$

$$= -\Gamma^a_{bc} + \frac{1}{2}(\Gamma^d_{bd}\delta^a_c + \Gamma^d_{cd}\delta^a_b) = \left[Q^{ad}_{be}\Gamma^e_{cd} + Q^{ad}_{ce}\Gamma^e_{bd}\right] \tag{2.9}$$

Note that it is symmetric in the lower two indices.

Next, the expressions for variation of the Einstein-Hilbert action in terms of the conjugate variables f^{ab} and N^c_{ab} becomes,

$$\delta(\sqrt{-g}R) = R_{ab}\delta f^{ab} + f^{ab}\delta R_{ab} = R_{ab}\delta f^{ab} - \partial_c[f^{ik}\delta N^c_{ik}], \tag{2.10}$$

In addition to simplifying the variation of the action, there are two main advantages of this conjugate pairs (f^{ab}, N^c_{ab}). The first one is that many known expressions and formula simplify considerably as expressed in terms of these conjugate variables. Secondly and more importantly, the *variations of these variables on a horizon leads to a direct thermodynamical interpretation* (see [11] for details).

2.2.3 *Einstein-Hilbert Action in Terms of the New Variables*

Using these new pair of canonically conjugate variables (f^{ab}, N^c_{ab}) we will try to rewrite the bulk part of the Lagrangian. Substituting the connections in Eq. (2.5) by N^c_{ab} from Eq. (2.9) we obtain the bulk part of the Lagrangian to be:

$$\sqrt{-g}L_{\rm quad} = g^{bd} N^i_{dj} N^k_{bl} \left[\delta^l_i \delta^j_k - \frac{1}{3} \delta^j_i \delta^l_k \right] = \frac{1}{2} N^c_{ab} \partial_c f^{ab} \tag{2.11}$$

with striking simplicity. Note that this exhibits the familiar '$p\dot{q}/2$' structure, which is a consequence of the fact that the Einstein-Hilbert Lagrangian is quadratic in $\dot{q}$. While for the surface term of the Einstein-Hilbert action we arrive at,

$$\sqrt{-g}L_{\rm sur} = \partial_c(\sqrt{-g}V^c) = -\partial_c \left[f^{ab} N^c_{ab} \right]. \tag{2.12}$$

The above relation defines the object V^c, such that,

$$\sqrt{-g}V^c = -f^{ab} N^c_{ab} \tag{2.13}$$

Various other geometrical relations connecting curvature tensor components to the new set of conjugate variables (f^{ab}, N^c_{ab}) have been derived and explored in [11].

2.2.3.1 Hamilton's Equations for General Relativity

In Lagrangian formulation of classical mechanics, variation of the action is carried out treating the dynamical variable q as independent, for the class of actions which do not depend on the higher time-derivatives of q. The momentum p on the other hand, is then defined as the partial derivative of the Lagrangian with respect to $\dot{q}$. In the Hamiltonian formulation one defines the Hamiltonian $H(q, p)$ through the following relation $L = p\dot{q} - H(q, p)$. In this case, fixing the variations of q at the boundary, yields the well-known Hamilton's equations, which is referred to as modified Hamilton's principle [20, 21].

Analogously there are two well-known variational principles in general relativity—one in which the variation of the gravitational action is carried out in terms of the variation of the metric. On the other hand, one can also consider the metric and the affine connection to be independent and hence varied separately, known as the Palatini variational principle [22]. We shall now outline a version of the Palatini variational principle in terms of the conjugate variables (f^{ab}, N^c_{ab}) for Einstein-Hilbert action following [11]. Using the variables (f^{ab}, N^c_{ab}), the gravitational Lagrangian density can be expressed as

$$\sqrt{-g}R = f^{ab} R_{ab} = f^{ab}(-\partial_c N^c_{ab} - N^c_{ad} N^d_{bc} + \frac{1}{3} N^c_{ac} N^d_{bd}) . \tag{2.14}$$

The variation of $\sqrt{-g}R$ with respect to N^c_{ab} with fixed f^{ab} is given by

$$\delta\left(\sqrt{-g}R\right)|_{f^{ab}} = \left[\partial_c f^{ab} - 2 f^{ad} N^b_{cd} + \frac{2}{3} f^{am} N^d_{dm} \delta^b_c \right] \delta N^c_{ab} - \partial_c (f^{ab} \delta N^c_{ab}) \tag{2.15}$$

Fixing N^a_{bc} at the boundary leads to the corresponding field equations, which can be obtained by equating the symmetrized coefficient of δN^c_{ab} to zero. These equations are

$$\partial_c f^{ab} = f^{ad} N^b_{cd} + f^{bd} N^a_{cd} - \frac{1}{3} f^{am} N^d_{dm} \delta^b_c - \frac{1}{3} f^{bm} N^d_{dm} \delta^a_c \qquad (2.16)$$

yielding the desired relation between N^c_{ab} with f^{ab} and its derivative. The connection of the above relation with the standard result $\nabla_c g^{ab} = 0$ obtained from Palatini variational principle has been explored in [11].

At this stage it is possible to obtain an analogue of the Hamilton's equation $\dot{q} = \partial H/\partial p$. The analogy can be made more accurate by introducing a "Hamiltonian" as,

$$\mathcal{H}_g = f^{ab}(N^c_{ad} N^d_{bc} - \frac{1}{3} N^c_{ac} N^d_{bd}) \qquad (2.17)$$

Comparing Eq. (2.17) with Eq. (2.11), we see that $\mathcal{H}_g = \sqrt{-g} L_{\text{quad}}$. Then, with the notional correspondence $f^{ab} \to q$ and $N^c_{ab} \to p$, one can immediately establish the following result:

$$\sqrt{-g}R \to -q\partial p - \mathcal{H}_g = \{p\partial q - \mathcal{H}_g\} - \partial(qp) = \sqrt{-g}\left(L_{\text{quad}} + L_{\text{sur}}\right) \qquad (2.18)$$

Thus, the quadratic Lagrangian density used to derive the field equations can be reinterpreted as a Hamiltonian density. Equation (2.16) can then be rewritten as a Hamilton's field equation,

$$\partial_c f^{ab} = \frac{\partial \mathcal{H}_g}{\partial N^c_{ab}} \qquad (2.19)$$

Proceeding by analogy with classical mechanics, one can also obtain the other Hamilton's equation of motion, viz., $\dot{p} = -\partial H/\partial q$. This can be achieved by considering the variation of the Einstein-Hilbert Lagrangian density with respect to f^{ab}, and can be expressed as,

$$\delta\left(\sqrt{-g}R\right)|_{N^c_{ab}} = R_{ab}\delta f^{ab} \qquad (2.20)$$

Hence the field equations associated with variation of f^{ab} is equivalent to $R_{ab} = 0$, the vacuum field equations of Einstein. Referring back to Eq. (2.17), we see that this equation can also be re-expressed as

$$\partial_c N^c_{ab} = -\frac{\partial \mathcal{H}_g}{\partial f^{ab}} \qquad (2.21)$$

yielding the second of the Hamilton's equations. However unlike the case in classical mechanics, where the momentum would be conserved in the absence of external forces, we observe that N^c_{ab} has the capability of driving its own change, thanks to the nonlinear nature of gravity.

2.2.3.2 Inclusion of Matter

The next natural step would be to consider the inclusion of the matter Lagrangian density $\sqrt{-g}L_{\rm m}$, which can be accomplished by introducing a total Hamiltonian as

$$\mathcal{H}_{\rm tot} = \mathcal{H}_g - \sqrt{-g}L_{\rm m} \tag{2.22}$$

If the first term $\mathcal{H}_g$ (incidentally this is equal to $\sqrt{-g}L_{\rm quad}$) be considered to be kinetic, then it is natural to think of $\sqrt{-g}L_{\rm m}$ as a potential term. We shall further assume that the matter Lagrangian density $\sqrt{-g}L_{\rm m}$ under consideration depends only on f^{ab} and not on N^c_{ab}. In such a case, Eq. (2.16) retains its original form. Expressing everything in terms of the metric and its derivative it becomes clear that our assumption is equivalent to the fact that $\sqrt{-g}L_{\rm m}$ does not depend on the derivatives of the metric. (This is similar in spirit to the case in classical mechanics when we consider velocity-independent potentials.)

Let us now provide the usual definition of the matter energy-momentum tensor as

$$T_{ab} = -\frac{2}{\sqrt{-g}}\frac{\partial(\sqrt{-g}L_{\rm m})}{\partial g^{ab}}\,, \tag{2.23}$$

from which one can obtain the following equality:

$$\frac{\partial\sqrt{-g}L_{\rm m}}{\partial f^{ab}} = -\frac{1}{2}\Big[T_{ab} - \frac{1}{2}T g_{ab}\Big] \equiv -\frac{1}{2}\overline{T}_{ab}\,, \tag{2.24}$$

where a new object $\overline{T}_{ab}$ has been defined, which bears to T_{ab} the same relation as G_{ab} bears to R_{ab}.

When matter is included the second equation of Hamiltonian, i.e., Eq. (2.21) gets modified to read

$$\partial_c N^c_{ab} = -\frac{\partial\mathcal{H}_g}{\partial f^{ab}} + \frac{\partial(\sqrt{-g}L_{\rm m})}{\partial f^{ab}} = -N^c_{ad}N^d_{bc} + \frac{1}{3}N^c_{ac}N^d_{bd} - \frac{1}{2}\overline{T}_{ab} \tag{2.25}$$

which is equivalent to the usual Einstein's field equations, $2R_{ab} = \bar{T}_{ab}$ [11]. Finally, we should mention that the surface variation contains only variations of the conjugate momentum N^c_{ab} and not of f^{ab}. Thus, we need to fix only the "momenta" N^c_{ab} at the boundary in order to obtain the correct field equations.

2.2.4 *Noether Current*

In this section, we shall relate the Noether current—arising from a differential geometric identity—with the variations of N^a_{bc} and f^{ab}. The conservation of J^a using differential geometric identity is straightforward and was first presented in [23], here we will briefly outline it.

Let us start with an arbitrary vector field v^a and decompose $\nabla_a v_b$ into symmetric and antisymmetric part as follows,

$$S_{ab} \equiv \nabla_a v_b + \nabla_b v_a; \qquad J_{ab} \equiv \nabla_a v_b - \nabla_b v_a \tag{2.26}$$

It is obvious from the antisymmetry of J^{ab}, that one can construct a conserved current J^a out of it by defining $J^a = \nabla_b J^{ab}$. The only remaining task is to express the conserved current in terms of the newly defined N^c_{ab} variable. This can be achieved by considering Lie derivative of the connection, which has the following expression [1],

$$\pounds_v \Gamma^a_{bc} = \nabla_b \nabla_c v^a + R^a{}_{cmb} v^m \tag{2.27}$$

From Eq. (2.9) one can relate Lie variation of connection to Lie variation of N^c_{ab}. This immediately leads to the following expression,

$$\pounds_v N^a_{bc} = \left[Q^{ad}_{be} \pounds_v \Gamma^e_{cd} + Q^{ad}_{ce} \pounds_v \Gamma^e_{bd}\right] \tag{2.28}$$

where Q^{ab}_{cd} is defined in Eq. (2.2). Now one can consider contraction of the above object with inverse metric g^{bc} leading to the following expression,

$$\begin{aligned} g^{bc} \pounds_v N^a_{bc} &= 2Q^{adc}{}_e \left(\pounds_v \Gamma^e_{cd}\right) = \left(g^{ac}\delta^d_e - g^{dc}\delta^a_e\right)\left(\nabla_c \nabla_d v^e + R^e{}_{dmc} v^m\right) \\ &= \nabla_b J^{ab} - 2R^a_m v^m \end{aligned} \tag{2.29}$$

which immediately leads to the expression for Noether current

$$J^a(v) = 2R^a_b v^b + g^{cd} \pounds_v N^a_{cd} \tag{2.30}$$

Thus, given a vector field v^a one can construct the Noether current by almost *trivial* differential geometric manipulations—which *happens* to be the Noether current arising from diffeomorphism invariance of Einstein-Hilbert Lagrangian. We will try to emphasize the physical significance of this result in later chapters.

2.2.5 Gravitational Momentum

Another structure we want to associate with an arbitrary vector field v^a is the gravitational four-momentum density $P^a(v)$, defined—in the context of general relativity–as

$$P^a(v) = -Rv^a - g^{ij} \pounds_v N^a_{ij} \tag{2.31}$$

From Eq. (2.30) we can substitute for the Lie variation term in Eq. (2.31), which relates the gravitational momentum and the Noether current as

$$J^a(v) = -P^a(v) + 2G^a_b v^b \tag{2.32}$$

The physical meaning of the gravitational momentum can be understood from the following result. (This was motivated and discussed in detail in [16, 23]. We will not repeat the motivation and logic behind this definition here). Consider the special case in which v^a is the velocity of an arbitrary observer, who will attribute to the matter, with energy momentum tensor T_{ab}, the momentum density $\mathcal{M}^a = -T^a_b v^b$. We would expect the total momentum associated with matter plus gravitation to be conserved [16] in nature, for all observers. This condition requires:

$$\begin{aligned} 0 = \nabla_a \left(P^a + \mathcal{M}^a\right) &= \nabla_a \left(-J^a + 2G^a_b v^b - T^a_b v^b\right) \\ &= \nabla_a \left(2G^a_b v^b - T^a_b v^b\right) = \left(2G^a_b - T^a_b\right) \nabla_a v^b \equiv \mathcal{S}^{ab} \nabla_a v_b \end{aligned} \tag{2.33}$$

where $\mathcal{S}^{ab} \equiv (2G^{ab} - T^{ab})$ is a symmetric tensor and in the last line we have used Bianchi identity and the fact that J^a and T^a_b are conserved. The above relation should hold for any normalised time-like vector field v^a, which requires $\mathcal{S}^{ab} = 0$, i.e., $G_{ab} = 8\pi T_{ab}$, which are the field equation for gravity. (This result should be obvious from the fact that $\nabla_a v_b$ can be chosen to be arbitrary at any event even for normalised timelike vector fields. A more formal proof, suggested by S. Date, goes as follows: Choose first v^a to be a normalized *geodesic* velocity field with $v^a v_a = -1$ and $v^a \nabla_a v^b = 0$. Then the most general $\mathcal{S}^{ab}$ which satisfies $\mathcal{S}^{ab} \nabla_a v^b = 0$ has the form $\mathcal{S}^{ab} = \alpha(X^a v^b + X^b v^a) + \beta v^a v^b$ with two arbitrary functions α and β and an arbitrary vector X^a which can be chosen without loss of generality to be purely spatial, i.e., $v_a X^a = 0$. Choose next the velocity field to be $u_a = -N\nabla_a t$ Using the form of $\mathcal{S}^{ab} \nabla_a u_b = 0$ leads to $\alpha = \beta = 0$. This immediately gives $\mathcal{S}^{ab} = 0$.) The gravitational momentum introduced above will find quiet a few interesting applications latter on.

2.3 Lanczos-Lovelock Gravity: A Brief Introduction

One key aspect for general relativity is that, it follows directly from the geometric properties of the Riemann curvature tensor—in particular the Bianchi identity plays a crucial role. Generalizing the curvature tensor by including higher curvature terms, if one imposes the criteria that the trace of its Bianchi derivative vanishes, leads to a divergence free second rank tensor. This tensor also uniquely leads to the Lanczos-Lovelock Lagrangian and agrees with the one obtained by variation of the Lanczos-Lovelock action [24]. In this section we will describe several algebraic and thermodynamic results related to Lanczos-Lovelock Lagrangian, which will be of extensive use in later chapters. The importance of Lanczos-Lovelock gravity is two-fold. It generalizes Einstein-Hilbert action by introducing higher curvature terms, which can arise naturally in strong gravity regime (incidentally, the first correction to Einstein-Hilbert action, known as Gauss-Bonnet action was obtained earlier from a string theoretic point of view [25]). Secondly, several thermodynamic features related to horizons in Einstein-Hilbert action carry over to Lanczos-Lovelock theories of gravity. In emergent gravity paradigm thermodynamics of horizons appear as a key input and hence there is a natural extension of this paradigm to Lanczos-Lovelock gravity. We will be mainly concerned about the thermodynamic aspects of Lanczos-Lovelock gravity in later chapters of this thesis.

2.3.1 How Does Lanczos-Lovelock Gravity Comes About?

In the first part of this chapter, we have been working with Einstein-Hilbert action in four-dimensional spacetime. A natural question to ask in that context is, why spacetime has four dimensions? Even though there is no satisfactory answer to that question, most of the high-energy physics community believes in existence of extra dimension, exploiting which one can answer some subtle questions, which cannot be answered in the standard four-dimensional models. One such problem has to do with the large difference between scale of weak interaction to that of strong interaction, known in the literature as the hierarchy problem. It turns out that extra dimensional models can provide a *natural* solution to this fine-tuning problem [26–28].

Another natural query being—what about gravity in these higher-dimensional scenarios? One can assume that gravitational action is still given by the Einstein-Hilbert Lagrangian, however there is one caveat. The effect of these extra dimensions can only be felt at very high energies and one can ask at such high energies how legitimate it is to use the Einstein-Hilbert Lagrangian. There is a general consensus in the community that at such high energies the Einstein-Hilbert Lagrangian would be supplemented by higher curvature terms. In principle there can be a huge number of such higher curvature corrections gravitational action can have, e.g., $f(R)$, $g(R_{ab}R^{ab})$ and so on. Thus we need some physical guiding principles to find out the most natural candidate.

2.3.1.1 A General Lagrangian

Before going into the guiding principles, let us start with an action, which is an arbitrary function of the curvature tensor $R^a{}_{bcd}$ and the metric g^{ab} in a D-dimensional spacetime. We will work with a single timelike coordinate, i.e., the total spacetime dimensions can be written as $1 + (D - 1)$. With imposition of all these conditions, the action functional for gravity turns out to be,

$$A = \int_{\mathcal{V}} d^D x \sqrt{-g} L \left(g^{ab}, R^a{}_{bcd}\right) \tag{2.34}$$

As we have already mentioned, the Lagrangian depends both on the curvature and the metric but not on the derivatives of the curvature. Given this Lagrangian, one can construct another fourth rank tensor P^{abcd}, obtained by taking derivative of L with respect to curvature tensor R_{abcd}. This tensor, which will turn out to be quiet significant for our later purpose, can be written as

$$P^{abcd} = \left(\frac{\partial L}{\partial R_{abcd}}\right)_{g_{ij}} \tag{2.35}$$

having all the algebraic symmetry properties of the curvature tensor. Note that for Einstein-Hilbert action the tensor P^{abcd} exactly coincides with the tensor Q^{abcd} defined in Eq. (2.2). An analogue of the Ricci tensor in general relativity can also be constructed by the definition

$$\mathcal{R}^{ab} \equiv P^{aijk} R^b{}_{ijk}. \tag{2.36}$$

It is a straightforward exercise to show that for Einstein-Hilbert action $\mathcal{R}_{ab}$ is indeed the Ricci tensor. The field equations for this gravity theory can be obtained by varying the action in Eq. (2.34) with respect to the metric g^{ab}, leading to the following result:

$$\begin{aligned}\delta A &= \delta \int_{\mathcal{V}} d^D x \sqrt{-g} L \\ &= \int_{\mathcal{V}} d^D x \sqrt{-g} E_{ab} \delta g^{ab} + \int_{\mathcal{V}} d^D x \sqrt{-g} \nabla_j \delta v^j\end{aligned} \tag{2.37}$$

Alike the case for Einstein-Hilbert action one obtains a bulk variation term, leading to field equations and a surface variation term. The term E_{ab} in the bulk variation is the field equation term and δv^a corresponds to the boundary term. They are given by the following expressions [1]

$$\begin{aligned}E_{ab} &\equiv \frac{1}{\sqrt{-g}} \left(\frac{\partial \sqrt{-g} L}{\partial g^{ab}}\right)_{R_{abcd}} - 2\nabla^m \nabla^n \left(\frac{\partial L}{R^{amnb}}\right) \\ &= \mathcal{R}_{ab} - \frac{1}{2} g_{ab} L - 2\nabla^m \nabla^n P_{amnb}\end{aligned} \tag{2.38}$$

$$\delta v^j = 2P^{ibjd} \nabla_b \delta g_{di} - 2\delta g_{di} \nabla_c P^{ijcd} \tag{2.39}$$

Note that for a general Lagrangian, P^{abcd}, being derivative of the Lagrangian with respect to R_{abcd} involves second order derivative of the metric through curvature tensor. Hence the term $\nabla^m \nabla^n P_{amnb}$ in the field equations contains fourth order derivatives of the metric. It is well known that field equations having more than second order derivatives of the metric (to be precise, time derivative) in general is plagued with various instabilities, e.g., Ostrogadski instability, existence of tachyonic modes etc.

2.3.1.2 The Lanczos-Lovelock Lagrangian

So far we have been discussing the most general Lagrangian constructed out of metric and curvature. However as we have just witnessed such Lagrangians in general have field equations, which involve fourth order derivatives of the metric leading to instabilities. To avoid all these instabilities and to obtain a physically well motivated theory we will work with *only* those gravitational theories, for which field equations contain at *most* second derivatives of the metric.

Therefore, we need to impose certain restrictions on the form of the Lagrangian, presented in Eq. (2.34) in order to yield second order field equations. From the general field equations in Eq. (2.38) to obtain second order field equations, we must impose an extra condition on P^{abcd}, such that

$$\nabla_a P^{abcd} = 0. \tag{2.40}$$

Thus the problem of finding an action functional leading to a second order field equation reduces to finding scalar functions of curvature and metric such that Eq. (2.40) is satisfied. Such a scalar indeed exists and is unique; it is given [1, 10, 18, 29, 30] by the Lanczos-Lovelock Lagrangian in D dimensions, as

$$L = \sum_m c_m L_m = \sum_m \frac{c_m}{m} \frac{\partial L_m}{\partial R_{abcd}} R_{abcd} = \sum_m \frac{c_m}{m} P^{abcd}_{(m)} R_{abcd} \tag{2.41}$$

This Lagrangian L_m being a homogeneous function of R_{abcd} of order m can also be written as $L_m = Q^{abcd}_{(m)} R_{abcd}$, which can be used to identify $P^{abcd}_{(m)} = m Q^{abcd}_{(m)}$. This relation can also be obtained using explicit expression for P^{ab}_{cd} in terms of the curvature tensor and L_m as

$$P^{ab}_{cd\ (m)} = \frac{\partial L_m}{\partial R^{cd}_{ab}} = m \delta^{aba_2b_2\ldots a_m b_m}_{cdc_2d_2\ldots c_m d_m} R^{c_2d_2}_{a_2b_2} \ldots R^{c_m d_m}_{a_m b_m} \equiv m Q^{ab}_{cd\ (m)} \tag{2.42}$$

This relation will be used extensively later. Like the Einstein-Hilbert Lagrangian, one can decompose the Lanczos-Lovelock Lagrangian in a quadratic part and a surface contribution. This decomposition can be written as,

$$\begin{aligned} \sqrt{-g}L &= \sqrt{-g}Q_a^{\ bcd} R^a_{\ bcd} \\ &= 2\sqrt{-g}Q_a^{\ bcd}\Gamma^a_{dk}\Gamma^k_{bc} + \partial_c\left[2\sqrt{-g}Q_a^{\ bcd}\Gamma^a_{bd}\right] \qquad (2.43) \\ &= \sqrt{-g}L_{\text{quad}} + \sqrt{-g}L_{\text{sur}} \qquad (2.44) \end{aligned}$$

where for m th order Lanczos-Lovelock Lagrangian $Q_a^{\ bcd} = Q_{a\ (m)}^{\ bcd}$, while for general Lagrangian $Q_a^{\ bcd} = \sum c_m Q_{a\ (m)}^{\ bcd}$. Further note that due to complete antisymmetry of the determinant tensor in a D dimensional space-time we have the following restriction $2m \leq D$, otherwise the determinant tensor would vanish identically. Lanczos-Lovelock models at dimension $D = 2m$ are known as critical dimensions for a given Lanczos-Lovelock term. In these critical dimensions the variation of the action functional reduces to a pure surface term [31, 32]. Further, the tensor $\mathcal{R}_{ab}$ defined in Eq. (2.36) in the context of Lanczos-Lovelock gravity is indeed symmetric; but the result is nontrivial to prove (for this result and other properties of these tensors, see [33]). So far we have been dealing with geometrical properties of the Lovelock action, but for the rest of the thesis some thermodynamic constructs would be useful. We will now briefly introduce these thermodynamic ingredients for Lanczos-Lovelock gravity.

2.3.2 Noether Current and Entropy for Lanczos-Lovelock Gravity

The Noether current for Lanczos-Lovelock gravity can also be derived manipulating several differential geometric identity. It coincides with the one derived from diffeomorphism invariance of the action and has the following expression [1, 29, 33, 34]:

$$J^a(\xi) \equiv \left(2E^{ab}\xi_b + L\xi^a + \delta_\xi v^a\right) \tag{2.45}$$

From the property of the Noether current $\nabla_a J^a = 0$, we can define an antisymmetric tensor referred to as *Noether Potential* by the condition, $J^a = \nabla_b J^{ab}$. Using Eq. (2.38) we can substitute for the boundary term, leading to an explicit form for both the Noether current and Potential. These general expressions can be found in [1]. In the context of Lanczos-Lovelock theories, where $\nabla_a P^{abcd} = 0$, they are given by

$$J^{ab} = 2P^{abcd}\nabla_c\xi_d \tag{2.46}$$

$$J^a = 2P^{abcd}\nabla_b\nabla_c\xi_d = 2\mathcal{R}^{ab}\xi_b + 2P_k^{\ ija}\pounds_\xi\Gamma^k_{ij} \tag{2.47}$$

with Γ^a_{bc} being the metric compatible connection.

The Noether current has a direct thermodynamic interpretation as well. For Lanczos-Lovelock theories of gravity, assuming the zeroth law, one can derive the first law [35], from which one can identify the entropy associated with Lanczos-Lovelock gravity. Also the Noether charge computed from the Noether current corresponding to diffeomorphism symmetry will play the role of entropy when evaluated on null surfaces. Further this entropy is refereed to as the Wald entropy associated with null surfaces in all Lanczos-Lovelock models. The corresponding entropy density (which, integrated over the horizon gives the entropy) is given by [35–44]

$$s = -\frac{1}{8}\sqrt{q}P^{abcd}\mu_{ab}\mu_{cd} \tag{2.48}$$

where $\mu_{ab} = u_a r_b - u_b r_a$ stands for the bi-normal, with the following normalization conditions, $u_a u^a = -1$ and $r_a r^a = 1$. Since the tensor P^{abcd} appears directly in the expression for Wald entropy, it is sometimes called the entropy tensor as well. The best way to get an intuitive feel for this result is to consider Einstein-Hilbert action, which can be obtained by considering $P^{abcd} = (1/2)(g^{ac}g^{bd} - g^{ad}g^{bc})$. For this case one will immediately obtain the entropy density to be $s = \sqrt{q}/4$, the standard Bekenstein entropy. We will use this expression for entropy density quiet frequently in later chapters.

2.4 Construction of Gaussian Null Coordinates

We will now elaborate on the construction of a suitable system of coordinates associated with an arbitrary null surface. This construction will be used extensively in later chapters. Let us consider the four-dimensional spacetime $V^4 = M^3 \times R$, where M^3 is a compact three-dimensional manifold. We will consider spacetimes to be time orientable with null embedded hypersurfaces, which are diffeomorphic to M^3 with closed null generators. We take $\mathcal{N}$ to be such a null hypersurface with null generator [45, 46]. On this null surface $\mathcal{N}$ we can introduce spacelike two surface with coordinates (x_1, x_2) defined on them. The null geodesics generating the null hypersurface $\mathcal{N}$ goes out of this spacelike two surface. Thus we can use these null generators to define coordinates on the null hypersurface. The intersection point of these null geodesics with the spacelike two surface can be determined by the coordinates (x_1, x_2), which then evolves along the null geodesic, which is parametrized by u, and label each point on the null hypersurface as (u, x_1, x_2). The above system of coordinates readily identify three basis vectors: (a) the tangent to the null geodesics, $\boldsymbol{\ell} = \partial/\partial u$, and (b) basis vectors on the two surface $\mathbf{e}_A = \partial/\partial_A$.

Having fixed the coordinates on the null surface $\mathcal{N}$ we now move out of the surface using another set of null generators with tangent k^a satisfying the following constraints: (a) $k_a k^a = 0$, (b) $e^a_A k_a = 0$ and finally $\ell_a k^a = -1$. These null geodesics are taken to intersect the null surface at coordinates (u, x_1, x_2) and then move out with affine parameter r, such that any point in the neighborhood of the null surface can be characterized by four coordinates (u, r, x_1, x_2). In this coordinate system, the null surface is given by the condition $r = 0$. This defines a coordinate system $\{u, r, x^A\}$ over the global manifold V^4. This system of coordinates are formed in a manner analogous with Gaussian Normal Coordinate and hence is referred to as *Gaussian Null Coordinates* (*GNC*).

Having set the full coordinate map near the null surface we now proceed to determine the metric elements in that region. Note that $\ell_a \ell^a = 0$ leads to $g_{uu} = 0$ on the null surface since $\boldsymbol{\ell} = \partial/\partial u$. We also note that the basis vectors e^a_A have to lie on the null surface implying $\ell_a e^a_A = 0$ on $\mathcal{N}$, which leads to $g_{uA} = 0$. Also the metric on the two-surface is given by $g_{AB} = e^a_A e^b_B g_{ab}$, which we denote by q_{AB}. We also need the criteria that q_{AB} is positive definite with finite determinant ensuring invertibility and non-degeneracy of the two metric. Thus the following metric components gets fixed to be:

$$\begin{aligned} g_{uu}|_{r=0} = g_{uA}|_{r=0} = 0; \\ g_{AB} = q_{AB} \end{aligned} \tag{2.49}$$

Let us now proceed to determine the other components of the metric. For this, we will use the vector $\mathbf{k} = -\partial/\partial r$ such that from $k^a k_a = 0$ we get $g_{rr} = 0$ throughout the spacetime manifold. Also from the criteria that the null geodesics are affinely parametrized by r we readily obtain $\partial_r g_{r\alpha} = 0$, where $\alpha = (u, x_1, x_2)$. Again, from the conditions $\ell^a k_a = -1$, we readily get $g_{ru} = 1$ and from $k_a e^a_A = 0$ we get $g_{rA} = 0$. From the criteria derived earlier showing $\partial_r g_{r\alpha} = 0$ we can conclude that the above two-metric coefficients are valid everywhere. Thus within the global region V^4 we can have smooth functions α and β_A such that, $\alpha \mid_{r=0}= (\partial g_{uu}/\partial r) \mid_{r=0}$ and $\beta_A \mid_{r=0}= (\partial g_{uA}/\partial r) \mid_{r=0}$. With these two identifications we have the following expression for the line element as

$$ds^2 = g_{ab}dx^a dx^b = -2r\alpha du^2 + 2du dr - 2r\beta_A du dx^A + q_{AB}dx^A dx^B \tag{2.50}$$

where q_{AB} is the two-dimensional metric representing the metric on the null surface. Note that the construction presented above is completely general; it can be applied in the neighborhood of any null hypersurface and, in particular, to the event horizon of a black hole and can also be generalized to D spacetime dimensions, with q_{AB} being the $(D-2)$-dimensional metric.

We will now introduce the time development vector ξ^a appropriate for this coordinate system as the one with the components $\xi^a = \delta^a_0$ in the GNC; that is:

$$\xi^a = (1, 0, 0, 0)\,; \qquad \xi_a = (-2r\alpha, 1, -r\beta_A) \tag{2.51}$$

It can be easily shown that ξ^a will be identical to the timelike Killing vector corresponding to the Rindler time coordinate if we rewrite the standard Rindler metric in the GNC form. Therefore, we can think of ξ^a as a natural generalization of the time development vector corresponding to the Rindler-like observers in the GNC; of course, it will not be a Killing vector in general. Since $\xi^2 = -2r\alpha$, we see that, in the $r \to 0$ limit, ξ^a becomes null. Given ξ^a, we can construct the four-velocity u_a for a comoving observer by dividing ξ^a by its norm $\sqrt{2r\alpha}$ obtaining:

$$u^i = \left(\frac{1}{\sqrt{2r\alpha}}, 0, 0, 0\right); \qquad u_i = \left(-\sqrt{2r\alpha}, \frac{1}{\sqrt{2r\alpha}}, \frac{-r\beta_A}{\sqrt{2r\alpha}}\right) \tag{2.52}$$

The form of u^i shows that the comoving observers can also be thought of as observers with $(r, x^A) =$ constant. This proves to be convenient for probing the properties of the null surface.

This four-velocity, has the four-acceleration $a^i = u^j \nabla_j u^i$. The magnitude of the acceleration $\sqrt{a_i a^i}$, multiplied by the redshift factor $\sqrt{2r\alpha}$, has a finite result in the null limit, (i.e., $r \to 0$ limit):

$$N a|_{r\to 0} = \left(\sqrt{2r\alpha}\right) a|_{r\to 0} = \alpha - \frac{\partial_u \alpha}{2\alpha} \tag{2.53}$$

When the acceleration α varies slowly in time (i.e., $\partial_u \alpha/\alpha^2 \ll 1$) the second term is negligible and $Na \to \alpha$. The redshifted Unruh-Davies [47, 48] temperature associated with the $r = 0$ surface, as measured by $(r, x^A) =$ constant observer is given [23] by Eq. (2.53). We will call this temperature as the "acceleration temperature".

We will next introduce the relevant null vectors associated with the GNC. Given the four-velocity u_a and four-acceleration a_i we can construct two null vectors $\bar{\ell}^i$ and $\bar{k}^i$ as:

$$\bar{\ell}^a = \frac{\sqrt{2r\alpha}}{2}\left(u^a + r^a\right); \qquad \bar{k}^a = \frac{1}{\sqrt{2r\alpha}}\left(u^a - r^a\right) \tag{2.54}$$

where r_i is the unit vector in the direction of the acceleration, i.e., $r_i = a_i/a$. These two vectors $\bar{\ell}^i$ and $\bar{k}^i$ satisfy: $\bar{\ell}^2 = 0$, $\bar{k}^2 = 0$ and $\bar{\ell}^a \bar{k}_a = -1$ and we have the following components on the null surface:

$$\bar{\ell}_i\Big|_{r\to 0} = (0, 1, 0, 0) \tag{2.55}$$

$$\bar{k}_i\Big|_{r\to 0} = \left(-1, \frac{q_{AB}a^A a^B}{4r\alpha a^2}, -\frac{a_A}{\sqrt{2r\alpha}\, a}\right) \tag{2.56}$$

Since we are essentially interested only in the $r \to 0$ limit, it is more convenient to work with a simpler vector field $\ell_i \equiv \nabla_i r$ *everywhere*, which reduces to this $\bar{\ell}_i$ on the null surface and defines the natural null normal to the $r = 0$ surface as a limiting case. Similarly, we can introduce another vector k_a in place of $\bar{k}_a$ to simplify the computations. Using the non-uniqueness in the definition of $\bar{k}_a$, we can change it to another vector k_a such that,

$$\bar{k}_a = k_a + A\ell_a + B_A e^A_a \tag{2.57}$$

where e^A_a are basis vectors on the null surface and $\ell_a = \bar{\ell}_a$. From the property $\ell_a e^a_A = 0$ and $\ell^2 = 0$ we get $\ell_a k^a = -1$, since $\ell_a \bar{k}^a = -1$. The condition $k^2 = 0$, leads to a condition between A and B_A as: $2A = q_{CD}B^C B^D$. Choosing $A = (q_{AB}a^A a^B/4r\alpha a^2)$ and $B_A = -(a_A/\sqrt{2r\alpha}\, a)$ leads to the simple form $k_a = (-1, 0, 0, 0)$. Thus the vector $\ell_a = \nabla_a r$ and the auxiliary $k_a = -\nabla_a u$ have the following components in the GNC:

$$\ell_a = (0, 1, 0, 0)\,, \qquad \ell^a = \left(1, 2r\alpha + r^2\beta^2, r\beta^A\right) \tag{2.58a}$$

$$k_a = (-1, 0, 0, 0)\,, \qquad k^a = (0, -1, 0, 0) \tag{2.58b}$$

Along with these two vectors we also have the vector ξ^a, which is the time development vector introduced earlier. Thus, through the GNC, we have introduced three vectors ℓ_a, k_a and ξ_a. The bi-normal associated with the null surface can be obtained in terms of ℓ_a and k_a as $\epsilon_{ab} = \ell_a k_b - \ell_b k_a$. It turns out that, for the $r = 0$ surface the non-affinity parameter, κ corresponding to the null normal ℓ^a is obtained from $\ell^b \nabla_b \ell^a = \kappa \ell^a$. Evaluation of which for the null normal in Eq. (2.58a) yields the non-affinity parameter to be: $\kappa = \alpha$. While the vector $k_a = -\nabla_a u$, is tangent to the ingoing null geodesic and is affinely parametrized, with affine parameter r. These null vectors will be used extensively as we go along the thesis.

2.5 Looking to the Future

In the previous sections we have discussed in detail, various properties of Einstein-Hilbert and Lanczos-Lovelock gravity. The obvious question to ask at this stage, whether, just like Einstein-Hilbert action, for Lanczos-Lovelock gravity as well, there exist some geometrical variables that can have nice thermodynamic features. We will try to answer this question in the later chapters. We will first introduce a new set of conjugate variables from differential geometric perspective, which can be used to describe Lanczos-Lovelock gravity in the next chapter. After identifying the correct geometric variables we will proceed to understand their thermodynamic importance as well and shall illustrate that variations of these variables are related to either variation of temperature or entropy associated with null surfaces in the subsequent chapters.

Also, we reiterate that in a recent work [23] it has been shown that evolution of spacetime can be interpreted as the difference between surface and bulk degrees of freedom, but in the context of Einstein-Hilbert action. In later chapters we will show that the same works for Lanczos-Lovelock gravity as well. Further, the notion of gravitational momentum have been defined for general relativity, whose thermodynamic significance as well as generalization to Lanczos-Lovelock gravity will be an interesting problem to look upon. All of these discussions will form an integral part of this thesis and will be dealt with greater detail in later chapters.

References

1. T. Padmanabhan, *Gravitation: Foundations and Frontiers* (Cambridge University Press, Cambridge, UK, 2010)
2. M.V. Ostrogradsky, Memoires de lAcademie Imperiale des Science de Saint Petersbourg, **4**, 385 (1850)

3. J. York, Role of conformal three geometry in the dynamics of gravitation. Phys. Rev. Lett. **28**, 1082–1085 (1972)
4. G. Gibbons, S. Hawking, Action integrals and partition functions in quantum gravity. Phys. Rev. D **15**, 2752–2756 (1977)
5. J. Charap, J. Nelson, Surface integrals and the gravitational action. J. Phys. A: Math. Gen. **16** 1661 (1983)
6. C. Lanczos, Z. Phys. **73**, 147 (1932)
7. C. Lanczos, Electricity as a natural property of Riemannian geometry. Rev. Mod. Phys. **39**, 716–736 (1932)
8. C. Lanczos, Z. Phys. **39**, 842 (1938)
9. C. Lanczos, A remarkable property of the Riemann-Christoffel tensor in four dimensions. Annals Math. **39**, 842–850 (1938)
10. D. Lovelock, The Einstein tensor and its generalizations. J. Math. Phys. **12**, 498–501 (1971)
11. K. Parattu, B.R. Majhi, T. Padmanabhan, Structure of the gravitational action and its relation with horizon thermodynamics and emergent gravity paradigm. Phys. Rev. D **87**, 124011 (Jun, 2013). arXiv:gr-qc/1303.1535 [gr-qc], doi:10.1103/PhysRevD.87.124011
12. A. Eddington, *The Mathematical Theory of Relativity*, 2nd edn. (Cambridge University Press, Cambridge, UK, 1924)
13. E. Schrodinger, *Space-Time Structure* (Cambridge Science Classics. Cambridge University Press, Cambridge, UK, 1950)
14. A. Einstein, B. Kaufman, A new form of the general relativistic field equations. Annals Math. **62**, 128–138 (1955)
15. A. Einstein, B. Kaufman, A new form of the general relativistic field equations. Annals Math. **62**, 128–138 (1955), http://www.jstor.org/stable/2007103
16. T. Padmanabhan, Momentum density of spacetime and the gravitational dynamics. arXiv:1506.03814 [gr-qc]
17. T. Padmanabhan, Holographic gravity and the surface term in the Einstein-Hilbert action. Braz. J. Phys. **35**, 362–372 (2005). arXiv:gr-qc/0412068 [gr-qc]
18. A. Mukhopadhyay, T. Padmanabhan, Holography of gravitational action functionals. Phys. Rev. D **74**, 124023 (2006). arXiv:hep-th/0608120
19. S. Kolekar, T. Padmanabhan, Holography in action. Phys. Rev. D **82**, 024036 (2010). arXiv:1005.0619 [gr-qc]
20. C.W. Misner, K.S. Thorne, J.A. Wheeler, *Gravitation*, 3rd edn. (W. H. Freeman and Company, 1973)
21. H. Goldstein, C. Poole, J. Safko, *Classical Mechanics*. 3rd edn. (Pearson Education, 2007)
22. A. Palatini, Deduzione invariantiva delle equazioni gravitazionali dal principio di hamilton. Rend. Circ. Mat. Palermo **43**, 203–212 (1919)
23. T. Padmanabhan, General relativity from a thermodynamic perspective. Gen. Rel. Grav. **46**, 1673 (2014). arXiv:1312.3253 [gr-qc]
24. N. Dadhich, Characterization of the Lovelock gravity by Bianchi derivative. Pramana. **74**, 875–882 (2010). arXiv:0802.3034 [gr-qc]
25. D.G. Boulware, S. Deser, String generated gravity models. Phys. Rev. Lett. **55**, 2656 (1985)
26. L. Randall, R. Sundrum, A Large mass hierarchy from a small extra dimension. Phys. Rev. Lett. **83**, 3370–3373 (1999). arXiv:hep-ph/9905221 [hep-ph]
27. P. Horava, E. Witten, Eleven-dimensional supergravity on a manifold with boundary. Nucl. Phys. B **475**, 94–114 (1996). arXiv:hep-th/9603142 [hep-th]
28. N. Arkani-Hamed, S. Dimopoulos, G. Dvali, The hierarchy problem and new dimensions at a millimeter. Phys. Lett. B **429**, 263–272 (1998). arXiv:hep-ph/9803315 [hep-ph]
29. T. Padmanabhan, D. Kothawala, Lanczos-Lovelock models of gravity. Phys. Rept. **531**, 115–171 (2013). arXiv:1302.2151 [gr-qc]
30. A. Paranjape, S. Sarkar, T. Padmanabhan, Thermodynamic route to field equations in Lancos-Lovelock gravity. Phys. Rev. D **74**, 104015 (2006). arXiv:hep-th/0607240 [hep-th]
31. A. Yale, T. Padmanabhan, Structure of Lanczos-Lovelock Lagrangians in Critical Dimensions. Gen. Rel. Grav. **43**, 1549–1570 (2011). arXiv:1008.5154 [gr-qc]

32. N. Kiriushcheva, S.V. Kuzmin, On Hamiltonian formulation of the Einstein-Hilbert action in two dimensions. Mod. Phys. Lett. A **21**, 899–906 (2006). arXiv:hep-th/0510260 [hep-th]
33. T. Padmanabhan, Some aspects of field equations in generalised theories of gravity. Phys. Rev. D **84**, 124041 (2011). arXiv:1109.3846 [gr-qc]
34. T. Padmanabhan, A physical interpretation of gravitational field equations. AIP Conf. Proc. **1241**, 93–108 (2010). arXiv:0911.1403 [gr-qc]
35. R.M. Wald, Black hole entropy is the Noether charge. Phys. Rev. D **48**, 3427–3431 (1993). arXiv:gr-qc/9307038 [gr-qc]
36. T. Padmanabhan, Thermodynamical aspects of gravity: new insights. Rept. Prog. Phys. **73**, 046901 (2010). arXiv:0911.5004 [gr-qc]
37. B.R. Majhi, T. Padmanabhan, Noether current, horizon virasoro algebra and entropy. Phys. Rev. D **85**, 084040 (2012). arXiv:1111.1809 [gr-qc]
38. V. Iyer, R.M. Wald, Some properties of Noether charge and a proposal for dynamical black hole entropy. Phys. Rev. D **50**, 846–864 (1994). arXiv:gr-qc/9403028 [gr-qc]
39. R.M. Wald, A. Zoupas, A general definition of 'conserved quantities' in general relativity and other theories of gravity. Phys. Rev. D **61**, 084027 (2000). arXiv:gr-qc/9911095 [gr-qc]
40. A. Strominger, C. Vafa, Microscopic origin of the Bekenstein-Hawking entropy. Phys. Lett. B **379**, 99–104 (1996). arXiv:hep-th/9601029 [hep-th]
41. A. Ashtekar, J. Baez, A. Corichi, K. Krasnov, Quantum geometry and black hole entropy. Phys. Rev. Lett. **80**, 904–907 (1998). arXiv:gr-qc/9710007 [gr-qc]
42. J.M. Garcia-Islas, BTZ black hole entropy: a spin foam model description. Class. Quant. Grav. **25**, 245001 (2008). arXiv:0804.2082 [gr-qc]
43. L. Bombelli, R.K. Koul, J. Lee, R.D. Sorkin, A quantum source of entropy for black holes. Phys. Rev. D **34**, 373–383 (1986)
44. B.R. Majhi, T. Padmanabhan, Thermality and heat content of horizons from infinitesimal coordinate transformations. arXiv:1302.1206 [gr-qc]
45. V. Moncrief, J. Isenberg, Symmetries of cosmological cauchy horizons. Commun. Math. Phys. **89**(3) 387–413 (1983). doi:10.1007/BF01214662
46. E.M. Morales, On a second law of black hole mechanics in a higher derivative theory of gravity (2008), http://www.theorie.physik.uni-goettingen.de/forschung/qft/theses/dipl/Morfa-Morales.pdf
47. P. Davies, Scalar particle production in Schwarzschild and Rindler metrics. J. Phys. A **8**, 609–616 (1975)
48. W. Unruh, Notes on black hole evaporation. Phys. Rev. D **14**, 870 (1976)

Chapter 3
Alternative Geometrical Variables in Lanczos-Lovelock Gravity

3.1 Introduction

The deep connection between gravitational dynamics and horizon thermodynamics makes it natural to think of spacetime as some kind of fluid with its thermodynamic properties arising from the dynamics of underlying "atoms of spacetime". This emergent gravity paradigm (see, for a review [1–3]) has received significant amount of support from later investigations as well [4–9]. This leads to two possible routes towards gravitational dynamics—(a) the conventional route, using the action functional and geometrical variables and (b) the thermodynamic route which uses suitably defined degrees of freedom.

In a recent work [10] two canonically conjugate variables

$$f^{ab} = \sqrt{-g}g^{ab}; \qquad N^c_{ab} = Q^{cp}_{aq}\Gamma^q_{bp} + Q^{cp}_{bq}\Gamma^q_{ap}; \qquad Q^{ab}_{cd} = \frac{1}{2}\left(\delta^a_c\delta^b_d - \delta^a_d\delta^b_c\right) \quad (3.1)$$

have been introduced in the context of general relativity and it has been shown that (the variation of) these quantities also possess simple thermodynamical interpretation in terms of $S\delta T$ and $T\delta S$ (see Chap. 2 for a more detailed description). The main purpose of this chapter is to identify one set of such variables in Lanczos-Lovelock gravity which satisfy the following conditions: (a) These variables reduce to the ones used in general relativity in $D = 4$, when the Lanczos-Lovelock model reduces to general relativity. (b) These variables can be used to probe the geometrical structure of Lanczos-Lovelock gravity.

The chapter is organized as follows: In Sect. 3.2 we consider possible generalization of these variables to Lanczos-Lovelock gravity. Then Sect. 3.3 describes gravity in terms of the newly introduced conjugate variables in both Einstein-Hilbert and Lanczos-Lovelock actions. Finally, we conclude with a discussion on our results. Some more details of the main calculations are presented in Appendix A.

S. Chakraborty, *Classical and Quantum Aspects of Gravity in Relation to the Emergent Paradigm*, Springer Theses, DOI 10.1007/978-3-319-63733-4_3

3.2 Possible Generalization of Conjugate Variables to Lanczos-Lovelock Gravity

In Einstein gravity, we have obtained a simpler description using the two conjugate variables, $f^{ab} = \sqrt{-g}g^{ab}$ and $N^c_{ab} = Q^{cd}_{ae}\Gamma^e_{bd} + Q^{cd}_{be}\Gamma^e_{ad}$. The most natural choice for the corresponding variables in Lanczos-Lovelock gravity will be the ones obtained by defining N^a_{bc} by the same relation, viz. by Eq. (2.9), as in general relativity but with Q^{ab}_{cd} given by that for the appropriate Lanczos-Lovelock model, viz. by Eq. (2.42). We will first explore this choice of variables and see whether they satisfy all the requirements.

Even in Lanczos-Lovelock theories, the decomposition of the Lagrangian into a bulk term and a surface term exists and has the identical expression as in Eq. (2.5) with the Q^{ab}_{cd} defined by Eq. (2.42). There are two desirable features we expect the variables to satisfy: First, the surface term should be expressible as, $-\partial(qp)$; second the quadratic part of the Lagrangian must be expressible as, $p\partial q$. Note that q and p are not absolute, since in the Hamiltonian formulation both are given equal weightage, and we can even interchange q and p. In Einstein-Hilbert action these relations are given by Eqs. (2.11) and (2.12) respectively.

With this motivation, let us examine whether these two variables, $\tilde{f}^{ab}$ which is same as the f^{ab} in Einstein-Hilbert action and $\tilde{N}^a_{bc}$ having the same expression as that of N^a_{bc} with Q^{ab}_{cd} corresponding to the one for the Lanczos-Lovelock model, satisfy our criteria. For that purpose we need to evaluate the following combinations, $\tilde{f}^{ab}\partial_c\tilde{N}^c_{ab}$ and $\tilde{N}^a_{bc}\partial_a\tilde{f}^{bc}$. These are evaluated explicitly in Appendix A.2 [see Eqs. (3.45) and (3.46) of Appendix A]. We state here the final results:

$$\sqrt{-g}L_{sur} = -\partial_c\left(\tilde{f}^{pq}\tilde{N}^c_{pq}\right) \tag{3.2}$$

$$\sqrt{-g}L_{quad} = \tilde{N}^c_{ab}\partial_c\tilde{f}^{ab} - \left(2\sqrt{-g}Q_p^{\ bqc}\Gamma^p_{qb}\Gamma^m_{cm} - 2\sqrt{-g}g^{bm}Q^{cq}_{ap}\Gamma^p_{qb}\Gamma^a_{cm}\right) \tag{3.3}$$

$$\sqrt{-g}Q_p^{\ qrs}R^p_{\ qrs} = -\tilde{f}^{ab}\partial_c\tilde{N}^c_{ab} - \left(2\sqrt{-g}Q_p^{\ bqc}\Gamma^p_{qb}\Gamma^m_{cm} - 2\sqrt{-g}g^{bm}Q^{cq}_{ap}\Gamma^p_{qb}\Gamma^a_{cm}\right) \tag{3.4}$$

Thus even though it leads to the proper surface term it does not simplify the other terms unlike in the Einstein-Hilbert case presented in Eqs. (2.11) and (2.12) respectively. Hence we conclude that these variables are not suitable.

The same conclusion can also be reached from the expression for the Noether current as well. As in the Einstein-Hilbert scenario where the last term becomes, $f^{ab}\pounds_\xi N^c_{ab}$, here also we would like the last term to be of the form $\tilde{f}^{pq}\pounds_\xi\tilde{N}^a_{pq}$. However this does not yield the correct Noether current. An extra term, depending on Lie variation of the entropy tensor P^{abcd}, comes into the picture. We have used several identities regarding Lie variation of P^{abcd} but none of these help to simplify the extra term. [Though these identities do not help, they are quiet interesting and have not been derived earlier; hence we present these relations in Appendix A.1]. In summary, these variables, though they are the simplest choice, do not fulfill the criteria we would like them to satisfy.

It is, however, possible to attack the problem from a different angle and obtain another set of variables that satisfies all our criteria. The clue comes from the Noether current itself. We know from Eq. (2.30) that in general relativity, the second term of the Noether current, given by $f^{ab}\pounds_\xi N^c_{ab}$, has the two variables $[f^{ab}, N^c_{ij}]$ on the two sides of $\pounds_\xi$ while in the mth order Lanczos-Lovelock model the corresponding term $P_k^{\ aji}\pounds_\xi\Gamma^k_{\ aj} = mQ_k^{\ aji}\pounds_\xi\Gamma^k_{\ aj}$ (see Eq. (2.47)) has the variables $[\Gamma^k_{\ aj}, Q_k^{\ aji}]$ on the two sides of $\pounds_\xi$. Taking a cue from this, let us define two variables:

$$\Gamma^a_{bc} = \frac{1}{2}g^{ad}\left(-\partial_d g_{bc} + \partial_b g_{cd} + \partial_c g_{bd}\right) \tag{3.5}$$

$$U_a^{\ bcd} = 2\sqrt{-g}\,Q_a^{\ bcd} \tag{3.6}$$

where Γ^a_{bc}, of course, is the standard connection and $Q_a^{\ bcd}$ is given by Eq. (2.42). (The factors are chosen to give the correct limit for general relativity when $m = 1$; we will see that these definitions work.)

Interestingly, these variables can be introduced in a somewhat different manner as well. Suppose we consider the Lanczos-Lovelock model Lagrangian which can be expressed entirely in terms of R^{im}_{kl} [with index placements (2,2)] plus Kronecker deltas. Since $R^{im}_{kl} = g^{mj}R^i_{\ jkl}$ it can also be expressed in terms of the variables $R^i_{\ jkl}$ [with index placements (1,3)] and the metric g^{ab}. On the other hand, $R^i_{\ jkl}$ [with index placements (1,3)] can be written entirely in terms of $\partial_l\Gamma^i_{\ jk}$ and $\Gamma^i_{\ jk}$ without the metric appearing anywhere. Therefore, one can think of the Lanczos-Lovelock Lagrangian, when expressed in terms of $R^i_{\ jkl}$ [with index placements (1,3)] and the metric g^{ab} as a functional of $[g^{ab}, \partial_l\Gamma^i_{\ jk}, \Gamma^i_{\ jk}]$. This suggests defining the "conjugate variable" to the connection:

$$mU_u^{\ vlw} = \frac{\partial\left(\sqrt{-g}L\right)}{\partial\left(\partial_l\Gamma^u_{vw}\right)} \tag{3.7}$$

The equality of the two sides, for $m-$th order Lanczos-Lovelock Lagrangian is easy to verify using Eq. (3.49). It can be proved that this quantity U_{abcd} has all the symmetries of the curvature tensor (essentially because the $\partial_l\Gamma^u_{vw}$ occurs in L only through the curvature tensor.) This suggests yet another meaning to the variables introduced in Eqs. (3.5) and (3.6).

Of course, this set of variables in Eqs. (3.5) and (3.6) works for the Einstein-Hilbert case as well. Now $U_u^{\ vlw}$ can be expressed entirely in terms of the metric and—in this sense—one can think of the metric (or rather the particular combination of metric components) as conjugate to connection (rather than the other way around!) even in the Einstein-Hilbert case. All the original relations for the Einstein-Hilbert action can be written in terms of these two variables instead of f^{ab} and N^a_{bc} as in Eqs. (3.47) and (3.48) matching exactly Eqs. (2.11) and (2.12). The reason has to do with the fact that for any covariant derivative operator $\hat{D}$ we have

$$
\begin{aligned}
f^{pq}\hat{D}N^a_{pq} &= 2\sqrt{-g}g^{pq}\hat{D}\left(Q^{ae}_{pd}\Gamma^d_{eq}\right) \\
&= 2\sqrt{-g}g^{pq}Q^{ae}_{pd}\hat{D}\Gamma^d_{eq} \\
&= 2\sqrt{-g}Q_d^{\ qea}\hat{D}\Gamma^d_{qe} = U_p^{\ qra}\hat{D}\Gamma^p_{qr}.
\end{aligned} \tag{3.8}
$$

This is possible only in the Einstein-Hilbert action where Q^{cd}_{ab} involves only the Kronecker deltas and hence can be pulled through any covariant derivative operator. Thus, in general, the pair $[\Gamma^a_{bc}, U_a^{\ pqr}]$ can also act as the conjugate variables, in Einstein gravity. Since $U_a^{\ pqr}$ can be entirely expressed in terms of the metric in general relativity, they are equivalent to f^{ab} and N^c_{ab} structurally. However in higher order Lanczos-Lovelock theories Q^{ab}_{cd} depends on curvature tensor and hence the above equivalence is broken. As we shall see it is better to work with the two variables as in Eqs. (3.5) and (3.6), as shown in Appendix A.2, Eqs. (3.47) and (3.48).

Finally, note that, by construction, the Noether current becomes in terms of these variables

$$
\begin{aligned}
\sqrt{-g}J^a &= m\left(2U_b^{\ pqr}R^a_{\ pqr}v^b + U_b^{\ cda}\mathcal{L}_\xi\Gamma^b_{cd}\right) \\
&= 2\sqrt{-g}\mathcal{R}^a_b v^b + mU_b^{\ cda}\mathcal{L}_\xi\Gamma^b_{cd}
\end{aligned} \tag{3.9}
$$

This is exactly of the same form as Eq. (2.30), the Noether current for Einstein-Hilbert action.

3.3 Describing Gravity in Terms of Conjugate Variables

In the previous section we have introduced two new variables, which reduce to the corresponding variables of general relativity in appropriate limit. Thus these variables satisfy our first requirement, i.e., they should reproduce the correct general relativistic limit. Having addressed the general relativity limit for these variables we now study whether the gravitational dynamics of Lanczos-Lovelock gravity can be described in terms of these variables.

3.3.1 Einstein-Hilbert Action with the New Set of Variables

It is important to first verify what happens to variational principle of general relativity when we use the two variables introduced in Sect. 3.2. For this purpose we shall start with the Einstein-Hilbert Lagrangian written as:

$$\begin{aligned} L = \sqrt{-g}R &= \sqrt{-g}\frac{1}{2}\left(g^{bd}\delta^c_a - g^{bc}\delta^d_a\right) R^a{}_{bcd} \\ &= \sqrt{-g}\left(g^{bd}\delta^c_a - g^{bc}\delta^d_a\right)\left(\partial_c\Gamma^a_{bd} - \Gamma^a_{dp}\Gamma^p_{bc}\right) \\ &\equiv U_a{}^{bcd}\left(\partial_c\Gamma^a_{bd} - \Gamma^a_{dp}\Gamma^p_{bc}\right) \end{aligned} \tag{3.10}$$

where we have defined the variable, $U_a{}^{bcd} = \sqrt{-g}\left(g^{bd}\delta^c_a - g^{bc}\delta^d_a\right)$ having all the symmetries of curvature tensor. Let us see what happens if we treat the variables $U_a{}^{bcd}$ and Γ^a_{bc} as independent as befitting conjugate variables. The variation of Γ^a_{bc} leads to:

$$\begin{aligned} \delta L\,|_{U_a{}^{bcd}} &= U_a{}^{bcd}\left(\partial_c\delta\Gamma^a_{bd} - \Gamma^p_{bc}\delta\Gamma^a_{dp} - \Gamma^a_{dp}\delta\Gamma^p_{bc}\right) \\ &= U_a{}^{bcd}\nabla_c\delta\Gamma^a_{bd} \end{aligned} \tag{3.11}$$

In order to get to the final expression we have used the fact that the Lagrangian is a scalar density and hence can be evaluated in a local inertial frame with $U_a{}^{bcd}$ and $\delta\Gamma^a_{bc}$ as tensors. (One can explicitly verify this result, even without using this trick.) Then we can rewrite Eq. (3.11) as:

$$\delta L\,|_{U_a{}^{bcd}} = \nabla_c\left(U_a{}^{bcd}\delta\Gamma^a_{bd}\right) - \delta\Gamma^a_{bd}\nabla_c U_a{}^{bcd} \tag{3.12}$$

When the above equation is integrated over the spacetime volume to obtain the variation δL of the action, the first term $\nabla_c\left(U_a{}^{bcd}\delta\Gamma^a_{bd}\right)$, being a total divergence contributes only on the surface and hence can be dropped. Thus the condition $\delta L = 0$ for arbitrary variations $\delta\Gamma^a_{bc}$ leads to:

$$\nabla_c U_a{}^{bcd} = 0 \tag{3.13}$$

From the definition of $U_a{}^{bcd}$ it is evident that Eq. (3.13) implies $\nabla_c g_{ab} = 0$, standard result from Palatini variation in Einstein-Hilbert action [11] [This result can also be obtained using Eq. (3.43) in Appendix A.2].

But we are led to a difficulty in this approach when we vary $U_a{}^{bcd}$ because the variation of $U_a{}^{bcd}$ leads to $R^a{}_{bcd} = 0$, i.e., flat spacetime! It is not possible to get Einstein's equations from *arbitrary* variations of $U_a{}^{bcd}$ since it has four indices, naturally leading to zero curvature tensor, equivalently flat spacetime. The reason for this disaster is simple. When we vary $U_a{}^{bcd}$ we are pretending that we are varying 20 independent components (because $U_a{}^{bcd}$ has the symmetries of curvature tensor); but we know that—since $U_a{}^{bcd}$ is completely determined by g_{ab} it really has only 10 independent components. So, in order to get correct field equations we need to restrict variations such that there are only 10 of them independent components in the variation $\delta U_a{}^{bcd}$. This is easy to achieve. Since an arbitrary symmetric second rank tensor S^p_q has 10 independent components, we can easily construct such a constrained variation by considering a subclass of $\delta U_a{}^{bcd}$ which is determined by the variations δS^p_q of an arbitrary second rank tensor. This leads us to consider variations of the form:

$$\delta U_p^{\ qrs} = \left(U_m^{\ qrs} \delta^n_p - \frac{1}{2} U_p^{\ qrs} \delta^n_m \right) \delta S^m_n \tag{3.14}$$

(In fact, it turns out that the above variation can be slightly generalized by introducing a sixth rank tensor $A_{pm}^{\ \ qrsn}$ satisfying the criteria $A_{pm}^{\ \ qrsn} R^p_{\ qrs} = 0$. We shall not consider these variations any more since they have no effect on the field equations.) With these restricted class of variations we arrive at:

$$\begin{aligned} \delta L \mid_{\Gamma^a_{bc}} &= \frac{1}{2} R^a_{\ bcd} \delta U_a^{\ bcd} \\ &= \frac{1}{2} R^p_{\ qrs} \left(U_m^{\ qrs} \delta^n_p - \frac{1}{2} \delta^n_m U_p^{\ qrs} \right) \delta S^m_n \end{aligned} \tag{3.15}$$

Then for arbitrary variations of the symmetric tensor S^a_b, we get the field equations:

$$R^p_{\ qrs} \left(U_m^{\ qrs} \delta^n_p - \frac{1}{2} \delta^n_m U_p^{\ qrs} \right) = 0 \tag{3.16}$$

To prove the equivalence with Einstein's equations we note that the following relations

$$R^p_{\ qrs} U_m^{\ qrs} = 2\sqrt{-g} R^p_m; \quad R^p_{\ qrs} U_p^{\ qrs} = 2\sqrt{-g} R \tag{3.17}$$

directly transform the field equations Eq. (3.16) to, $G_{ab} = 0$, the source-free Einstein's equations. (We have not included matter fields to our system about which we shall comment later.) In the next section we will show the validity of the above formalism for Lanczos-Lovelock gravity.

3.3.2 *Generalization to Lanczos-Lovelock Gravity*

In the case of Lanczos-Lovelock models the appropriate Lagrangian to consider for our purpose is:

$$L = U_a^{\ bcd} \left(\partial_c \Gamma^a_{bd} - \Gamma^a_{md} \Gamma^m_{bc} \right) \tag{3.18}$$

which, using Eq. (3.6), can be identified with the Lanczos-Lovelock Lagrangian. The variation of the above Lagrangian with respect to Γ^a_{bc} leads to

$$\nabla_c U_a^{\ bcd} = 0 \tag{3.19}$$

as in the Einstein-Hilbert scenario (see Eq. (3.13)). This condition is equivalent to the criteria that in Lanczos-Lovelock gravity $\nabla_c P^{abcd} = 0$. *This result is quiet remarkable*, since the criterion that the field equation should be of second order in the dynamical variable gets into picture automatically from variation of the Lagrangian. In fact, this condition has another aspect to it. In general, when we study the metric formulation we treat the Lagrangian with g_{ab} as the independent variable (with connections given in terms of the metric) while in the Palatini formulation, we treat both the metric and the connections as independent and their variation leads to the relation between them *and* the field equations. For an arbitrary Lagrangian the metric and Palatini variation do not coincide [12]. However if the condition $\nabla_c\left(\partial L/\partial R_{abcd}\right)=0$ is satisfied then both the metric and Palatini formulations coincide. This is identical to the condition presented in Eq. (2.40) and it is interesting to see this condition emerging from a variation here.

Next we need to vary the Lagrangian with respect to $U_a^{\ bcd}$. Arbitrary variation of $U_a^{\ bcd}$ treating all the 20 components independent will lead to trouble, just as in general relativity. Since the Lagrangian in Eq. (3.18) can equivalently be written as $L=(1/2)U_a^{\ bcd}R^a_{\ bcd}$, such that for arbitrary variation of $U_a^{\ bcd}$ we get $R^a_{\ bcd}=0$, i.e. flat spacetime solution—just as in general relativity. In order to get the field equations we need to again consider only a subclass of variations as we did in the Einstein-Hilbert scenario to derive the equations of motion in Eq. (3.16). Here again we need to assume that not all the independent components of $U_a^{\ bcd}$ are contributing to the variation but only 10 degrees of freedom, which can be encoded by a symmetric second rank part with arbitrary variation. This amounts to taking:

$$\delta U_p^{\ qrs}=\left(U_m^{\ qrs}\delta^n_p-\frac{1}{2}U_p^{\ qrs}\delta^n_m\right)\delta S^m_n \tag{3.20}$$

(Here also we can introduce an additional sixth rank tensor as we did after Eq. (3.14). However as far as the field equations are concerned it has no effect and thus will not be considered any more.) With these restricted class of variations the Lagrangian variation leads to:

$$\delta L\mid_{\Gamma^a_{bc}}=\left(mU_a^{\ pqr}\delta^b_s-\frac{1}{2}\delta^b_aU_s^{\ pqr}\right)R^s_{\ pqr}\delta S^a_b \tag{3.21}$$

where δS^a_b is variation of an arbitrary symmetric second rank tensor and the factor m comes from the fact that we are considering mth order Lanczos-Lovelock Lagrangian. When the variation δS^a_b is considered arbitrary the field equations turn out to be

$$\left(mU_a^{\ pqr}\delta^b_s-\frac{1}{2}\delta^b_aU_s^{\ pqr}\right)R^s_{\ pqr}=0. \tag{3.22}$$

To show that the above field equations are indeed identical to the field equations in Lanczos-Lovelock gravity we just use Eq. (3.6) to substitute for $U_a^{\ bcd}$ leading to:

$$\begin{aligned}0&=mQ_a^{\ pqr}R^b_{\ pqr}-\frac{1}{2}\delta^b_aQ_s^{\ pqr}R^s_{\ pqr}\\&=\mathcal{R}^a_b-\frac{1}{2}\delta^a_bL\end{aligned} \tag{3.23}$$

which is the Lanczos-Lovelock field equations. Thus we observe that these two variables satisfy all the criteria that conjugate variables should.

The above result is derived for mth order Lanczos-Lovelock Lagrangian and can be easily generalized to general Lanczos-Lovelock Lagrangian $L = \sum_m c_m L^{(m)}$. Then the above variation of Lanczos-Lovelock Lagrangian leads to the following expression:

$$\begin{aligned}\delta\left(\sqrt{-g}L\right)|_{\Gamma^a_{bc}} &= \sum_m c_m \delta\left(\sqrt{-g}L^{(m)}\right)|_{\Gamma^a_{bc}} \\ &= \left\{\left(\sum_m c_m m U_a^{\ pqr}\right)\delta^b_s - \frac{1}{2}\delta^b_a\left(\sum_m c_m U_s^{\ pqr}\right)\right\} R^s_{\ pqr}\delta S^a_b \end{aligned} \tag{3.24}$$

For arbitrary variation of the symmetric tensor S^a_b the field equations can be obtained as:

$$\left\{\left(\sum_m c_m m U_a^{\ pqr}\right)\delta^b_s - \frac{1}{2}\delta^b_a\left(\sum_m c_m U_s^{\ pqr}\right)\right\} R^s_{\ pqr} = 0 \tag{3.25}$$

Note that with the following relations

$$\sum_m c_m m U_a^{\ pqr} = \frac{\partial\sqrt{-g}L}{\partial R^a_{\ pqr}} = \sqrt{-g}P_a^{\ pqr} \tag{3.26}$$

$$\sum_m c_m U_s^{\ pqr} R^s_{\ pqr} = \sqrt{-g}L \tag{3.27}$$

the above Eq. (3.25) becomes equivalent to

$$\mathcal{R}^b_a - \frac{1}{2}\delta^b_a L = 0 \tag{3.28}$$

which is the most general Lanczos-Lovelock field equations.

In this approach, we arrive at the condition that needs to be imposed in order to get second order field equations, directly from a variational principle along with the field equations themselves. The price we pay is the following: (a) We need to restrict the form of the variations, the physical meaning of which is unclear. (b) The inclusion of matter in this scheme is difficult. Usually, the energy momentum tensor comes from the variation of the matter Lagrangian with respect to the metric alone and since we have not included the metric in our formulation it is not clear how to include matter. These issues require further investigation.

3.4 Concluding Remarks

The link between the standard approach to gravity and the thermodynamical one motivates us to look for geometrical variables in which the expression for action simplifies and which will have direct thermodynamical interpretation. This goal was achieved for the Einstein-Hilbert action recently [10] by introducing canonically conjugate variables as $f^{ab} = \sqrt{-g}g^{ab}$ and the corresponding momenta N^{c}_{ab}. In terms of these variables, the surface term turns out to have the structure $-\partial(qp)$ and their variations have direct thermodynamic interpretation.

We have demonstrated that such variables exist in Lanczos-Lovelock gravity as well. The variables in the case of Lanczos-Lovelock models have the following properties: (a) These variables reduce to the ones used in general relativity in $D = 4$ when the Lanczos-Lovelock model reduces to general relativity. (b) These variables can describe the gravitational dynamics of Lanczos-Lovelock action as well. These variables, by themselves, are interesting and deserves further study. For example, we found that they can be thought of as connections and conjugate momenta associated with connections in a formal sense. However to be of significant importance in describing Lanczos-Lovelock models of gravity, the geometrical properties are not sufficient, the variables (or their variations) need to have thermodynamic interpretations as well. This is what we take up in the next chapter.

Appendix

In this appendix, we shall present the supplementary material for this chapter.

A.1 Identities Regarding Lie Variation of P^{abcd}

In this section we shall derive some identities related to Lie variation of the entropy tensor, P^{abcd}. For that purpose we first consider Lie variation of the Lagrangian treated as a scalar function of the metric g_{ab} and R_{abcd} leading to

$$\begin{aligned}
\pounds_{\xi} L\left(g_{ab}, R_{ijkl}\right) &= \xi^{m}\nabla_{m} L\left(g_{ab}, R_{ijkl}\right) \\
&= \frac{\partial L}{\partial g_{ab}}\xi^{m}\nabla_{m} g_{ab} + \frac{\partial L}{\partial R_{ijkl}}\xi^{m}\nabla_{m} R_{ijkl} \\
&= P^{ijkl}\xi^{m}\nabla_{m} R_{ijkl}
\end{aligned} \tag{3.29}$$

where we have used the fact that covariant derivative of metric tensor vanishes. Then for the Lagrangian which is homogeneous function of degree m we get

$$\pounds_\xi L = \xi^m \nabla_m \left(\frac{1}{m} P^{ijkl} R_{ijkl} \right) \tag{3.30}$$

Then using Eq. (3.29) we readily obtain

$$R_{abcd} \xi^m \nabla_m P^{abcd} = (m-1) P^{abcd} \xi^m \nabla_m R_{abcd} \tag{3.31}$$

We also have the following relation:

$$\begin{aligned} P^{ijkl} \pounds_\xi R_{ijkl} &= P^{ijkl} \Big(\xi^m \nabla_m R_{ijkl} + R_{ajkl} \nabla_i \xi^a + R_{iakl} \nabla_j \xi^a \\ &\quad + R_{ijal} \nabla_k \xi^a + R_{ijka} \nabla_l \xi^a \Big) \\ &= P^{ijkl} \xi^m \nabla_m R_{ijkl} + 4 \nabla_i \xi_m R^i{}_{jkl} P^{mjkl} \\ &= P^{ijkl} \xi^m \nabla_m R_{ijkl} + 4 \nabla_i \xi_m \mathcal{R}^{im} \end{aligned} \tag{3.32}$$

Again we can also write, $m \pounds_\xi L = P^{ijkl} \pounds_\xi R_{ijkl} + R_{ijkl} \pounds_\xi P^{ijkl}$. Then we obtain:

$$\begin{aligned} R_{ijkl} \pounds_\xi P^{ijkl} &= m \pounds_\xi L - P^{ijkl} \pounds_\xi R_{ijkl} \\ &= m \pounds_\xi L - P^{ijkl} \xi^m \nabla_m R_{ijkl} - 4 \nabla_i \xi_m \mathcal{R}^{im} \\ &= (m-1) P^{abcd} \xi^m \nabla_m R_{abcd} - 4 \nabla_i \xi_m \mathcal{R}^{im} \end{aligned} \tag{3.33}$$

This equation can also be casted in a different form as

$$R_{abcd} \left(\pounds_\xi P^{abcd} - \xi^m \nabla_m P^{abcd} \right) = -4 \nabla_i \xi_m \mathcal{R}^{im} \tag{3.34}$$

Now we can rewrite the metric as a function of g_{ab} and $R^a{}_{bcd}$, in which case the Lie variation leads to

$$\pounds_\xi L \left(g_{ij}, R^a{}_{bcd} \right) = P_i{}^{jkl} \xi^m \nabla_m R^i{}_{jkl} \tag{3.35}$$

With the Lagrangian as homogeneous function of curvature tensor to mth order leads to

$$R^a{}_{bcd} \pounds_\xi P_a{}^{bcd} = (m-1) P_i{}^{jkl} \xi^m \nabla_m R^i_{jkl} \tag{3.36}$$

Then we arrive at the following identity

$$P^{abcd} \pounds_\xi \left(g_{am} R^m_{\ bcd} \right) = P_a{}^{bcd} \xi^m \nabla_m R^a{}_{bcd} + 4 \nabla_i \xi_m \mathcal{R}^{im} \tag{3.37}$$

or

$$\begin{aligned} P_a{}^{bcd} \pounds_\xi R^a{}_{bcd} &= P_a{}^{bcd} \xi^m \nabla_m R^a{}_{bcd} + 4 \nabla_i \xi_m \mathcal{R}^{im} - \mathcal{R}^{am} \mathcal{L}_\xi g_{am} \\ &= P_a{}^{bcd} \xi^m \nabla_m R^a{}_{bcd} + 2 \nabla_i \xi_m \mathcal{R}^{im} \end{aligned} \tag{3.38}$$

This leads to the following relation:

$$R^{a}{}_{bcd}\left(£_{\xi}P_{a}{}^{bcd}-\xi^{m}\nabla_{m}P_{a}{}^{bcd}\right)=-2\nabla_{i}\xi_{m}\mathcal{R}^{im} \tag{3.39}$$

If we proceed along the same lines we readily obtain another such relation given as:

$$R^{ij}_{kl}\left(£_{\xi}P^{kl}_{ij}-\xi^{m}\nabla_{m}P^{kl}_{ij}\right)=0 \tag{3.40}$$

These relations illustrate the Lie variation of P^{abcd} when contracted with the curvature tensor.

A.2 Derivation of Various Identities Used in Text

We will consider the $p\partial q$ and $q\partial p$ structure arising from the identification of $\tilde{f}^{ab}$ as coordinate and $\tilde{N}^{c}_{ab}$ as momentum in Lanczos-Lovelock gravity. For the calculation, the following identity will be used here and there:

$$\begin{aligned}0=\nabla_{c}Q^{cd}_{ab}&=\partial_{c}Q^{cd}_{ab}+\Gamma^{c}_{ck}Q^{kd}_{ab}\\&\quad-\Gamma^{k}_{ca}Q^{cd}_{kb}-\Gamma^{k}_{cb}Q^{cd}_{ak}+\Gamma^{d}_{ck}Q^{ck}_{ab}\end{aligned} \tag{3.41}$$

However Q^{cd}_{ab} being antisymmetric in (c,d) while Γ^{c}_{ab} being symmetric in (a,b) the last term in the above expansion vanishes. Thus ordinary derivative of the quantity Q^{cd}_{ab} has the following expression

$$\partial_{c}Q^{cd}_{ab}=-\Gamma^{c}_{ck}Q^{kd}_{ab}+\Gamma^{k}_{ca}Q^{cd}_{kb}+\Gamma^{k}_{cb}Q^{cd}_{ak} \tag{3.42}$$

Note that we can include $\sqrt{-g}$ in the above expression leading to:

$$\partial_{c}\left(\sqrt{-g}Q_{a}{}^{bcd}\right)=\left(\sqrt{-g}Q_{p}{}^{bcd}\right)\Gamma^{p}_{ac}-\left(\sqrt{-g}Q_{a}^{pcd}\right)\Gamma^{b}_{cp} \tag{3.43}$$

Thus we get the following expression from Eq. (3.42):

$$\begin{aligned}\partial_{c}\tilde{N}^{c}_{ab}&=\partial_{c}\left[Q^{cq}_{bp}\Gamma^{p}_{aq}+Q^{cq}_{ap}\Gamma^{p}_{bq}\right]\\&=\left(\partial_{c}Q^{cq}_{bp}\right)\Gamma^{p}_{aq}+Q^{cq}_{bp}\partial_{c}\Gamma^{p}_{aq}+\left(\partial_{c}Q^{cq}_{ap}\right)\Gamma^{p}_{bq}+Q^{cq}_{ap}\partial_{c}\Gamma^{p}_{bq}\\&=Q^{cq}_{kp}\Gamma^{k}_{cb}\Gamma^{p}_{aq}+Q^{cq}_{bk}\Gamma^{k}_{cp}\Gamma^{p}_{aq}-Q^{kq}_{bp}\Gamma^{p}_{aq}\Gamma^{c}_{ck}+Q^{cq}_{kp}\Gamma^{k}_{ca}\Gamma^{p}_{bq}\\&\quad+Q^{cq}_{ak}\Gamma^{k}_{cp}\Gamma^{p}_{bq}-Q^{kq}_{ap}\Gamma^{c}_{ck}\Gamma^{p}_{bq}+Q^{cq}_{bp}\partial_{c}\Gamma^{p}_{aq}+Q^{cq}_{ap}\partial_{c}\Gamma^{p}_{bq}\end{aligned} \tag{3.44}$$

where in order to arrive at the last equality Eq. (3.42) has been used. Now contracting the above expression with $\tilde{f}^{ab}$ we readily obtain

$$
\begin{aligned}
\tilde{f}^{ab}\partial_c \tilde{N}^c_{ab} &= \sqrt{-g}g^{ab}\left[2Q^{cq}_{ap}\partial_c \Gamma^p_{bq} + 2Q^{cq}_{kp}\Gamma^k_{ca}\Gamma^p_{bq} + 2Q^{cq}_{ak}\Gamma^k_{cp}\Gamma^p_{bq} - 2Q^{kq}_{ap}\Gamma^c_{ck}\Gamma^p_{bq}\right] \\
&= -2\sqrt{-g}Q_p^{\ bcq}\left(\partial_c \Gamma^p_{bq} + \Gamma^p_{ck}\Gamma^k_{bq}\right) \\
&+ 2\sqrt{-g}Q_p^{\ bkq}\Gamma^c_{ck}\Gamma^p_{bq} + 2\sqrt{-g}g^{ab}Q^{cq}_{kp}\Gamma^k_{ca}\Gamma^p_{bq} \\
&= -\sqrt{-g}Q_p^{\ bqc}R^p_{\ bqc} + 2\sqrt{-g}Q_p^{\ bkq}\Gamma^c_{ck}\Gamma^p_{bq} + 2\sqrt{-g}g^{ab}Q^{cq}_{kp}\Gamma^k_{ca}\Gamma^p_{bq} \quad (3.45)
\end{aligned}
$$

Note that in the Einstein-Hilbert limit the last two terms adds up to yield $-\sqrt{-g}L_{quad}$. Then consider the other combination which can be expressed as

$$
\begin{aligned}
\tilde{N}^c_{ab}\partial_c \tilde{f}^{ab} &= \left(Q^{cq}_{ap}\Gamma^p_{qb} + Q^{cq}_{bp}\Gamma^p_{qa}\right)\partial_c\left(\sqrt{-g}g^{ab}\right) \\
&= \sqrt{-g}\left(Q^{cq}_{ap}\Gamma^p_{qb} + Q^{cq}_{bp}\Gamma^p_{qa}\right)\left(\partial_c g^{ab} + g^{ab}\Gamma^p_{cp}\right) \\
&= 2\sqrt{-g}Q^{cq}_{ap}\Gamma^p_{qb}\partial_c g^{ab} + 2\sqrt{-g}Q_p^{\ bqc}\Gamma^p_{qb}\Gamma^m_{cm} \\
&= 2\sqrt{-g}Q_p^{\ bcq}\Gamma^l_{bc}\Gamma^p_{ql} + 2\sqrt{-g}Q_p^{\ bqc}\Gamma^p_{qb}\Gamma^m_{cm} - 2\sqrt{-g}g^{bm}Q^{cq}_{ap}\Gamma^p_{qb}\Gamma^a_{cm} \\
&= \sqrt{-g}L_{quad} + 2\sqrt{-g}Q_p^{\ bqc}\Gamma^p_{qb}\Gamma^m_{cm} - 2\sqrt{-g}g^{bm}Q^{cq}_{ap}\Gamma^p_{qb}\Gamma^a_{cm}. \quad (3.46)
\end{aligned}
$$

In the Einstein-Hilbert limit the above term leads to $2\sqrt{-g}L_{quad}$. Next we will derive similar relations which actually behaves as conjugate variables, with the identification, $p \equiv 2\sqrt{-g}Q_a^{\ bcd}$ and $q \equiv \Gamma^a_{bc}$. Then the respective $p\partial q$ and $q\partial p$ expressions are given in the following results:

$$
\begin{aligned}
2\sqrt{-g}Q_e^{\ bdc}\partial_c \Gamma^e_{bd} &= \sqrt{-g}Q_e^{\ bdc}\left(\partial_c \Gamma^e_{bd} - \partial_d \Gamma^e_{bc}\right) \\
&= \sqrt{-g}Q_e^{\ bdc}R^e_{\ bcd} - 2\sqrt{-g}Q_e^{\ bdc}\Gamma^e_{mc}\Gamma^m_{bd} \\
&= -\sqrt{-g}Q_e^{\ abc}R^e_{\ abc} - \sqrt{-g}L_{quad} \quad (3.47)
\end{aligned}
$$

and

$$
\begin{aligned}
\Gamma^d_{be}\partial_c\left(2\sqrt{-g}Q_d^{\ bec}\right) &= 2\sqrt{-g}\Gamma^d_{be}\partial_c Q_d^{\ bec} + 2\Gamma^d_{be}Q_d^{\ bec}\partial_c\sqrt{-g} \\
&= 2\sqrt{-g}\Gamma^d_{be}\left(\Gamma^a_{cd}Q_a^{\ bec} - \Gamma^b_{ca}Q_d^{\ aec} - \Gamma^c_{ca}Q_d^{\ bea}\right) \\
&+ 2\Gamma^d_{be}Q_d^{\ bec}\partial_c\sqrt{-g} \\
&= 2\sqrt{-g}L_{quad} \quad (3.48)
\end{aligned}
$$

Finally we will explicitly demonstrate the result of one derivative used in the text for Lanczos-Lovelock Lagrangian:

$$
\begin{aligned}
\frac{\partial\left(\sqrt{-g}L\right)}{\partial\left(\partial_l \Gamma^u_{vw}\right)} &= m\sqrt{-g}\delta^{aba_2b_2\ldots a_mb_m}_{cdc_2d_2\ldots c_md_m}\frac{\partial R^{cd}_{ab}}{\partial\left(\partial_l\Gamma^u_{vw}\right)}R^{c_2d_2}_{a_2b_2}\ldots R^{c_md_m}_{a_mb_m} \\
&= m\sqrt{-g}\delta^{aba_2b_2\ldots a_mb_m}_{cdc_2d_2\ldots c_md_m}\left[g^{dp}\delta^c_u\delta^v_p\left(\delta^l_a\delta^w_b - \delta^l_b\delta^w_a\right)\right]R^{c_2d_2}_{a_2b_2}\ldots R^{c_md_m}_{a_mb_m} \\
&= 2m\sqrt{-g}g^{dv}Q^{lw}_{ud} = mU_u^{\ vlw} \quad (3.49)
\end{aligned}
$$

References

1. T. Padmanabhan, Gravity and the thermodynamics of horizons. Phys. Rep. **406**, 49–125 (2005). arXiv:gr-qc/0311036 [gr-qc]
2. R.M. Wald, The thermodynamics of black holes. Living Rev. Relativ. **4**, 6 (2001). arXiv:gr-qc/9912119 [gr-qc]
3. T. Padmanabhan, Thermodynamical aspects of gravity: new insights. Rep. Prog. Phys. **73**, 046901 (2010). arXiv:0911.5004 [gr-qc]
4. T. Padmanabhan, Classical and quantum thermodynamics of horizons in spherically symmetric space-times. Class. Quantum Gravity **19**, 5387–5408 (2002). arXiv:gr-qc/0204019 [gr-qc]
5. R.-G. Cai, S.P. Kim, First law of thermodynamics and Friedmann equations of Friedmann-Robertson-Walker universe. JHEP **0502**, 050 (2005). arXiv:hep-th/0501055 [hep-th]
6. A. Paranjape, S. Sarkar, T. Padmanabhan, Thermodynamic route to field equations in Lancos-Lovelock gravity. Phys. Rev. D **74**, 104015 (2006). arXiv:hep-th/0607240 [hep-th]
7. T. Padmanabhan, A. Paranjape, Entropy of null surfaces and dynamics of spacetime. Phys. Rev. D **75**, 064004 (2007). arXiv:gr-qc/0701003 [gr-qc]
8. S. Kolekar, T. Padmanabhan, Action principle for the fluid-gravity correspondence and emergent gravity. Phys. Rev. D **85**, 024004 (2012). arXiv:1109.5353 [gr-qc]
9. T. Padmanabhan, General relativity from a thermodynamic perspective. Gen. Relativ. Gravit. **46**, 1673 (2014). arXiv:1312.3253 [gr-qc]
10. K. Parattu, B.R. Majhi, T. Padmanabhan, Structure of the gravitational action and its relation with horizon thermodynamics and emergent gravity paradigm. Phys. Rev. D **87**, 124011 (2013). arXiv:gr-qc/1303.1535 [gr-qc], doi:10.1103/PhysRevD.87.124011
11. T. Padmanabhan, *Gravitation: Foundations and Frontiers* (Cambridge University Press, Cambridge, UK, 2010)
12. Q. Exirifard, M.M. Sheikh-Jabbari, Lovelock gravity at the crossroads of Palatini and metric formulations. Phys. Lett. B **661**, 158–161 (2008). arXiv:0705.1879 [hep-th]

Part III
Thermodynamics, Gravity and Null Surfaces

Chapter 4
Thermodynamic Interpretation of Geometrical Variables

4.1 Introduction

We have two possible routes towards explaining the gravitational dynamics. One is the conventional route, using the action functional and geometrical variables. The other is the thermodynamic route which uses suitably defined degrees of freedom, heat content etc. The link between these two routes, obviously, is provided by the action functional itself, which—as is known from previous investigations [1–7]—has both dynamical and thermodynamical interpretation. This is a peculiar aspect of gravitational action functionals, not shared by other theories which possess no thermodynamic or emergent interpretation and hence is worth probing deeply.

A first step in this direction was taken in [8] by introducing two new canonically conjugate variables, as described in Chap. 2 in the context of general relativity. These quantities possess simple thermodynamical interpretation in terms of $S\delta T$ and $T\delta S$. In Chap. 3 we have identified two such geometrical variables in Lanczos-Lovelock gravity. These variables reduce to the ones used in general relativity in $D = 4$ when the Lanczos-Lovelock model reduces to general relativity and they describe the geometrical aspects of Lanczos-Lovelock gravity correctly. The remaining feature that needs to be shown corresponds to—the variation of these quantities should correspond to $S\delta T$ and $T\delta S$ respectively where S is now the correct Wald entropy of the Lanczos-Lovelock model. In this chapter we shall show that this can indeed be done.

The chapter is organized as follows: Sect. 4.2 presents thermodynamic quantities for a general static spacetime and then it is generalized to arbitrary null surface constructed in the context of Lanczos-Lovelock gravity. This explicitly shows the thermodynamic nature of these geometrical variables. We finally conclude with a discussion on our results.

S. Chakraborty, *Classical and Quantum Aspects of Gravity in Relation to the Emergent Paradigm*, Springer Theses, DOI 10.1007/978-3-319-63733-4_4

4.2 Thermodynamics Related to Lanczos-Lovelock Action

In this section, we will show that the variables Γ^c_{ab} and $U_a^{\ bcd}$ introduced in Chap. 3 are closely related to the thermodynamic properties of null surfaces. The variations $\Gamma^a_{pq}\delta U_a^{\ pqr}$ and $U_a^{\ pqr}\delta\Gamma^a_{pq}$ obtained from the conjugate variables have direct thermodynamic interpretation associated with them. This result was obtained earlier in [8] in the context of general relativity, but here we shall describe the corresponding results for the Lanczos-Lovelock models. We shall first illustrate the results for a general static spacetime and then generalize these results to the arbitrary null surface constructed in Sect. 2.4.

The surface term in Lanczos-Lovelock Lagrangian, discussed in Chap. 2, can be written as:

$$A_{\text{sur}} = -\int d^D x \partial_c \left(2\sqrt{-g}\, Q_a^{\ bdc}\Gamma^a_{bd}\right) \equiv -\int d^D x \partial_c \left(\sqrt{-g}\, V^c\right) \tag{4.1}$$

Then under infinitesimal variation, the surface term variation can be subdivided into two parts

$$\delta\left(\sqrt{-g}\, V^c\right) = U_a^{\ bdc}\delta\Gamma^a_{bd} + \Gamma^a_{bd}\delta U_a^{\ bdc} \tag{4.2}$$

where one term involves variation of connections, while the other one is quiet complex and involves variation of both the metric and the entropy tensor. Hence the variation of the surface term can be written by introducing the surface Hamiltonian $H_{\text{sur}} = (-\partial A_{\text{sur}}/\partial t)$ as follows (introducing the 16π factor again):

$$\begin{aligned}
\delta H_{\text{sur}} &= \frac{1}{16\pi}\int d^{D-2}x\; n_c \delta\left(\sqrt{-g}\, V^c\right) \\
&= \frac{1}{16\pi}\int d^{D-2}x\; n_c U_a^{\ bdc}\delta\Gamma^a_{bd} + \frac{1}{16\pi}\int d^{D-2}x\; n_c \Gamma^a_{bd}\delta U_a^{\ bdc} \\
&= \delta H^{(1)}_{\text{sur}} + \delta H^{(2)}_{\text{sur}}
\end{aligned} \tag{4.3}$$

Also the variation of the connection due to infinitesimal change of metric $g_{ab} \to g_{ab} + h_{ab}$ corresponds to,

$$\delta\Gamma^p_{qr} = \frac{1}{2}h^{pa}\left(-\partial_a g_{qr} + \partial_q g_{ar} + \partial_r g_{aq}\right) + \frac{1}{2}g^{pa}\left(-\partial_a h_{qr} + \partial_q h_{ar} + \partial_r h_{aq}\right) \tag{4.4}$$

Using these results we can calculate the variation explicitly for different metrics. We will show that, the first term in Eq. (4.3) leads to $s\delta T$ while the second term leads to $T\delta s$, where s is the Wald entropy density.

4.2.1 A General Static Spacetime

We will first prove these relations for a general static spacetime and shall subsequently discuss them in the context of arbitrary null surface constructed in Chap. 2. An arbitrary static D-dimensional spacetime with horizon can be described by the following line element [9],

$$ds^2 = -N^2dt^2 + dn^2 + \sigma_{AB}dy^A dy^B \tag{4.5}$$

In the above line element, n represents spatial direction, normal to the $(D-2)$-dimensional hypersurface with σ_{AB} being transverse metric on the surface. Let $\boldsymbol{\xi} = \partial/\partial t$ be a timelike Killing vector field, with Killing horizon being located at, $N^2 \rightarrow 0$. The coordinate system is chosen in such a way that $n = 0$ on the Killing horizon. Then, in the near horizon regime, the following expansions of N and σ_{AB} are valid [9]:

$$\begin{aligned} N &= \kappa(x^A)n + \mathcal{O}(n^3) \\ \sigma_{AB} &= [\sigma_H(y)]_{AB} + \frac{1}{2}[\sigma_2(y)]_{AB}\, n^2 + \mathcal{O}(n^3) \end{aligned} \tag{4.6}$$

In the above expression κ is the local gravitational acceleration defined as: $\kappa = \partial_n N$ and in the $n \rightarrow 0$ limit, $\kappa \rightarrow \kappa_H$, satisfying all the standard properties of surface gravity. Also κ/N represents the normal component of the four acceleration of an observer at fixed (n, x^A). Throughout the calculation we shall evaluate quantities on a $n =$ constant surface and then take the $n \rightarrow 0$ limit.

For the line element presented in Eq. (4.5) we find that (in the relevant $n \rightarrow 0$ limit) only three components of connection are non zero. These are : $\Gamma^n_{tt} = N\partial_n N$, $\Gamma^t_{nt} = \partial_n N/N$ and Γ^A_{BC}. Another two expressions we need for the calculation are the $(D-2)$ dimensional surface element and Wald entropy for Lanczos-Lovelock theories. From Chap. 2 we have the following expression for the Wald entropy on the static horizon located at $n = 0$ as,

$$S = \frac{1}{2}\int d^{D-2}x\sqrt{\sigma}P^{nt}_{nt} \equiv \int d^{D-2}x s \tag{4.7}$$

where $s = (1/2)\sqrt{\sigma}P^{nt}_{nt}$ is the entropy density of the $n = 0$ surface. Next we evaluate the variation $U_a^{\ bdn}\delta\Gamma^a_{bd}$ which turns out to be:

$$\begin{aligned} U_a^{\ bdn}\delta\Gamma^a_{bd}\ |_H &= 2\sqrt{-g}Q_a^{\ bdn}\delta\Gamma^a_{bd} \\ &= 2N^2\sqrt{\sigma}Q_n^{\ ttn}\delta\left(\partial_n N\right) + 2\sqrt{\sigma}Q_t^{\ ntn}\delta\left(\partial_n N\right) \\ &= 4\sqrt{\sigma}Q^{nt}_{nt}\delta\kappa \end{aligned} \tag{4.8}$$

Along the similar lines we can compute the other part of the variation, $\Gamma^a_{bd}\delta U_a^{\ bdn}$ which can be written as:

$$\Gamma^a_{bd}\delta U_a^{\ bdn}\mid_H = 4\kappa\delta\left(\sqrt{\sigma}Q^{nt}_{nt}\right) \tag{4.9}$$

From Eq. (4.7) we get the expression for entropy and the temperature associated with the Killing horizon corresponds to $\kappa/2\pi$. With these two identifications we find:

$$\frac{1}{16\pi}\int d^{D-2}x\ \Gamma^a_{bd}\delta(mU_a^{\ bdn}) = \frac{1}{2}\int d^{D-2}x\ \frac{\kappa}{2\pi}\ \delta\left(\sqrt{\sigma}P^{nt}_{nt}\right) = \int d^{D-2}x T\delta s \tag{4.10}$$

$$\frac{1}{16\pi}\int d^{D-2}x\ (mU_a^{\ bdn})\delta\Gamma^a_{bd} = \frac{1}{2}\int d^{D-2}x\sqrt{\sigma}P^{nt}_{nt}\delta\left(\frac{\kappa}{2\pi}\right) = \int d^{D-2}x\ s\delta T \tag{4.11}$$

In the above expressions s represents the entropy density of the horizon, which reduces to $\sqrt{\sigma}/4$ in the Einstein-Hilbert limit. The above results show that the two terms in Eq. (4.3) leads to $T\delta s$ and $s\delta T$ respectively, with T being $\kappa/2\pi$, the temperature associated with the Killing horizon. Also we observe that these are identical to those obtained in Einstein-Hilbert scenario as in [8]. Thus we have natural generalization of the results in general relativity to all Lanczos-Lovelock models if we use the variables $U_a^{\ bcd}$ and Γ^a_{bc} introduced in Sect. 3.2.

4.2.2 *Generalization to Arbitrary Null Surface*

Having discussed the thermodynamic interpretation of the two variables for a general static spacetime, we will now extend the result to an arbitrary null surface. The metric near the null surface has been constructed in Sect. 2.4 and we shall use that metric to evaluate various quantities of interest. As in the case of static situation, here also we shall calculate all the quantities on a r = constant surface and then shall take the limit $r \to 0$ to retrieve the null surface. For that purpose we start with normal to the r = constant surface and then take the null limit. With proper choice of l_a and k_a [10] the Wald entropy turns out to be:

$$S = \frac{1}{2}\int d^{D-2}x\sqrt{q}P^{ur}_{ur} \tag{4.12}$$

Let us next calculate the surface Hamiltonian $H_{\rm sur} = -(\partial A_{\rm sur}/\partial u)$ that comes from the r = constant surface leading to (introducing 16π factor),

$$H_{\rm sur} = \frac{1}{8\pi}\int d^{D-2}x\sqrt{q}Q_a^{\ bdr}\Gamma^a_{bd} \tag{4.13}$$

Using the connection components which will remain nonzero in the $r \to 0$ limit, the surface Hamiltonian turns out to be,

$$H_{\rm sur} = \frac{1}{8\pi}\int d^{D-2}x\sqrt{q}\Big[2\alpha Q^{ur}_{ur} + 2\beta_A Q^{Ar}_{ur} + q^{BD}\hat{\Gamma}^A_{BC}Q^{Cr}_{AD} - q^{AC}\partial_r q_{AB}Q^{Br}_{uC}$$
$$+\frac{1}{2}\left\{-\partial_u q_{AB}Q_r{}^{ABr} + \partial_u q_{BC}\left(Q^{CBur} + Q^{CuBr}\right)\right\}\Big] \tag{4.14}$$

In the Einstein-Hilbert limit only the Q^{ur}_{ur} term contributes leading to the thermodynamic interpretation. Now using Eq. (2.42) all these Q^{ab}_{cd} terms can be calculated. They lead to:

$$Q^{Ar}_{ur} = \delta^{AruBPQ...}_{urCDRS...}R^{CD}_{uB}R^{RS}_{PQ}\cdots \tag{4.15}$$
$$Q^{Cr}_{AD} = \delta^{CrMNuL...}_{ADurPQ...}R^{ur}_{MN}R^{PQ}_{uL}\ldots + \delta^{CruMPQ...}_{ADurRS...}R^{ur}_{uM}R^{RS}_{PQ}\cdots$$
$$+\,\delta^{CrMPuKUV...}_{ADuQLJrW...}R^{uQ}_{MP}R^{LJ}_{uK}R^{rW}_{UV}\ldots + \delta^{CrMPuLUV...}_{ADuQrKXY...}R^{uQ}_{MP}R^{rK}_{uL}R^{XY}_{UV}\cdots$$
$$+\,\delta^{CruPMNUV...}_{ADuQrLXY...}R^{uQ}_{uP}R^{rL}_{MN}R^{XY}_{UV}\cdots \tag{4.16}$$
$$Q^{Br}_{uC} = \delta^{BruA...}_{uCrD...}R^{rD}_{uA}\ldots + \delta^{BruPJK...}_{uCQRrM...}R^{QR}_{uP}R^{rM}_{JK}\cdots \tag{4.17}$$
$$Q^{Br}_{rA} = \delta^{BruCMN...}_{rAuDPQ...}R^{uD}_{uC}R^{PQ}_{MN}\ldots + \delta^{BruCMN...}_{rAEFuR...}R^{EF}_{uC}R^{uR}_{MN}\cdots \tag{4.18}$$
$$Q^{CB}_{ru} = \delta^{CBruMN...}_{ruPQRS...}R^{PQ}_{ru}R^{RS}_{MN}\ldots + \delta^{CBrPuM...}_{ruQRNS...}R^{MN}_{rP}R^{NS}_{uM}\cdots \tag{4.19}$$

In arriving at the above results we have used the fact that the determinant tensor is antisymmetric in any two indices. Thus all the remaining terms, in the above expressions, contain only the components of the curvature tensor with indices depending on coordinates on the null surface. They are all fully characterized by the $(D-2)$ metric q_{AB}. In the null limit we have:

$$R^{CQ}_{uB} = q^{QP}R^C{}_{PuB};\quad R^{rM}_{JK} = q^{MN}R^r{}_{NJK};\quad R^{rD}_{uC} = -q^{AD}R^r{}_{ACu}; \tag{4.20}$$

Having obtained all the components of the surface term we will now consider the thermodynamic interpretation.

Thermodynamic interpretation can be given provided some additional conditions are imposed. We can use two types of conditions; either on the variation (with the metric remaining arbitrary) or on the metric (keeping the variations arbitrary). As in the case of general relativity [8], it is preferable to impose the conditions the variations but keeping the background metric arbitrary. From the result we had in the Einstein-Hilbert action we would expect these conditions to include $\delta\left(q^{CA}\partial_i q_{AB}\right) = 0$, $\hat{D}_M\delta\alpha = 0$ and $\partial_i\delta q_{AB} = 0$, where $\hat{D}_M$ stands for two dimensional covariant derivative. It turns out that, along with these conditions we also need to set $\delta\beta_A = 0$. With these conditions we get the only non-zero contributing term as:

$$H_{\rm sur} = \frac{1}{4\pi}\int d^{D-2}x\,\sqrt{q}\alpha Q^{ur}_{ur} = \frac{1}{m}\int d^{D-2}x\,\left(\frac{\alpha}{2\pi}\right)\left(\frac{1}{2}\sqrt{\sigma}P^{ur}_{ur}\right) = \int d^{D-2}xTs \tag{4.21}$$

Hence the variation of the surface Hamiltonian leads to:

$$\delta H_{\text{sur}} = -\frac{1}{4\pi}\int d^{D-2}x\left[\delta\left(\sqrt{q}Q^{ur}_{ur}\right)\alpha + \sqrt{q}Q^{ur}_{ur}\delta\alpha\right] \tag{4.22}$$

As well as we have the following expression from Eq. (4.3),

$$\delta H^{(1)}_{\text{sur}} = \frac{1}{16\pi}\int d^{D-2}x\; n_c U_a^{\ bdc}\delta\Gamma^a_{bd} = \frac{1}{4\pi}\int d^{D-2}x\sqrt{q}Q^{ur}_{ur}\delta\alpha \tag{4.23}$$

This along with Eqs. (4.22) and (4.3) leads to,

$$\delta H^{(2)}_{\text{sur}} = \frac{1}{16\pi}\int d^{D-1}x\; n_c\Gamma^a_{bd}\delta U_a^{\ bdc} = \frac{1}{4\pi}\int d^{D-1}x\;\alpha\delta\left(\sqrt{q}Q^{ur}_{ur}\right) \tag{4.24}$$

Then Eqs. (4.23) and (4.24) can also be interpreted in terms of entropy density $s = (m/2)\sqrt{q}Q^{ur}_{ur}$ leading to:

$$\begin{aligned}\delta H^{(1)}_{\text{sur}} &= \int d^{D-2}x\frac{\sqrt{q}Q^{ur}_{ur}}{2}\delta\left(\frac{\alpha}{2\pi}\right)\\ &= \frac{1}{m}\int d^{D-2}x\; s\delta T\end{aligned} \tag{4.25}$$

$$\begin{aligned}\delta H^{(2)}_{\text{sur}} &= \int d^{D-2}x\;\frac{\alpha}{2\pi}\delta\left(\frac{\sqrt{q}Q^{ur}_{ur}}{2}\right)\\ &= \frac{1}{m}\int d^{D-2}x\; T\delta s\end{aligned} \tag{4.26}$$

The above results hold for mth order Lanczos-Lovelock Lagrangian, which can be generalized in a straight forward manner to a general Lanczos-Lovelock Lagrangian made out of sum of individual Lanczos-Lovelock Lagrangians of different order.

For the sake of completeness, we mention that the same results can also be obtained for arbitrary variations if we impose a constraint on the background metric. This constraint is identical in general relativity and in Lanczos-Lovelock models and is given by $\partial_u q_{AB} = 0$ and $\hat{D}_M\alpha = \partial_u(\beta_M/2)$. As all the relevant curvature tensor components vanish, when we impose the condition $\partial_u q_{AB} = 0$, leaving only Q^{ur}_{ur} term nonzero. This again leads to the standard expressions as in Eqs. (4.25) and (4.26).

4.3 Concluding Remarks

The link between the standard approach to gravity and the thermodynamical one is provided by the action principle of gravity. Previous studies have shown that these actions have several peculiar features and—under suitable conditions—a thermodynamical interpretation. This motivates us to look for geometrical variables in which the expression for action simplifies and which will have direct thermodynamical interpretation. This goal was achieved recently [8] for the Einstein-Hilbert action by introducing canonically conjugate variables. Variations of these variables turned out to have direct thermodynamic interpretation.

It has been noticed in the past that virtually every result involving the thermodynamical interpretation of gravity, which was valid for general relativity, could be generalized to Lanczos-Lovelock models. We have shown that this fact holds for the above result as well. In Chap. 3 we could introduce two suitable variables in the case of Lanczos-Lovelock models describing all the geometric properties consistently. In this chapter we have shown that the variation of these quantities correspond to $s\delta T$ and $T\delta s$ where s is now the correct Wald entropy density of the Lanczos-Lovelock model. This result holds rather trivially on any static (but not necessarily spherically symmetric or matter-free) horizon and—more importantly—on any arbitrary null surface acting as local Rindler horizon. Since local Rindler structures can be imposed on any event, this shows that, around any event, certain geometric variables can be attributed thermodynamical significance.

The analysis once again confirms that the thermodynamic interpretation goes far deeper than general relativity and is definitely telling us something nontrivial about the structure of the spacetime. We note that the nature of Wald entropy density in Lanczos-Lovelock models is far more complicated than a simple constant (1/4) in general relativity; yet, everything works out exactly as expected. The action principle somehow encodes the information about horizon thermodynamics, which is a key result in emergent gravity paradigm.

References

1. A. Mukhopadhyay, T. Padmanabhan, Holography of gravitational action functionals. Phys. Rev. D **74**, 124023 (2006). arXiv:hep-th/0608120
2. S. Kolekar, D. Kothawala, T. Padmanabhan, Two aspects of black hole entropy in Lanczos-Lovelock models of gravity. Phys. Rev. D **85**, 064031 (2012). arXiv:1111.0973 [gr-qc]
3. T. Padmanabhan, Holographic gravity and the surface term in the Einstein-Hilbert action. Braz. J. Phys. **35**, 362–372 (2005). arXiv:gr-qc/0412068 [gr-qc]
4. T. Padmanabhan, Equipartition of energy in the horizon degrees of freedom and the emergence of gravity. Mod. Phys. Lett. A **25**, 1129–1136 (2010). arXiv:0912.3165 [gr-qc]
5. T. Padmanabhan, Entropy density of spacetime and the Navier-Stokes fluid dynamics of null surfaces. Phys. Rev. D **83**, 044048 (2011). arXiv:1012.0119 [gr-qc]
6. S. Kolekar, T. Padmanabhan, Action principle for the fluid-gravity correspondence and emergent gravity. Phys. Rev. D **85**, 024004 (2012). arXiv:1109.5353 [gr-qc]
7. T. Padmanabhan, Surface density of spacetime degrees of freedom from equipartition law in theories of gravity. Phys. Rev. D **81**, 124040 (2010). arXiv:1003.5665 [gr-qc]
8. K. Parattu, B.R. Majhi, T. Padmanabhan, Structure of the gravitational action and its relation with horizon thermodynamics and emergent gravity paradigm. Phys. Rev. D **87**, 124011 (2013). arXiv:gr-qc/1303.1535 [gr-qc], doi:10.1103/PhysRevD.87.124011
9. A. Medved, D. Martin, M. Visser, Dirty black holes: space-time geometry and near horizon symmetries. Class. Quantum. Gravity **21**, 3111–3126 (2004). arXiv:gr-qc/0402069 [gr-qc]
10. K. Parattu, S. Chakraborty, B.R. Majhi, T. Padmanabhan, Null Surfaces: Counter-term for the Action Principle and the Characterization of the Gravitational Degrees of Freedom. arXiv:1501.01053 [gr-qc]

Chapter 5
Spacetime Evolution and Equipartition in Lanczos-Lovelock Gravity

5.1 Introduction

Recently, in [1] the connection between gravity and thermodynamics have been explored at a significantly deeper level albeit in the context of general relativity. In this work [1] it was explicitly demonstrated that, in the context of general relativity, the following results hold:

1. The total Noether charge in a three-volume $\mathcal{V}$, related to the time evolution vector field, can be interpreted as the heat content of the boundary $\partial\mathcal{V}$ of the three-volume:

$$\int_{\mathcal{V}} d^3x\sqrt{h}u_a J^a\left(\xi\right) = \epsilon\int_{\partial\mathcal{V}} \frac{\sqrt{\sigma}d^2x}{4}T_{\rm loc} = \epsilon\int_{\partial\mathcal{V}} d^2x T_{\rm loc}s \tag{5.1}$$

where σ is the determinant of the two-dimensional metric on $\partial\mathcal{V}$, with $\sqrt{\sigma}/4$ being the entropy density and $T_{\rm loc}$ corresponds to the local Unruh-Davies temperature. Further, $\xi^a = Nu^a$ is the time evolution vector corresponding to observers having four-velocity $u_a = -N\nabla_a t$, that is the normal to the t = constant surface. Also, ϵ is a numerical factor, which is $+1$ if the acceleration of u_a and the normal to $\partial\mathcal{V}$ have identical directions, otherwise it is -1. The above relation provides a "holographic" result connecting the bulk and boundary variables. (Even projection of the Noether current along the acceleration has thermodynamics interpretation, see Appendix A).
2. The time evolution of the spacetime itself can be described in an elegant manner by the equation:

$$\int_{\mathcal{V}} \frac{d^3x}{8\pi}h_{ab}\pounds_{\xi}p^{ab} = \epsilon\frac{1}{2}k_B T_{\rm avg}(N_{\rm bulk} - N_{\rm sur}) \tag{5.2}$$

where h_{ab} is the induced metric on the t = constant surface and p^{ab} being its conjugate momentum. Further, one have the following identification: $h_{ab}\pounds_{\xi}p^{ab} =$

S. Chakraborty, *Classical and Quantum Aspects of Gravity in Relation to the Emergent Paradigm*, Springer Theses, DOI 10.1007/978-3-319-63733-4_5

$\sqrt{h} u_a g^{ij} \pounds_\xi N^a_{ij}$. The $N_{\rm sur}$ and $N_{\rm bulk}$ are the degrees of freedom in the surface and bulk of a 3-dimensional region $\mathcal{V}$ and $T_{\rm avg}$ is the average Davies-Unruh temperature associated with the boundary. The surface degrees of freedom $N_{\rm sur}$ depends on the area of the two-surface, while $N_{\rm bulk}$ depends on the average temperature $T_{\rm avg}$ and Komar energy density $\rho_{\rm Komar}$ [1]. (The parameter $\epsilon = \pm 1$ ensures that the $N_{\rm bulk}$ is positive even when Komar energy turns negative.) This equation depicts that the rate of change of gravitational momentum is driven by the departure from holographic equipartition, measured by $(N_{\rm bulk} - N_{\rm sur})$. The metric will be time independent in the chosen foliation if and only if $N_{\rm sur} = N_{\rm bulk}$, which is true for all static geometries. The validity of Eq. (5.2) for all observers (i.e., foliations) essentially implies the validity of Einstein's uations. In short, *deviation from holographic equipartition leads to the time evolution of the spacetime metric.*
3. For suitably defined gravitational momentum related to a specific time evolution vector field, total energy (gravity+matter) in a bulk volume $\mathcal{V}$ turns out to be the heat density of the surface $\partial\mathcal{V}$.
4. For a bulk region bounded by null surfaces, the total Noether charge within that spacetime region is related to 'heating' of the boundary surface.

In this chapter we will mainly investigate whether the first two results can be generalized to Lanczos-Lovelock models. The generalization of the other two results to Lanczos-Lovelock gravity will be discussed in the later chapters. The importance of this result stems from the fact that the expression for horizon entropy in general relativity is rather trivial and is just a quarter of horizon area. While in Lanczos-Lovelock models, the corresponding expression is much more complex which, in turn, modifies the expression for $N_{\rm sur}$. It is, therefore, not clear a priori whether our results—interpretation of Noether charge and Eq. (5.2)—will generalize to Lanczos-Lovelock models of gravity. We will show in this chapter that, these results indeed possess a natural generalization to Lanczos-Lovelock gravity as well.

The rest of the chapter is organised as follows: In Sect. 5.2 we provide some comprehensive examples to illustrate the validity of the above results in the context of general relativity. In Sect. 5.3 we generalise all the results associated with spacelike surfaces to Lanczos-Lovelock models of gravity. Subsequently, in Sect. 5.3.1 we relate the Noether charge to the surface heat content in the Lanczos-Lovelock models and finally in Sect. 5.3.2 we derive the evolution equation in terms of surface and bulk degrees of freedom. The last section summarises the conclusions.

5.2 Warm Up: Some Illustrative Examples in General Relativity

An important aspect of the dynamical evolution equation derived in [1] is the following: while it is covariant, it is foliation dependent through the normal to the $t =$ constant hypersurface. For example, even in a static spacetime (which possesses a timelike Killing vector field) the *non-static* observers will perceive a time-dependent metric and hence departure from the holographic equipartition (so that both sides of Eq. (5.2) are nonzero), while static observers (with velocities along the Killing direction) will perceive a time-independent metric and hence holographic equipartition will hold for them (with both sides of Eq. (5.2) being zero). This contrast is most striking when one studies two natural class of observers in a static spacetime. The first set corresponds to observers with four-velocities along the timelike Killing vector, having a nonzero acceleration. In the frame of this observer the metric components are independent of time and the left hand side of Eq. (5.2) vanishes leading to holographic equipartition $N_{\rm sur} = N_{\rm bulk}$. But we know that *any* spacetime metric can also be expressed in the synchronous frame coordinates with the line element:

$$ds^2 = -d\tau^2 + g_{\alpha\beta}dx^\alpha dx^\beta \tag{5.3}$$

In the synchronous frame the observers at x^α = constant are comoving with the following four velocity: $u_a = (-1, 0, 0, 0)$. Obviously, the comoving observer is not accelerating, (i.e., the curves x^α = constant are geodesics) and hence the local Davies-Unruh temperature associated with these observers will vanish. We want to consider Eq. (5.2) in two such coordinate systems in order to clarify some of the issues.

Let us begin with the synchronous frame in which $T_{\rm avg} \to 0$ implying $T_{\rm avg}N_{\rm sur} \to 0$ with $T_{\rm avg}N_{\rm bulk}$ remaining finite, so that Eq. (5.2) reduces to the following form:

$$\frac{1}{8\pi}\int_{\mathcal{R}} d^3x\sqrt{h}u_a g^{ij}\pounds_\xi N^a_{ij} = -\frac{\epsilon}{2}T_{\rm avg}N_{\rm bulk} = -\int_{\mathcal{R}} d^3x\sqrt{h}\rho_{\rm Komar} \tag{5.4}$$

The quantity $u_a g^{ij}\pounds_\xi N^a_{ij}$ in such an arbitrary synchronous frame takes the following form:

$$\begin{aligned}\sqrt{h}u_a g^{ij}\pounds_\xi N^a_{ij} &= 2\sqrt{h}\left(K_{ab}K^{ab} - u^a\nabla_a K\right)\\ &= \sqrt{h}\left(g^{\alpha\beta}\partial_\tau^2 g_{\alpha\beta} + \frac{1}{2}\partial_\tau g^{\alpha\beta}\partial_\tau g_{\alpha\beta}\right)\end{aligned} \tag{5.5}$$

where we have used Eq. (157) of [1]. It can be shown that equating this expression to $-16\pi\bar{T}_{ab}u^a u^b$ reproduces the standard time-time component of Einstein's equation in the synchronous frame showing the correctness of our result.

As an explicit example, consider the Friedmann universe for which $g_{\alpha\beta} = a^2(t)\delta_{\alpha\beta}$ leading to the following expressions:

$$\partial_\tau g_{\alpha\beta} = 2a\dot{a}\delta_{\alpha\beta};\quad \partial_\tau^2 g_{\alpha\beta} = \left(2\dot{a}^2 + 2a\ddot{a}\right)\delta_{\alpha\beta};\quad \partial_\tau g^{\alpha\beta} = -2\frac{\dot{a}}{a^3}\delta^{\alpha\beta} \tag{5.6}$$

along with $\bar{T}_{ab}u^a u^b = (1/2)(\rho + 3p)$. On substitution of Eq. (5.6), in Eq. (5.5) we arrive at the following expression for the time evolution of the scale factor (or, the Hubble parameter):

$$\frac{\ddot{a}}{a} = -\frac{4\pi}{3}\left(\rho + 3p\right). \tag{5.7}$$

The above equation when supplemented by the equation of state of the perfect fluid leads to the standard results. Thus in Friedmann universe dynamical evolution of the spacetime leads to dynamical evolution equation of the scale factor sourced by the Komar energy density. Before proceeding further it is worthwhile to clarify the following point: In the case of Friedmann universe, one can *also* obtain [2] the following result

$$\frac{dV}{dt} = N_{sur} - \sum \epsilon N_{bulk} \tag{5.8}$$

with $V = (4\pi/3)H^{-3}$ being the areal volume of the Hubble radius sphere if we define the degrees of freedom using the temperature $T \equiv H/2\pi$. (The ϵ factor has to chosen for each bulk component appropriately in order to keep all N_{bulk} positive definite as indicated by the summation; see [2] for a detailed discussion). Though this is also equivalent to Einstein's equations, it is structurally quite different from the evolution equation as in Eq. (5.2) (and should not be confused with it) for the following reasons: (a) The left hand sides of Eqs. (5.2) and (5.8) are completely different. (b) The placement of ϵ-s are different in the right hand sides of both Eqs. (5.2) and (5.8). (c) One uses Friedmann time coordinate in the left hand side of Eq. (5.8) while attributing a temperature $T \equiv H/2\pi$ to define the degrees of freedom. (d) Most importantly, Eq. (5.8) holds only for Friedmann universe while Eq. (5.2) is completely general and holds for any spacetime.

Coming back to the consequences of Eq. (5.2), since this result is true for any Friedmann universe, it will also be true for the de Sitter spacetime written in synchronous (Friedmann) coordinates. The de Sitter metric, as seen by comoving observers has an explicit time dependence $a(t) \propto \exp(Ht)$ while for these observers the perceived Davies-Unruh temperature identically vanishes. Nevertheless, Eq. (5.2) will of course provide the correct evolution equation. On the other hand, de Sitter spacetime can also be expressed in static coordinates as described by the line element:

$$ds^2 = -\left(1 - \frac{r^2}{l^2}\right)dt^2 + \frac{dr^2}{\left(1 - \frac{r^2}{l^2}\right)} + r^2\left(d\theta^2 + \sin^2\theta d\phi^2\right) \tag{5.9}$$

The observers with $x^\alpha =$ constant in this coordinate system are clearly not geodesic observers. They have the following expressions for their four velocity and four acceleration respectively:

$$u_a = \sqrt{\left(1 - \frac{r^2}{l^2}\right)}(-1, 0, 0, 0) \tag{5.10}$$

$$a^i = \left(0, -(r/l^2), 0, 0\right) \tag{5.11}$$

Let us see what happens when we use this four velocity u_a, defining the foliation. In this case, the acceleration a^i and the normal r_i are directed opposite to each other since r_i is the outward directed normal. (Note that in the de Sitter spacetime the free-falling observers are moving outwards and *with respect to them* the static observers are moving *inwards* i.e., opposite to the outward pointing normal.) Hence in this situation we require $\epsilon = -1$. The magnitude of the acceleration being

$$a = \frac{r}{l^2}\frac{1}{\sqrt{\left(1-\frac{r^2}{l^2}\right)}} \tag{5.12}$$

which is obtained from Eq. (5.11). Thus the local Davies-Unruh temperature becomes,

$$T_{\rm loc} = \frac{Na}{2\pi} = \frac{r}{2\pi l^2} = T_{\rm avg} \tag{5.13}$$

Since the spacetime is static ξ_i is a time-like Killing vector and hence the Lie derivative of the connection present in Eq. (5.2) vanishes. Therefore, in this foliation, holographic equipartition must hold. To verify this explicitly, we start by calculating the surface degrees of freedom which turns out to be:

$$N_{\rm sur} \equiv A = \int_{\partial\mathcal{R}} \sqrt{\sigma} d^2x = 4\pi r^2 \tag{5.14}$$

Again the bulk degrees of freedom become,

$$N_{\rm bulk} = \frac{\epsilon}{(1/2)T_{\rm avg}} \int d^3x\sqrt{h}\rho_{\rm Komar} = 4\pi\frac{\frac{8\pi}{3}\rho r^3}{rl^{-2}} \tag{5.15}$$

Note that the ϵ factor in the definition of the bulk degrees of freedom, keeps it positive, even though the Komar energy density in this case is negative. Then in de Sitter spacetime we can introduce $8\pi\rho = (3/l^2)$ from which we readily observe that:

$$N_{\rm bulk} = (8\pi\rho)(l^2/3)4\pi r^2 = 4\pi r^2 = N_{\rm sur} \tag{5.16}$$

Hence for de Sitter spacetime in static coordinates the holographic equipartition does hold as expected. (Alternatively, setting $N_{\rm bulk} = N_{\rm sur}$ will naturally lead to the correct identification of l in the metric with matter energy density by $8\pi\rho = (3/l^2)$.)

One can easily verify, by explicit computation, how these results generalize to any static, spherically symmetric spacetime, with the line element:

$$ds^2 = -f(r)dt^2 + \frac{dr^2}{f(r)} + r^2 d\Omega^2 \tag{5.17}$$

covering several interesting metrics with horizons. In this static coordinates the holographic equipartition holds, as one can easily check. A more interesting situation arises in the case of geodesic observers in the synchronous frame. To illustrate this, we start with a coordinate transformation: $(t, r, \theta, \phi) \rightarrow (\tau, R, \theta, \phi)$ in which the variables are related by the following equations:

$$dt = dR - \frac{1}{\sqrt{1-f(r)}}\frac{dr}{f(r)} \tag{5.18}$$

$$dR = d\tau + \frac{dr}{\sqrt{1-f(r)}} \tag{5.19}$$

In terms of these newly defined variables the static, spherically symmetric line element reduces to the synchronous form:

$$ds^2 = -d\tau^2 + [1-f(r)]\,dR^2 + r^2 d\Omega^2 \tag{5.20}$$

The comoving observers, having four velocities $u_a = (-1, 0, 0, 0)$ are geodesic observers with zero acceleration and thus the local Davies-Unruh temperature vanishes for them. One can then use Eqs. (5.4) and (5.5) to describe the evolution and the relevant derivatives necessary for this computation are:

$$\begin{aligned}
&\partial_\tau g_{RR} = -f'(r)\dot{r};\ \ \partial_\tau^2 g_{RR} = -f'(r)\ddot{r} - f''(r)\dot{r}^2;\ \ \partial_\tau g^{RR} = \frac{f'(r)\dot{r}}{[1-f(r)]^2} \\
&\partial_\tau g_{\theta\theta} = 2r\dot{r};\ \ \partial_\tau^2 g_{\theta\theta} = 2r\ddot{r} + 2\dot{r}^2;\ \ \partial_\tau g^{\theta\theta} = -\frac{2\dot{r}}{r^3} \\
&\partial_\tau g_{\phi\phi} = 2r\dot{r}\sin^2\theta;\ \ \partial_\tau^2 g_{\theta\theta} = \left(2r\ddot{r} + 2\dot{r}^2\right)\sin^2\theta;\ \ \partial_\tau g^{\theta\theta} = -\frac{2\dot{r}}{r^3}\frac{1}{\sin^2\theta}
\end{aligned} \tag{5.21}$$

On substitution of these derivatives in Eq. (5.5) we obtain the following differential equation satisfied by the unknown function $f(r)$:

$$f''(r) + \frac{2f'(r)}{r} = 16\pi\bar{T}_{\tau\tau} = -16\pi\bar{T}^0_0 \tag{5.22}$$

One can easily verify that this is the correct field equation in this case (see e.g., page 302 of [3]). For example, if we consider the metric of a charged particle with $\bar{T}_{\tau\tau} = Q^2/8\pi r^4$, the above equation can be solved to yield $f(r) = 1-(2M/r)+(Q^2/r^2)$, which, of course, is the Reissner-Nordström metric. The above description being covariant but foliation dependent, is actually a very desirable and inevitable feature from the thermodynamical point of view [4, 5].

5.3 Generalization to Lanczos-Lovelock Gravity

In the previous section we have given explicit examples in the context of Einstein-Hilbert action, that the departure from holographic equipartition leads to the dynamics of the spacetime and have also illustrated that in static spacetime the surface degrees of freedom equals the bulk degrees of freedom. We will now generalize the above description to Lanczos-Lovelock theories of gravity. The basic setup of Lanczos-Lovelock gravity has already been discussed in Chap. 2 and shall not be repeated again in this chapter.

5.3.1 *Heat Content of Spacetime in Lanczos-Lovelock Gravity*

We will work with the same spacetime foliations as introduced in [1] and thus will use the vectors u^a, ξ^a introduced earlier. We begin by connecting the Noether charge in a spatial volume to the heat content of the boundary. To do this we will start by relating the Noether current for a vector q_a to that of another vector $f(x)q_a = v_a$ for any arbitrary function $f(x)$ in the context of Lanczos-Lovelock gravity. From Appendix A.3, using Eq. (5.59) the desired relation can be obtained as:

$$\left\{q_a J^a(fq) - f q_a J^a(q)\right\} = \nabla_b \left(2P^{abcd} q_a q_d \nabla_c f\right) \tag{5.23}$$

The usefulness of the above equation originates from the fact that if $q_a = \nabla_a \phi$ then its Noether current vanishes and hence the Noether current for $v_a = f(x)q_a$ acquires a particularly simple form. Applying the above result for the vector field ξ^a (see Eq. (5.67)) we obtain the simple relation:

$$u_a J^a(\xi) = 2D_\alpha \left(N\chi^\alpha\right) \tag{5.24}$$

where we have introduced a new vector field χ^a which is given by (see Eq. (5.60))

$$\chi^a = -2P^{abcd} u_b u_d a_c \tag{5.25}$$

and satisfies the condition $u_a \chi^a = 0$ (so that it is a spatial vector). Further it also have the property: $D_i \chi^i = \nabla_i \chi^i - a_i \chi^i$. We can now integrate Eq. (5.24) over $(D-1)$ dimensional volume bounded by $N =$ constant surface within $t =$ constant hypersurface leading to (introducing the factor of 16π judiciously):

$$\int_{\mathcal{V}} d^{D-1}x \sqrt{h} u^a J_a(\xi) = \int_{\partial\mathcal{V}} \frac{d^{D-2}x\sqrt{\sigma}}{8\pi} N r_\alpha \chi^\alpha \tag{5.26}$$

As in the case of Einstein-Hilbert Action, here also the vector r_α is the unit normal to the $N =$ constant hypersurface. This vector is either parallel or anti-parallel to the acceleration vector such that $r_\alpha = \epsilon a_\alpha/a$, where $\epsilon = +1$ implies parallel to acceleration and vice-versa. With this convention, we obtain the following result from the vector field χ_α:

$$\sqrt{\sigma}\frac{Nr_\alpha \chi^\alpha}{8\pi} = \epsilon\left(\frac{Na}{2\pi}\right)\left(\frac{1}{2}\sqrt{\sigma}P^{\alpha bd\beta}r_\alpha u_b u_d r_\beta\right) \tag{5.27}$$

The term in the bracket is closely related to the entropy density of the $(D-2)$-surface in Lanczos-Lovelock gravity, defined in Eq. (2.48) and ultimately leading to,

$$s = -\frac{1}{8}\sqrt{\sigma}P^{abcd}\mu_{ab}\mu_{cd} = \frac{1}{2}\sqrt{\sigma}P^{\alpha bd\beta}r_\alpha u_b u_d r_\beta. \tag{5.28}$$

Using this expression for entropy density in Eq. (5.27) we immediately obtain

$$\sqrt{\sigma}\frac{Nr_\alpha \chi^\alpha}{8\pi} = \epsilon T_{\rm loc}s \tag{5.29}$$

where $T_{\rm loc} = Na/2\pi$ corresponds to the redshifted local Unruh-Davies temperature as measured by the observers moving normal to $t =$ constant surface, with respect to the local vacuum of freely falling observers. We thus see that the results in general relativity do have a natural generalization to Lanczos-Lovelock models. With all these results, Eq. 5.26 finally reduces to:

$$\int_{\mathcal{V}} d^{D-1}x\sqrt{h}u^a J_a(\xi) = \epsilon\int_{\partial\mathcal{V}} d^{D-2}x\ T_{loc}s. \tag{5.30}$$

Thus *in Lanczos-Lovelock gravity as well the Noether charge in a bulk spatial region is equal to the surface heat content of the boundary*. The similar result derived for Einstein-Hilbert action can be thought of as a special case of the more general Lanczos-Lovelock gravity; this implies that the connection between the bulk Noether charge and the surface heat content goes way beyond the Einstein-Hilbert action. The result *is nontrivial because the expression for entropy density in the general Lanczos-Lovelock models is significantly differenent* in comparison with general relativity in which it is just one quarter per unit area.

5.3.2 *Evolution Equation of Spacetime in Lanczos-Lovelock Gravity*

Let us next generalize our result as presented in Eq. (5.2) for all Lanczos-Lovelock models, thereby obtaining the dynamical evolution as due to deviation from holographic equipartition. We start by substituting the Noether current expression for ξ^a as presented in Eq. (2.47) to Eq. (5.24) which leads to the following result:

$$2u_a P_i^{\ jka} \pounds_\xi \Gamma^i_{jk} = D_\alpha \left(2N\chi^\alpha\right) - 2N\mathcal{R}_{ab}u^a u^b \tag{5.31}$$

Let us first consider the pure Lanczos-Lovelock theory (i.e., with a single Lovelock polynomial) with the mth order Lanczos-Lovelock Lagrangian. (We shall consider the generalization to Lanczos-Lovelock models with a sum of Lagrangians, however at the end.) Contracting the field equation $\mathcal{R}_{ab} - (1/2)g_{ab}L = 8\pi T_{ab}$ for the mth order Lanczos-Lovelock gravity with g^{ab} we get $L = [8\pi] / [m - (D/2)]\, T$, where D is the space-time dimension. Therefore the gravitational field equations can also be rewritten as:

$$\mathcal{R}_{ab} = 8\pi \left(T_{ab} - \frac{1}{2}\frac{1}{(D/2) - m} g_{ab}T\right) \equiv 8\pi \bar{T}_{ab} \tag{5.32}$$

Using this and integrating Eq. (5.31) over $(D-1)$ dimensional volume we obtain

$$\int_{\mathcal{V}} \frac{d^{D-1}x\sqrt{h}}{8\pi} 2u_a P_i^{\ jka} \pounds_\xi \Gamma^i_{jk} = \int_{\partial\mathcal{V}} \frac{d^{D-2}x\sqrt{\sigma}}{4\pi} N\chi^\alpha r_\alpha - \int_{\mathcal{V}} d^{D-1}x\sqrt{h}2N\bar{T}_{ab}u^a u^b. \tag{5.33}$$

As before, the r_α is the normal to the $N =$ constant surfaces within any $t =$ constant surface and is either parallel or anti-parallel to the acceleration. The energy momentum term can therefore be written in an identical fashion by using the Komar energy density, defined as: $\rho_{\mathrm{Komar}} = 2N\bar{T}_{ab}u^a u^b$. We can proceed using Eq. (5.27), which on substitution into Eq. (5.33) ultimately leads to:

$$\int_{\mathcal{V}} \frac{d^{D-1}x\sqrt{h}}{8\pi} 2u_a P_i^{\ jka} \pounds_\xi \Gamma^i_{jk} = -2\epsilon \int_{\partial\mathcal{V}} d^{D-2}x\sqrt{\sigma} P^{\alpha b\beta d} r_\alpha u_b r_\beta u_d \left(\frac{1}{2}T_{\mathrm{loc}}\right) - \int_{\mathcal{V}} d^{D-1}x\sqrt{h}\rho_{\mathrm{Komar}} \tag{5.34}$$

Rest of the analysis requires proper definition of N_{sur}, N_{bulk} etc. in the context of Lanczos-Lovelock gravity, which we do in analogy with the case of general relativity. The number of surface degrees of freedom is defined as four times the entropy in Lanczos-Lovelock theory following general relativity:

$$N_{\mathrm{sur}} \equiv 4S = 2\int_{\partial\mathcal{V}} d^{D-2}x\sqrt{\sigma} P^{abd\beta} r_\alpha u_b u_d r_\beta \tag{5.35}$$

The average temperature is also properly defined using the surface degrees of freedom as the local weights, ensuring that the total heat content can be reproduced:

$$\frac{1}{2}N_{\mathrm{sur}}k_B T_{\mathrm{avg}} = \frac{1}{2}\int dN_{\mathrm{sur}}k_B T_{\mathrm{loc}}; \qquad T_{\mathrm{avg}}S = \int T_{\mathrm{loc}}dS. \tag{5.36}$$

This result can be written, more explicitly as:

$$T_{\rm avg} = \frac{\int_{\partial\mathcal{V}} d^{D-2}x\sqrt{\sigma}P^{ab\beta d}r_\alpha u_b r_\beta u_d T_{\rm loc}}{\int_{\partial\mathcal{V}} d^{D-2}x\sqrt{\sigma}P^{ab\beta d}r_\alpha u_b r_\beta u_d}$$
$$= \frac{1}{S}\int dS\, T_{\rm loc} = \frac{1}{N_{\rm sur}}\int dN_{\rm sur} T_{\rm loc} \tag{5.37}$$

Once $T_{\rm avg}$ has been defined, the number of bulk degrees of freedom is given by the corresponding equipartition value:

$$N_{\rm bulk} = \frac{\epsilon}{(1/2)T_{\rm avg}}\int_{\mathcal{V}} d^{D-1}x\sqrt{h}\rho_{\rm Komar} \tag{5.38}$$

with ϵ included (as in general relativity), to ensure that $N_{\rm bulk}$ always stays positive. Inserting Eqs. (5.35), (5.37) and (5.38) in Eq. (5.34) we immediately find that the dynamical evolution of the spacetime in Lanczos-Lovelock gravity is determined by the following relation:

$$\int_{\mathcal{V}} \frac{d^{D-1}x\sqrt{h}}{8\pi} 2u_a P_i^{\ jka} \pounds_\xi \Gamma^i_{jk} = \epsilon\left(\frac{1}{2}T_{\rm avg}\right)(N_{\rm sur} - N_{\rm bulk}) \tag{5.39}$$

which is a direct generalization of the corresponding result for general relativity.

For a static spacetime the Lie variation of the connection vanishes since ξ^a becomes a time-like Killing vector field. This immediately leads to the holographic equipartition in the context of Lanczos-Lovelock gravity:

$$N_{\rm sur} = N_{\rm bulk} \tag{5.40}$$

(for an earlier derivation of this result see [6].) When the foliation leads to a time dependent metric, the departure from holographic equipartition drives the dynamical evolution of the metric through the Lie derivative term on the left hand side of Eq. (5.39).

The above result was derived for mth order Lanczos-Lovelock Lagrangian and the definitions of $\bar{T}_{ab}$, $\rho_{\rm Komar}$ and $N_{\rm bulk}$ introduce the m dependence though the expression for $\mathcal{R}_{ab}$ in Eq. (5.32). If, instead, we consider a Lanczos-Lovelock Lagrangian, which is made out of a sum of Langrangians with different m, then the field equations $\mathcal{R}_{ab} - (1/2)g_{ab}L = 8\pi T_{ab}$ on contraction with g^{ab} lead to the result:

$$\sum_m c_m\,[m - (D/2)]\,L_{(m)} = 8\pi T \tag{5.41}$$

which in general cannot be solved in closed form for L in terms of T. However, one can take care of this issue by redefining $\rho_{\rm Komar}$ and $N_{\rm bulk}$ formally in terms of the Lovelock Ricci $\mathcal{R}_{ab}$. That is, we define the Komar energy density as: $\rho = 2N(\mathcal{R}_{ab}/8\pi)u^a u^b$. With this definition of Komar energy, the bulk degrees of freedom reduces to the following form:

$$N_{\text{bulk}} = \frac{\epsilon}{(1/2)T_{\text{avg}}} \int_{\mathcal{R}} d^{D-1}x\sqrt{h}\rho \tag{5.42}$$

Then we again obtain a similar result:

$$\int_{\mathcal{V}} \frac{d^{D-1}x\sqrt{h}}{8\pi} 2u_a P_i^{\ jka} \pounds_\xi \Gamma^i_{jk} = \epsilon \left(\frac{1}{2}T_{\text{avg}}\right)(N_{\text{sur}} - N_{\text{bulk}}) \tag{5.43}$$

with the understanding that, for a given model, one should re-express the variables in terms of T_{ab}.

The above results provide a direct connection between the evolution of spacetime and departure from holographic equipartition. The results also encode the holographic nature of gravity by introducing naturally defined bulk and surface degrees of freedom. The difference between the description of spacetime evolution along these lines and that of standard gravitational field equations $\mathcal{R}_{ab} - (1/2)g_{ab}L = 8\pi T_{ab}$ is the following: For the standard gravitational field equations the left hand side *does not* have a clear physical meaning. There is also no distinction between static and dynamic spacetimes and hence the standard treatment cannot answer the following question: what drives the time-dependence of the spacetime metric? The answer is obviously not T_{ab} since one can obtain time dependent solutions even with $T_{ab} = 0$ while static solutions with $T_{ab} \neq 0$. In contrast the evolution depicted in Eq. (5.43) addresses all these issues and we have a natural separation between the static and the time-evolving metrics via holographic equipartition. When the surface and bulk degrees of freedom are unequal, it results in a departure from holographic equipartition and drives the time-dependence of the metric. Thus the driving force behind dynamical evolution of any spacetime is essentially the departure from holographic equipartition, providing a physically transparent statement about spacetime dynamics.

5.4 Discussion

Our aim in this chapter was to consider the relationship between the Noether current and gravitational dynamics in a general covariant manner. Noether currents can be thought of as originating from mathematical identities in differential geometry, having *no connection to the diffeomorphism invariance of the gravitational action* [1]. This result holds not only in Einstein-Hilbert Action but also in Lanczos-Lovelock gravity (see, for example Appendix A.1).

Even though such conserved currents can be associated with any vector field, the time development vectors are always special, motivating to introduce the vector ξ^a in the spacetime. The vector ξ^a is parallel to the velocity vector u^a for fundamental observers and represents proper time flow normal to t = constant surface. To our expectation the Noether charge and current associated with this vector have elegant and physically interesting thermodynamic interpretation. In particular we showed that, for the vector field ξ^a in Lanczos-Lovelock gravity in arbitrary spacetime dimensions, the *total Noether charge* in any bulk spatial volume $\mathcal{V}$, bounded by constant lapse surface, equals *the heat content of the boundary surface*. Also the equipartition energy of the surface equals twice the Noether charge within the volume. While defining the heat content, we have used the notion of local Unruh-Davies temperature and the Wald entropy for Lanczos-Lovelock gravity. This result holds for Lanczos-Lovelock gravity of all orders and does not rely on some specific criteria, e.g., existence of Killing vectors.

The above identification allow us to study holographic equipartition for static spacetimes and relate the time evolution of the spacetime metric as due to departure from the holographic equipartition. With a suitable and natural definition for the degrees of freedom in the surface as well as in the bulk, we find that for static spacetimes (described in the natural foliation) the surface and the bulk degrees of freedom are equal in number yielding holographic equipartition. It is the departure from this holographic equipartition that drives spacetime evolution and results in time dependent situation. This result holds not only for Einstein-Hilbert action but also in Lanczos-Lovelock gravity.

All the results derived above are generally covariant but they do depend on the choice of foliation. This implies that these results depend intrinsically on observers and their acceleration. This is an inevitable feature since the Davies-Unruh temperature is intrinsically observer dependent. As the dynamical evolution is connected to thermodynamic concepts in this approach, different observers *must* perceive the dynamical evolution differently. For example, the de Sitter spacetime is time dependent when written in synchronous frame, but becomes time independent in static spherically symmetric coordinate. Our description adapts naturally to both these different situations.

Appendix A

In this appendix, we shall present the supplementary material for this chapter.

A.1 Derivation of Noether Current from Differential Identities in Lanczos-Lovelock Gravity

In this section the Noether current for Lanczos-Lovelock gravity will be derived starting from identities in differential geometry without using any diffeomorphism invariance of action principles. The conceptual importance of this approach has already been emphasized in [1], in the context of Einstein gravity, and we shall generalize the result for Lanczos-Lovelock models. We start with the fact that the covariant derivative of any vector field can be decomposed into a symmetric and an antisymmetric part. From the antisymmetric part we can define another antisymmetric tensor field as,

$$16\pi J^{aj} = 2P^{ajki}\nabla_k v_i = P^{ajki}\left(\nabla_k v_i - \nabla_i v_k\right) \tag{5.44}$$

It is evident from the antisymmetry of P^{abcd} that a conserved current exists such that, $J^a = \nabla_j J^{aj}$. We recall the identities:

$$\left(\nabla_j \nabla_k - \nabla_k \nabla_j\right) v^i = R^i_{\ cjk} v^c \tag{5.45}$$

and,

$$\mathcal{L}_v \Gamma^i_{jk} = \nabla_j \nabla_k v^i - R^i_{\ kjm} v^m \tag{5.46}$$

and use them in the definition in Eq. (2.36) to get:

$$\begin{aligned}
\mathcal{R}^{ab} v_b &= P^{aijk} R^b_{\ ijk} v_b = -P^{aijk}\left(\nabla_j \nabla_k - \nabla_k \nabla_j\right) v_i \\
&= P^{aijk}\nabla_k \nabla_j v_i + \left(P^{akij} + P^{ajki}\right)\nabla_j \nabla_k v_i \\
&= P^{aijk}\nabla_k \nabla_j v_i + P^{akij}\nabla_j \nabla_k v_i + \nabla_j\left(P^{ajki}\nabla_k v_i\right)
\end{aligned} \tag{5.47}$$

where in the second line we have used the identity, $P^{a(bcd)} = 0$. Then from Eq. (5.44) we obtain:

$$\begin{aligned}
16\pi J^a &= 2\mathcal{R}^{ab} v_b - 2P^{aijk}\nabla_k \nabla_j v_i - 2P^{akij}\nabla_j \nabla_k v_i \\
&= 2\mathcal{R}^{ab} v_b + 2P_i^{\ ajk}\nabla_k \nabla_j v^i - 2P_i^{\ jak}\nabla_j \nabla_k v^i \\
&= 2\mathcal{R}^{ab} v_b + 2P_i^{\ ajk}\left(\mathcal{L}_v \Gamma^i_{kj} + R^i_{\ jkm} v^m\right) - 2P_i^{\ jak}\left(\mathcal{L}_v \Gamma^i_{jk} + R^i_{\ kjm} v^m\right) \\
&= 2\mathcal{R}^{ab} v_b + 2P_i^{\ jka}\mathcal{L}_v \Gamma^i_{jk}
\end{aligned} \tag{5.48}$$

while arriving at the third line we have used Eq. (5.46) and for the last line we have used the fact that, $P^{ijak} R_{ikjm} = P^{akij} R_{ikjm} = -P^{kaij} R_{ikjm} = P^{kaij} R_{kijm}$. Thus Eq. (2.47) can be derived without any reference to the diffeomorphism invariance of the gravitational action, using only the identities in differential geometry and various symmetry properties.

A.2 Projection of Noether Current Along Acceleration and Newtonian Limit

The analysis in [1] uses the fact that $u_a J^a(\xi)$ is a 3-divergence, so that the spatial volume integral of $u_a J^a(\xi)$ can be converted to a surface integral. In this case, it is natural to interpret $u_a J^a(\xi)$ as a spatial density, viz., charge per unit volume of space. It turns out that similar results can be obtained even for the component of $J^a(\xi)$ in the direction of the normal to the equipotential surface along the following lines. It can be easily shown that

$$\hat{a}_p J^p(\xi) = -\left(g^{ij} - \hat{a}^i \hat{a}^j\right) \nabla_i \left(2Nau_j\right) = -g_\perp^{ij} \nabla_i \left(2Nau_j\right) \tag{5.49}$$

where the tensor $g_\perp^{ij}$ acts as a projection tensor transverse to the unit vector $\hat{a}^i$. However in order to define a surface covariant derivative we need $\hat{a}^i$ to foliate the spacetime, which in turn implies $u^i \nabla_i N = 0$. In this case Eq. (5.49) can be written as $\hat{a}_p J^p(\xi) = -\mathcal{D}_i(2Nau^i)$, where $\mathcal{D}_i$ is the covariant derivative operator corresponding to the induced metric $g_\perp^{ij}$ on the $N =$ constant surfaces with normal $\hat{a}_i$. (When $u^i \nabla_i N = 0$, we have $a_j = \nabla_j \ln N$.) To obtain an integral version of this result, let us transform from the original (t, x^α) coordinates to a new coordinate system (t, N, x^A) using N itself as a "radial" coordinate. In this coordinate we have $a_i \propto \delta_i^N$ and thus $u^N = 0$, thanks to the relation $u^i a_i = 0$. Thus $\mathcal{D}_i(2Nau^i)$ will transform into $\mathcal{D}_{\bar{\alpha}}(2Nau^{\bar{\alpha}})$, where $\bar{\alpha}$ stands for the set of coordinates (t, x^A) on the N = constant surface. Integrating both sides $\hat{a}_p J^p(\xi) = -\mathcal{D}_{\bar{\alpha}}(2Nau^{\bar{\alpha}})$, over the N = constant surface will now lead to the result (with restoration of $1/16\pi$ factor):

$$\int d^2x dt \sqrt{-g_\perp}\, \hat{a}_p J^p(\xi) = -\int d^2x \sqrt{q} N \left(\frac{Na}{8\pi}\right) u^t = \int d^2x \left(\frac{Na}{2\pi}\right)\left(\frac{\sqrt{q}}{4}\right)\Bigg|_{t_2}^{t_1} \tag{5.50}$$

where we have used the standard result $\sqrt{-g_\perp} = N\sqrt{q}$, with q being the determinant of the two-dimensional hypersurface. The right hand side can be thought of as the difference in the heat content $Q(t_2) - Q(t_1)$ between the two surfaces $t = t_2$ and $t = t_1$ where:

$$Q(t) \equiv \int d^2x \left(\frac{Na}{2\pi}\right)\left(\frac{\sqrt{q}}{4}\right) = \int d^2x (Ts) \tag{5.51}$$

This looks very similar to the result we obtained in the case of the integral over $u_i J^i$ earlier (see Eq. (5.1) with $a^\alpha = r^\alpha$ on the N = constant surface), but there is a difference in the interpretation of the left hand side. While $u_i J^i$ can be thought of as the charge density per unit *spatial* volume, the quantity $\hat{a}_p J^p(\xi)$ represents the *flux* of Noether current through a time-like surface; therefore, $\hat{a}_p J^p(\xi)$ should be thought of as a current per unit area per unit time. We will see later that the flux of Noether current through null surfaces leads to a very similar result.

We conclude this section with a discussion of the Newtonian limit of general relativity using the Noether current which has some amusing features. The Newtonian limit is obtained by setting $N^2 = 1 + 2\phi$, $g_{0\alpha} = 0$ and $g_{\alpha\beta} = \delta_{\alpha\beta}$, where ϕ is the Newtonian potential [3]. Then the acceleration of the fundamental observers turn out to be $a_\alpha = \partial_\alpha \phi$. Since the spatial section of the spacetime is flat, the extrinsic curvature identically vanishes and so does the Lie variation term. Also $2\bar{T}_{ab}u^a u^b = \rho_{\rm Komar} = \rho$, which immediately leads to (with G inserted, see Eq. (5.2)):

$$\nabla^2\phi = 4\pi G\rho \tag{5.52}$$

the correct Newtonian limit. The same can also be obtained using the four velocity u_a. The Noether charge associated with u_a turns out to have the following expression [1]

$$D_\alpha a^\alpha = 16\pi u_a J^a(u) = 16\pi \bar{T}_{ab}u^a u^b + u_a g^{bc}\pounds_u N^a_{bc} \tag{5.53}$$

In the Newtonian limit the following results hold $2\bar{T}_{ab}u^a u^b = \rho$ and $u_a g^{bc}\pounds_u N^a_{bc} = -D_\alpha a^\alpha$ (which follows from the Newtonian limit of the result $u_a g^{bc}\pounds_\xi N^a_{bc} = ND_\alpha a^\alpha + 2a^\alpha D_\alpha N - Nu_a g^{ij}\pounds_u N^a_{ij}$ and the fact that in spacetime with flat spatial section the term $u_a g^{bc}\pounds_\xi N^a_{bc}$ identically vanishes). This immediately leads to Eq. (5.52).

We also see that the Noether charge is positive as long as $\rho > 0$ in the Newtonian limit. In fact, the Noether charge contained inside any equipotential surface is always a positive definite quantity as long as r^α and a^α point in the same direction (which happens when $\bar{T}_{ab}u^a u^b > 0$). To prove this we can integrate the Noether charge over a small region on a $t = \text{constant}$ hypersurface to obtain,

$$\begin{aligned}\int_{t=\text{constant}} d^3x\sqrt{h}u_a J^a(u) &= \frac{1}{8\pi}\int_{N,t=\text{constant}} d^2x\sqrt{q}2Na^\alpha r_\alpha \\ &= \int_{N,t=\text{constant}} d^2x\left(\frac{Na}{2\pi}\right)\left(\frac{\sqrt{q}}{4}\right)\end{aligned} \tag{5.54}$$

Since ρ is positive definite in this case the fundamental observers are accelerating outwards and thus $r_\alpha a^\alpha = a$. The temperature as measured by these fundamental observers is a positive definite quantity and so is the entropy density and hence the positivity of Noether charge follows.

A.3 Identities Regarding Noether Current in Lanczos-Lovelock Action

The Noether potential J^{ab} is antisymmetric in (a, b) and from its expression given by Eq. (2.46) it is evident that $J^{ab}(q)$ would identically vanish for $q_a = \nabla_a\phi$. We will use the above fact in order to obtain a relation between the Noether current for two vector fields q_a and v_a connected by $v_a = f(x)q_a$. This result, in the case of general relativity is detailed in [1]. While in the case of Lanczos-Lovelock gravity the Noether current for a vector field $v_a = f(x)q_a$ can be decomposed as:

$$\begin{aligned}16\pi J^{ab}(v) &= 2P^{abcd}\nabla_c(fq_d) \\ &= 2P^{abcd}q_d\nabla_c f + 2fP^{abcd}\nabla_c q_d\end{aligned} \tag{5.55}$$

Then the corresponding Noether current has the following expression:

$$\begin{aligned}16\pi J^a(v) &= 2P^{abcd}\nabla_b\left(q_d\nabla_c f\right) + 2P^{abcd}\nabla_b\left(f\nabla_c q_d\right)\\ &= 2P^{abcd}q_d\nabla_b\nabla_c f + 2P^{abcd}\nabla_c f\nabla_b q_d\\ &+ 2P^{abcd}\nabla_b f\nabla_c q_d + 2fP^{abcd}\nabla_b\nabla_c q_d\end{aligned} \tag{5.56}$$

From the above equation we readily arrive at:

$$\begin{aligned}16\pi\left\{J^a(v) - fJ^a(q)\right\} &= 2P^{abcd}q_d\nabla_b\nabla_c f + 2P^{abcd}\nabla_c f\nabla_b q_d + 2P^{abcd}\nabla_b f\nabla_c q_d\\ &= P^{abcd}\nabla_b A_{cd} + 16\pi J^{ab}(q)\nabla_b f\end{aligned} \tag{5.57}$$

where we have defined the antisymmetric tensor A_{cd} as $A_{cd} = q_d\nabla_c f - q_c\nabla_d f$. Now consider the following result: $q_a\nabla_b A_{cd} = \nabla_b\left(q_a A_{cd}\right) - A_{cd}\nabla_b q_a$ which leads to:

$$\begin{aligned}P^{abcd}q_a\nabla_b A_{cd} &= \nabla_b\left(P^{abcd}q_a A_{cd}\right) - 2P^{abcd}q_d\nabla_c f\nabla_b q_a\\ &= \nabla_b\left(P^{abcd}q_a A_{cd}\right) - 16\pi q_a J^{ab}\left(q\right)\nabla_b f\end{aligned} \tag{5.58}$$

Then Eq. (5.57) can be rewritten in the following manner:

$$\begin{aligned}16\pi\left\{q_a J^a(fq) - fq_a J^a(q)\right\} &= 16\pi J^{ab}(q)\nabla_b f q_a + \nabla_b\left(P^{abcd}q_a A_{cd}\right) - 16\pi q_a J^{ab}(q)\nabla_b f\\ &= \nabla_b\left(2P^{abcd}q_a q_d\nabla_c f\right)\end{aligned} \tag{5.59}$$

It can be easily verified that in the Einstein-Hilbert limit $P^{abcd} = Q^{abcd} = (1/2)\left(g^{ac}g^{bd} - g^{ad}g^{bc}\right)$, under which the above equation reduces to the respective one in general relativity.

Applying the above equation to $u_a = -N\nabla_a t$ with $q_a = \nabla_a t = -u_a/N$ and $f = -N$ we arrive at:

$$16\pi u_a J^a\left(u\right) = 2N\nabla_b\left(P^{abcd}u_a u_d\frac{\nabla_c N}{N^2}\right) \tag{5.60}$$

In order to proceed we define a new vector field such that:

$$\begin{aligned}\chi^a &= -2P^{abcd}u_b u_d\frac{\nabla_c N}{N}\\ &= -2P^{abcd}u_b u_d\left(a_c - \frac{1}{N}u_c u^j\nabla_j N\right)\\ &= -2P^{abcd}u_b a_c u_d\end{aligned} \tag{5.61}$$

Note that in the Einstein-Hilbert limit this vector reduces to the acceleration four vector as follows:

$$\chi^a = -2P^{abcd}u_b a_c u_d = -\left(g^{ac}g^{bd} - g^{ad}g^{bc}\right)u_b a_c u_d = -u^b u_b a^a + u^b a_b u^a = a^a \tag{5.62}$$

Also just as in the case of acceleration for the vector χ^a as well we have:

$$u_a \chi^a = -2a P^{ab\beta d} u_a u_b r_\beta u_d = 0 \tag{5.63}$$

where antisymmetry of P^{abcd} in the first two components has been used. We can also have the following relation for the vector field χ^a:

$$N a_b \chi^b = \chi^b \nabla_b N + \chi^b u_b u^j \nabla_j N = \chi^b \nabla_b N \tag{5.64}$$

where we have used the relation $u_a \chi^a = 0$ from Eq. (5.63). Thus Eq. (5.60) can be written in terms of the newly defined vector field χ^a in the following way:

$$\begin{aligned} 16\pi u_a J^a(u) &= N\nabla_b\left(\frac{\chi^b}{N}\right) \\ &= \nabla_b \chi^b - \frac{\nabla_b N}{N}\chi^b \\ &= D_\alpha \chi^\alpha \end{aligned} \tag{5.65}$$

The last relation follows from the fact that:

$$D_\alpha \chi^\alpha = D_b \chi^b = \nabla_b \chi^b - a_b \chi^b = \nabla_b \chi^b - \frac{\nabla_b N}{N}\chi^b \tag{5.66}$$

Then it is straightforward to get the Noether current for ξ^a by using $q_a = u_a$ and $f = N$ in Eq. (5.59) with Eq. (5.65) as:

$$\begin{aligned} 16\pi u^a J_a(\xi) &= 16\pi N u_a J^a(u) + \nabla_b\left(N\chi^b\right) \\ &= N D_\alpha \chi^\alpha + \nabla_b\left(N\chi^b\right) \\ &= D_\alpha\left(2N\chi^\alpha\right) \end{aligned} \tag{5.67}$$

Here also we have used the following identity:

$$\begin{aligned} D_\alpha\left(N\chi^\alpha\right) &= \left(g^{ij} + u^i u^j\right)\nabla_i\left(N\chi_j\right) \\ &= \nabla_i\left(N\chi^i\right) + u^i u^j \nabla_i\left(N\chi_j\right) \\ &= N\nabla_i \chi^i + N\chi^i a_i - N\chi^j\left(u^i \nabla_i u_j\right) \\ &= N\nabla_i \chi^i \end{aligned} \tag{5.68}$$

Thus we have derived the desired relation for the Noether current of the vector field ξ_a and it turns out to have identical structure as that of Einstein-Hilbert action with χ^a playing the role of four acceleration.

References

1. T. Padmanabhan, General relativity from a thermodynamic perspective. Gen. Relativ. Gravit. **46**, 1673 (2014). arXiv:1312.3253 [gr-qc]
2. T. Padmanabhan, Emergent perspective of gravity and dark energy. Res. Astron. Astrophys. **12**, 891–916 (2012). arXiv:1207.0505 [astro-ph.CO]
3. T. Padmanabhan, *Gravitation: Foundations and Frontiers* (Cambridge University Press, Cambridge, UK, 2010)
4. T. Padmanabhan, Structural aspects of gravitational dynamics and the emergent perspective of gravity. AIP Conf. Proc. **1483**, 212–238 (2012). arXiv:1208.1375 [hep-th]
5. T. Padmanabhan, A physical interpretation of gravitational field equations. AIP Conf. Proc. **1241**, 93–108 (2010). arXiv:0911.1403 [gr-qc]
6. T. Padmanabhan, Surface density of spacetime degrees of freedom from equipartition law in theories of gravity. Phys. Rev. D **81**, 124040 (2010). arXiv:1003.5665 [gr-qc]

Chapter 6
Lanczos-Lovelock Gravity from a Thermodynamic Perspective

6.1 Introduction

The four results enforcing the connection between gravity and thermodynamics in the context of general relativity [1] has been introduced in the previous chapter. The first two results, related to Noether charge and spacetime evolution, have already been generalized to Lanczos-Lovelock gravity and have been presented in Chap. 5, while in this chapter we will show that the other two can also be generalized to hold in Lanczos-Lovelock models of gravity.

Besides the above, we will also explore another very important and curious connection between gravitational dynamics and horizon thermodynamics. This originates from the fact that field equations for gravity near a horizon in both static [2] and spherically symmetric spacetime [3] can be written as a thermodynamics identity. In this chapter, we will try to generalize this result in Lanczos-Lovelock gravity for an arbitrary spacetime with a null surface, which is neither static nor spherically symmetric.

The chapter is organized as follows: First in Sect. 6.2 we will derive the equivalence of gravitational field equations with a thermodynamic identity for an arbitrary null surface, which then is applied in Sect. 6.3 to static spacetime and spherically symmetric spacetime to match with earlier results. Furthermore, in Sect. 6.4 we have presented the results related to Lanczos-Lovelock gravity, involving gravitational momentum and null surfaces. Finally, we have concluded with a discussion on our results. All the relevant derivations are summarized in Appendix A.

6.2 Thermodynamic Identity from Gravitational Field Equations

An arbitrary spacetime with a null surface can always be parametrized using Gaussian Null Coordinates (henceforth refereed to as GNC) which can be constructed in analogy with Gaussian normal coordinates (the construction and other properties have

S. Chakraborty, *Classical and Quantum Aspects of Gravity in Relation to the Emergent Paradigm*, Springer Theses, DOI 10.1007/978-3-319-63733-4_6

been detailed in Sect. 2.4). The GNC line element contains $(D-2)(D-1)/2$ independent parameters in the $(D-2)$-dimensional metric q_{AB}, $(D-2)$ independent parameters in β_A and finally one independent parameter α. All of them are dependent on the coordinates (u, r, x^A). The surface $r = 0$ is the null surface under our consideration. In GNC coordinates we will now consider the near null surface behavior of gravitational field equations in the mth order Lanczos-Lovelock gravity. As in Einstein gravity [4] in this case as well we will start with a subclass of the GNC parametrization in order to bring out the physics involved. For that as in the case of Einstein-Hilbert action, in Lanczos-Lovelock gravity as well we impose two additional requirements: $\beta_A|_{r=0} = 0$ and hypersurface orthogonality for time-like unit vector u_a constructed from ξ_a. This immediately leads to $\partial_A \alpha|_{r=0} = 0$ (see, e.g., [4]). Hence these two conditions imply that α should be independent of transverse coordinates. This can be thought of as an *extension of the zeroth law of black hole thermodynamics for an arbitrary null surface in Lanczos-Lovelock theories of gravity*. On using this condition we arrive at the following expression for $T^{ab}\ell_a k_b$ (which equals T^r_r in the null limit; also we have introduced the 16π factor) as (see Eq. (6.46) in Appendix A.1):

$$
\begin{aligned}
T^r_r &= \frac{1}{8\pi} E^r_r \\
&= \frac{m}{8}\frac{1}{2^{m-1}}\left(\frac{\alpha}{2\pi}\right)\left(\delta^{PA_1B_1\ldots A_{m-1}B_{m-1}}_{QC_1D_1\ldots C_{m-1}D_{m-1}} R^{C_1D_1}_{A_1B_1}\ldots R^{C_{m-1}D_{m-1}}_{A_{m-1}B_{m-1}}\right)\left(q^{QE}\partial_r q_{PE}\right) \\
&- \frac{1}{16\pi}\frac{1}{2^m}\delta^{A_1B_1\ldots A_mB_m}_{C_1D_1\ldots C_mD_m} R^{C_1D_1}_{A_1B_1}\ldots R^{C_mD_m}_{A_mB_m} \\
&- \frac{m}{8\pi}\frac{1}{2^m}\Big\{\delta^{PA_1B_1\ldots A_{m-1}B_{m-1}}_{QC_1D_1\ldots C_{m-1}D_{m-1}}\Big[-\frac{1}{2}q^{QE}\partial_u\partial_r q_{PE} + \frac{1}{4}\left(q^{QE}\partial_r q_{PF}\right)\left(q^{FL}\partial_u q_{EL}\right)\Big] \\
&\times R^{C_1D_1}_{A_1B_1}\ldots R^{C_{m-1}D_{m-1}}_{A_{m-1}B_{m-1}} \\
&+ 2(m-1)\delta^{PA_1B_1\ldots A_{m-1}B_{m-1}}_{QC_1D_1\ldots C_{m-1}D_{m-1}} R^{QC_1}_{uP} R^{uD_1}_{A_1B_1}\ldots R^{C_{m-1}D_{m-1}}_{A_{m-1}B_{m-1}}\Big\}
\end{aligned} \tag{6.1}
$$

Let us now consider the Einstein-Hilbert limit of the above equation, which can be obtained by substituting $m = 1$ in the above equation. This leads to:

$$
T^r_r = -\delta^A_B R^{uB}_{uA} - \frac{1}{4}\delta^{AB}_{CD} R^{CD}_{AB} \tag{6.2}
$$

which exactly coincides with the expression obtained in [4]. Hence our general result reduces to the corresponding one in Einstein-Hilbert action under appropriate limit.

Now let us multiply Eq. (6.1) with the virtual displacement along k^a (which is parametrized by the affine parameter $\bar{\lambda}$), which is $\delta\bar{\lambda}$, and $\sqrt{q}$, where q is the determinant of the transverse metric leading to:

$$
\begin{aligned}
T^r_r\delta\bar{\lambda}\sqrt{q} &= \frac{m}{8}\frac{1}{2^{m-1}}\sqrt{q}\left(\frac{\alpha}{2\pi}\right)\left(\delta^{PA_1B_1\ldots A_{m-1}B_{m-1}}_{QC_1D_1\ldots C_{m-1}D_{m-1}}R^{C_1D_1}_{A_1B_1}\ldots R^{C_{m-1}D_{m-1}}_{A_{m-1}B_{m-1}}\right)\left(q^{QE}\delta_{\bar{\lambda}}q_{PE}\right) \\
&- \delta\bar{\lambda}\sqrt{q}\Big\{\frac{1}{16\pi}\frac{1}{2^m}\delta^{A_1B_1\ldots A_mB_m}_{C_1D_1\ldots C_mD_m}R^{C_1D_1}_{A_1B_1}\ldots R^{C_mD_m}_{A_mB_m} + \frac{m}{8\pi}\frac{1}{2^{m-1}}\delta^{PA_1B_1\ldots A_{m-1}B_{m-1}}_{QC_1D_1\ldots C_{m-1}D_{m-1}} \\
&\times\Big[-\frac{1}{2}q^{QE}\partial_u\partial_r q_{PE} + \frac{1}{4}\left(q^{QE}\partial_r q_{PF}\right)\left(q^{FL}\partial_u q_{EL}\right)\Big]R^{C_1D_1}_{A_1B_1}\ldots R^{C_{m-1}D_{m-1}}_{A_{m-1}B_{m-1}} \\
&+ \frac{m(m-1)}{8\pi}\frac{1}{2^{m-1}}\delta^{PA_1B_1\ldots A_{m-1}B_{m-1}}_{QC_1D_1\ldots C_{m-1}D_{m-1}}R^{QC_1}_{uP}R^{uD_1}_{A_1B_1}\ldots R^{C_{m-1}D_{m-1}}_{A_{m-1}B_{m-1}}\Big\}
\end{aligned} \tag{6.3}
$$

Now $\alpha/2\pi$ can be interpreted as the temperature associated with the null surface, and using $\delta_{\bar{\lambda}}s$ from Eq. (6.50) we can integrate the above equation over $(D-2)$ dimensional null surface yielding (the most general expression has been provided in Appendix A.1):

$$
\begin{aligned}
\int d\Sigma T^r_r\delta\bar{\lambda} &= T\delta_{\bar{\lambda}}\int d^{D-2}x\, s - \int d\Sigma\delta\bar{\lambda}\Big[\left(\frac{1}{16\pi}\frac{1}{2^m}\delta^{A_1B_1\ldots A_mB_m}_{C_1D_1\ldots C_mD_m}R^{C_1D_1}_{A_1B_1}\ldots R^{C_mD_m}_{A_mB_m}\right) \\
&+ \Big\{\frac{m}{8\pi}\frac{1}{2^{m-1}}\delta^{PA_1B_1\ldots A_{m-1}B_{m-1}}_{QC_1D_1\ldots C_{m-1}D_{m-1}}\Big[-\frac{1}{2}q^{QE}\partial_u\partial_r q_{PE} \\
&+ \frac{1}{4}\left(q^{QE}\partial_r q_{PF}\right)\left(q^{FL}\partial_u q_{EL}\right)\Big]\times R^{C_1D_1}_{A_1B_1}\ldots R^{C_{m-1}D_{m-1}}_{A_{m-1}B_{m-1}} \\
&+ \frac{m(m-1)}{8\pi}\frac{1}{2^{m-1}}\delta^{PA_1B_1\ldots A_{m-1}B_{m-1}}_{QC_1D_1\ldots C_{m-1}D_{m-1}}R^{QC_1}_{uP}R^{uD_1}_{A_1B_1}\ldots R^{C_{m-1}D_{m-1}}_{A_{m-1}B_{m-1}}\Big\}\Big]
\end{aligned} \tag{6.4}
$$

where $d\Sigma = d^{D-2}x\sqrt{q}$. To bring out the physics presented in Eq. (6.4) we introduce the concept of transverse metric $g^{\perp}_{ab}$ and a work function [5, 6]. Let us start by considering u_a to be a normalized timelike vector and another normalized but spacelike vector r_a. They are related to the null vectors (ℓ_a, k_a) by the following relations: $u_a = (1/2A)\ell_a + Ak_a$ and $r_a = (1/2A)\ell_a - Ak_a$, where A is an arbitrary function. Given this setup the transverse metric is defined as $g^{\perp}_{ab} = u_a u_b - r_a r_b = \ell_a k_b + \ell_b k_a$. Using this transverse metric the work function is defined [5, 6] to be $P = (1/2)T_{ab}g^{ab}_{\perp} = T_{ab}\ell^a k^b$. In the adapted coordinate system on the null surface using the null vectors from Eq. (2.58a) and Eq. (2.58b) we have $P = T^r_r$. As an aside we would like to mention that in the case of spherically symmetric spacetime, P will be the transverse pressure. However, in this chapter we will not bother to illustrate the physical meaning of P which can be obtained from [5, 6].

Given this physical input we can rewrite Eq. (6.4) in the following form:

$$
\bar{F}\delta\bar{\lambda} = T\delta_{\bar{\lambda}}S - \delta_{\bar{\lambda}}E \tag{6.5}
$$

This exactly coincides with the conventional first law of thermodynamics (not the first law for black hole mechanics but that in conventional thermodynamics; for some related works in black hole thermodynamics see [7–10]), provided: (i) we identify the quantity S to be the entropy of the null surface in Lanczos-Lovelock gravity and this *exactly* matches with existing expression for entropy in Lanczos-Lovelock gravity [11–13]. (ii) We interpret $\bar{F}$ to be the average force over the null surface which is defined as the integral of the work function over the null surface as

$$\bar{F} = \int d^{D-2}x\sqrt{q}P \tag{6.6}$$

Finally, (iii) we should identify the second term on the right hand side of the Eq. (6.5) as variation of an energy as the null surface is moved by an affine parameter distance $\delta\bar{\lambda}$. The energy variation due to motion of the null surface has the following expression:

$$\begin{aligned}
\delta_{\bar{\lambda}}E = \delta\bar{\lambda}\int d\Sigma\Big\{&\frac{1}{16\pi}\frac{1}{2^m}\delta^{A_1B_1\ldots A_mB_m}_{C_1D_1\ldots C_mD_m}R^{C_1D_1}_{A_1B_1}\ldots R^{C_mD_m}_{A_mB_m} \\
&+\frac{m}{8\pi}\frac{1}{2^{m-1}}\delta^{PA_1B_1\ldots A_{m-1}B_{m-1}}_{QC_1D_1\ldots C_{m-1}D_{m-1}}\Big[-\frac{1}{2}q^{QE}\partial_u\partial_r q_{PE}+\frac{1}{4}\left(q^{QE}\partial_r q_{PF}\right)\left(q^{FL}\partial_u q_{EL}\right)\Big] \\
&\times R^{C_1D_1}_{A_1B_1}\ldots R^{C_{m-1}D_{m-1}}_{A_{m-1}B_{m-1}} \\
&+\frac{m(m-1)}{8\pi}\frac{1}{2^{m-1}}\delta^{PA_1B_1\ldots A_{m-1}B_{m-1}}_{QC_1D_1\ldots C_{m-1}D_{m-1}}R^{QC_1}_{uP}R^{uD_1}_{A_1B_1}\ldots R^{C_{m-1}D_{m-1}}_{A_{m-1}B_{m-1}}\Big\}
\end{aligned} \tag{6.7}$$

In the above expression for energy the $\bar{\lambda}$ integral must be done *after* the above integral has been performed, which in turn tells us that the detailed form of the expression inside bracket is required to get E explicitly.

The thermodynamic identity obtained in Eq. (6.5) can be better explained if written in the following fashion:

$$\delta_{\bar{\lambda}}E = T\delta_{\bar{\lambda}}S - \bar{F}\delta\bar{\lambda} \tag{6.8}$$

This equation in this context can be interpreted as: Work done due to infinitesimal virtual displacement from $r = 0$ to $r = \delta\bar{\lambda}$ of the null surface, subtracted from the heat energy, i.e., temperature times entropy change, equals to the energy engulfed during this process. It is to be noted that in the general relativistic limit the last term in the energy expression would be absent and the second term in it leads to time rate of change of transverse area. Hence the energy expression for general relativity can be obtained by taking suitable limit of the above energy expression.

6.3 Applications

In the previous section we have derived the equivalence of gravitational field equations in Lanczos-Lovelock gravity to the thermodynamic identity $P\delta_{\bar{\lambda}}V = T\delta_{\bar{\lambda}}S - \delta_{\bar{\lambda}}E$ near an arbitrary null surface. In this section we will illustrate two applications of our general result: First, the case of an arbitrary static spacetime (see [2]). Second, the spherically symmetric spacetime (see [3]). The discussion will be brief since the details are sketched in the references cited above.

6.3.1 Stationary Spacetime

A spacetime will be called stationary, when we impose Killing conditions on the time evolution vector field. In GNC the most natural time evolution vector field corresponds to: $\boldsymbol{\xi} = \partial/\partial u$. Imposing Killing condition on this vector demands all the metric components, namely α, β_A and q_{AB} to be independent of the u coordinate (see [4]). Thus imposing this condition the energy expression as given in Eq. (6.7) reduces to the following form (identifying $\bar{\lambda} = r$):

$$\begin{aligned}\delta_{\bar{\lambda}} E &= \delta\bar{\lambda} \int d\Sigma \left\{ \frac{1}{16\pi} \frac{1}{2^m} \delta^{A_1 B_1 \ldots A_m B_m}_{C_1 D_1 \ldots C_m D_m} R^{C_1 D_1}_{A_1 B_1} \ldots R^{C_m D_m}_{A_m B_m} \right\} \\ &= \delta r \int d\Sigma \ L^{(D-2)} \end{aligned} \tag{6.9}$$

which immediately leads to the following differential equation for energy:

$$\frac{\partial E}{\partial r} = \int d\Sigma \ L^{(D-2)} \tag{6.10}$$

This exactly matches with the expression given in [2]. For $m = 1$ and $D = 4$ this reduces to the expression obtained for Einstein's gravity.

In order to achieve staticity we must impose hypersurface orthogonality on the time evolution vector ξ^a (or, equivalently on the four velocity constructed out of it). This requires $\partial_A \alpha|_{r=0}$ to vanish [4]. However from Eq. (6.9) it is evident that this leads to no modification to our energy expression.

This is an interesting result. It shows that in arriving at this relation we have used two assumptions, viz, (a) $\beta_A = 0$ on the null surface and (b) the spacetime is stationary. Hence the above result does not require spacetime to be static. Thus starting from the thermodynamic identity $T\delta_{\lambda} S = \delta_{\lambda} E + P\delta_{\bar{\lambda}} V$ for arbitrary null surface we have shown that it holds for arbitrary static and stationary spacetimes as well. We will now take up the case for spherically symmetric but time dependent spacetime.

6.3.2 Spherically Symmetric Spacetime

We will finish by considering another application of our result: a spherically symmetric but *not* necessarily static spacetime. GNC metric can be expressed in a spherically symmetric form by choosing the transverse coordinates x^A to be the angular coordinates and enforcing $(D-2)$-sphere geometry on the $(u, r) =$ constant surface. Then we have the following restrictions on the GNC parameters, namely, $\partial_A \alpha = 0$, $\beta_A = 0$ and $q_{AB} = f(u, r) d\Omega^2_{(D-2)}$. When these conditions are imposed the line element takes the following form:

$$ds^2 = -2r\alpha(r, u)du^2 + 2dudr + f(u, r)d\Omega^2_{(D-2)} \tag{6.11}$$

We will define the radial coordinate [4] as: $R(r, u) = \sqrt{f(r, u)}$. Then making a Taylor series expansion about $r = 0$, we get $R(r, u) = R_H(u) + rg(r, u)$. Hence the null surface has a radius $R_H(u)$, which can change with u. This clearly shows that we have spherical symmetry but have retained time dependence. Hence by imposing spherical symmetry the energy satisfies a partial differential equation with the following form:

$$\begin{aligned}\frac{\partial E}{\partial \bar{\lambda}} &= \int d\Sigma \Big\{ \frac{1}{16\pi} \frac{1}{2^m} \delta^{A_1 B_1 \ldots A_m B_m}_{C_1 D_1 \ldots C_m D_m} R^{C_1 D_1}_{A_1 B_1} \cdots R^{C_m D_m}_{A_m B_m} \\ &+ \frac{m}{8\pi} \frac{1}{2^{m-1}} \delta^{P A_1 B_1 \ldots A_{m-1} B_{m-1}}_{Q C_1 D_1 \ldots C_{m-1} D_{m-1}} \Big[-\frac{1}{2} q^{QE} \partial_u \partial_r q_{PE} + \frac{1}{4} \left(q^{QE} \partial_r q_{PF}\right) \left(q^{FL} \partial_u q_{EL}\right) \Big] \\ &\times R^{C_1 D_1}_{A_1 B_1} \cdots R^{C_{m-1} D_{m-1}}_{A_{m-1} B_{m-1}} \\ &+ \frac{m(m-1)}{8\pi} \frac{1}{2^{m-1}} \delta^{P A_1 B_1 \ldots A_{m-1} B_{m-1}}_{Q C_1 D_1 \ldots C_{m-1} D_{m-1}} R^{Q C_1}_{u P} R^{u D_1}_{A_1 B_1} \cdots R^{C_{m-1} D_{m-1}}_{A_{m-1} B_{m-1}} \Big\} \end{aligned} \tag{6.12}$$

A much simpler form can be derived in which the 2-sphere line element is just $(r + R_H)^2 d\Omega^2_{(D-2)}$, where the radial coordinate at the null surface R_H is a constant. Then trading off r in favor of R the line element becomes

$$ds^2 = -2\,(R - R_H)\,\alpha(R, u) du^2 + 2du dR + R^2 d\Omega^2_{(D-2)} \tag{6.13}$$

Then using the result that $\partial_u R_H = 0$, the differential equation for energy can be integrated leading to,

$$\begin{aligned} E(R, u) &= \int d\bar{\lambda} \int d\Sigma\; \mathcal{L}^{(D-2)} + X(u) \\ &= \frac{1}{16\pi} A_{D-2} R^{D-(2m+1)} \prod_{j=2}^{2m} (D - j) + X(u) \end{aligned} \tag{6.14}$$

where $X(u)$ is an arbitrary function appearing as a "constant" of integration and A_{D-2} originates from the differential volume element. Having introduced the radial coordinate R, we can replace $\bar{\lambda}$ by R, since $R - R_H = r$, which coincides with the defining equation for $\bar{\lambda}$. As we move along ingoing radial lines, which are also ingoing radial null geodesic $-\partial/\partial r$, we will gradually hit the center of the $(D-2)$-spheres (assuming that it exists). The affine parameter at the center would be $R = 0$. Then Eq. (6.14) at the center turns out to be:

$$E(R = 0, u) = X(u) \tag{6.15}$$

Since the center is a single point it is natural to associate zero energy with it, which determines the arbitrary function to be $X(u) = 0$. Thus substituting this result in Eq. (6.14) and evaluating on the null surface, we obtain the energy associated with the null surface in a spherically symmetric spacetime to be given by

$$E = \frac{1}{16\pi} A_{D-2} R_H^{D-(2m+1)} \prod_{j=2}^{2m} (D-j) \tag{6.16}$$

This again matches exactly with the result obtained in [3]. Another important thing to note is the following: the expression for energy, first obtained for static spherically symmetric configuration *also* holds for time dependent situation with *only* spherical symmetry assumed.

6.4 Thermodynamic Interpretations in Lanczos-Lovelock Gravity

As we have mentioned earlier, in this chapter we will be dealing exclusively with Lanczos-Lovelock gravity. We will provide a generalization of a four vector from Einstein gravity to Lanczos-Lovelock gravity, which will carry the notion of gravitational momentum. Then we will concentrate on variation of this momentum and its meaning in thermodynamic language. Finally we discuss the null surfaces in the context of Lanczos-Lovelock gravity.

6.4.1 Bulk Gravitational Dynamics and Its Relation to Surface Thermodynamics in Lanczos-Lovelock Gravity

In [1] it has been illustrated that, total energy of matter and gravity equals the surface heat content in the context of general relativity. To prove this, a suitably defined gravitational four-momentum P^a has been used such that when integrated over a $t =$ constant surface with proper integration measure it leads to a notion of gravitational energy. The notion of energy is quiet ambiguous in the sense of observer dependence. For example, even in special relativity the energy of a particle with four momentum $\mathbf{p}$ as measured by an observer with four velocity $\mathbf{u}$ is: $E = -\mathbf{u}.\mathbf{p}$. This immediately suggests that we should use identical trick to identify the energy by contracting a suitably defined four momentum P^a with the four velocity $u_a = -N\nabla_a t$, normal to $t =$ constant surface. We will first briefly describe the situation for general relativity and shall generalize subsequently to Lanczos-Lovelock gravity.

Bulk Energy Versus Surface Heat Energy The Einstein-Hilbert action can be written explicitly in terms of two canonically conjugate variables, namely, $f^{ab} = \sqrt{-g}g^{ab}$ and $N^c_{ab} = Q^{cp}_{aq}\Gamma^q_{bp} + Q^{cp}_{bq}\Gamma^q_{ap}$ [14]. It turns out that the Einstein-Hilbert action can be interpreted as a momentum space action with f^{ab} as the coordinate and N^c_{ab} as its conjugate momentum [14]. Using these two variables a natural definition of gravitational momentum can be provided as (Note that the negative sign in front of P^a is just a convention, such that $-u_a P^a$ relates to the energy.):

$$-P^a(q) = g^{ij}\pounds_q N^a_{ij} + q^a L \tag{6.17}$$

where q^a is an arbitrary vector field. This follows from the result that Hamiltonian can be written as $H = q\dot{p} + L$ with the identification of p as N^c_{ab} and q as g^{ab} (or f^{ab} if we consider $-\sqrt{-g}P^a$). However in order to generalize the above result to Lanczos-Lovelock gravity, we should rewrite the above expression in a slightly modified form, such that: $g^{ij}\pounds_q N^a_{ij} = 2g^{ij}\pounds_q(Q^{ap}_{im}\Gamma^m_{jp}) = 2Q_m^{\ jpa}\pounds_q\Gamma^m_{jp}$. Thus the above action can also be interpreted with Γ^m_{jp} and $Q_m^{\ jpa}$ as conjugate variables. Surprisingly, this is also a valid pair of conjugate variables as illustrated in [15] and can be generalized readily to Lanczos-Lovelock gravity. However in this case we need to interpret Γ^m_{jp} as the momenta and $Q_m^{\ jpa}$ as the coordinate, which is true since Γ^m_{jp} has 40 independent degrees of freedom while $Q_m^{\ jpa}$ as constructed out of the metric only has 10 independent degrees of freedom (a detailed discussion has been presented in [15]). The above setup can be generalized in a natural fashion to Lanczos-Lovelock gravity following [15] and leads to:

$$-\sqrt{-g}P^a(q) = 2\sqrt{-g}P_p^{\ qra}\pounds_q\Gamma^p_{qr} + \sqrt{-g}Lq^a \tag{6.18}$$

The physical structure of this momentum can be understood in greater detail by using the Noether current. Writing the corresponding expression for Noether current explicitly in the case of Lanczos-Lovelock gravity and then simplifying we obtain (see Appendix A.2.1):

$$J^a(q) = \nabla_b J^{ab}(q) = 2E^a_b q^b - P^a(q) \tag{6.19}$$

This leads to another definition for the momentum which will turn out to be quiet useful and can be given as,

$$-P^a(q) = \nabla_b J^{ab}(q) - 2E^a_b q^b \tag{6.20}$$

Then divergence of the momentum has the following expression:

$$\nabla_a P^a(q) = 2E^a_b \nabla_a q^b \tag{6.21}$$

in arriving at the above expression we have used two results, Noether potential J^{ab} being antisymmetric and E_{ab} satisfying Bianchi identity. From the expression it is evident that if we enforce field equations for pure gravity, which amounts to: $E_{ab} = 0$, the momentum becomes divergence free. Also appearance of Noether current explicitly in this expression shows intimate connection of Noether current with energy in all Lanczos-Lovelock models of gravity.

So far the results are completely general, holding for any vector field q^a. Now we will specialize to the vector field ξ^a and will show that it leads to several remarkable results. First we start with momentum for ξ^a and its contraction with u_a leading to (see Appendix A.2.1):

$$-u_a P^a(\xi) = D_\alpha \left(2N\chi^\alpha\right) - 2N E^{ab} u_a u_b \tag{6.22}$$

where the vector χ^a is defined in Eq. (5.25). Now using the equation of motion i.e. $2E_{ab} = T_{ab}$ and integrating the above expression on a t = constant surface bounded by N = constant surface we get the following expression (16π factor has been introduced):

$$\int_{\mathcal{R}} d^{D-1}x\sqrt{h}u_a \left\{-P^a(\xi) + N T^a_b u^b\right\} = \int_{\partial\mathcal{R}} d^{D-2}x \; T_{\rm loc} s \tag{6.23}$$

The expression on the left hand side represents total, i.e., matter energy plus gravitational energy in a bulk region and the right hand side represents the heat content of the boundary surface. The temperature as usual is given by: $Na/2\pi$, i.e., the red-shifted Unruh-Davies temperature and s is the Wald entropy density associated with the boundary surface. The right hand side of the above expression can also be identified as half of the equipartition energy of the boundary surface. Hence *the bulk energy originating from both gravity and matter is equal to the surface heat content.*

Variation of Gravitational Energy From the above paragraph it is clear that the momentum P^a and the corresponding gravitational energy $-u_a P^a(\xi)$ have very interesting thermodynamic properties. In this light, it seems natural to consider variation of the above momentum under various physical processes, e.g., how it changes due to processes acting on the boundary. We will first work with the arbitrary vector field q^a and then we will specialize to the choice: $q^a = \xi^a$. To our surprise just as in general relativity even in Lanczos-Lovelock gravity variation of the gravitational momentum is connected to symplectic structures [16, 17].

Thus by variation of $-\sqrt{-g}P^a(q)$ and manipulating the terms carefully we obtain the symplectic structure as (see Appendix A.2.1)

$$-\delta\left(\sqrt{-g}P^a\right) = \sqrt{-g}E_{pq}\delta g^{pq}q^a + \sqrt{-g}\omega^a + \partial_c\left(2\sqrt{-g}P_p^{\ qr[a}q^{c]}\delta\Gamma^p_{qr}\right) \tag{6.24}$$

where the symplectic form ω^a has the following expression:

$$\sqrt{-g}\omega^a\left(\delta, \pounds_q\right) = \delta\left(2\sqrt{-g}P_p^{\ qra}\right)\pounds_q\Gamma^p_{qr} - \pounds_q\left(2\sqrt{-g}P_p^{\ qra}\right)\delta\Gamma^p_{qr} \tag{6.25}$$

This expression is true for any arbitrary vector field q^a and involves one arbitrary variation and another Lie variation along q^a.

Having obtained the general result we will now specialize to the vector field ξ^a. Then we can use the above formalism in order to obtain the change in gravitational energy of the system due to its evolution, which is related to Lie derivative along ξ^a. This can be achieved using a simple trick. First with the help of Eq. (6.25) we can compute the object $\delta[-\sqrt{h}u_a P^a(\xi)]$. The variation has the following expression (see Appendix A.2.1):

$$-\delta\left[\sqrt{h}u_a P^a(\xi)\right] + \sqrt{-g}E_{pq}\delta g^{pq} = \sqrt{h}u_a\omega^a + \partial_c\left[2\sqrt{-g}h^c_a P_p^{\ qra}\delta\Gamma^p_{qr}\right] \quad (6.26)$$

This holds for an arbitrary variation, however what we are interested in is when the above variation is due to a diffeomorphism along ξ^a. Again from Appendix A.2.1 using the field equation $2E_{ab} = T_{ab}$ we arrive at:

$$\begin{aligned}
\pounds_\xi \mathcal{H}_{grav} &= \pounds_\xi\left(-\int_{\mathcal{R}} d^{D-1}x\sqrt{h}u_a P^a(\xi)\right) \\
&= \int_{\partial\mathcal{R}} d^{D-2}x\sqrt{q}Nr_a\left(T^{cd}\xi_d + 2P_p^{\ qra}\pounds_\xi\Gamma^p_{qr}\right) \qquad (6.27)
\end{aligned}$$

where r_a is the unit normal to $N =$ constant surface within the $t =$ constant surface and $\mathcal{H}_{grav}$ stands for integral of $-\sqrt{h}u_a P^a(\xi)$ over the $(D-1)$-dimensional volume. Hence change of energy in the bulk is directly related to boundary effects (with the assumption that $T^t_c\xi^c = 0$). Among the two terms present on the right hand side, the first term is due to flow of matter energy across the boundary and the second term is related to our old friend $P_p^{\ qra}\pounds_\xi\Gamma^p_{qr}$. For pure gravity, i.e., $T_{ab} = 0$, the above result takes a much simpler form as:

$$\pounds_\xi\mathcal{H}_{grav} = \int_{\partial\mathcal{R}} d^{D-2}x\sqrt{q}Nr_a 2P_p^{\ qra}\pounds_\xi\Gamma^p_{qr} \qquad (6.28)$$

This can have direct influence on gravity wave propagation, i.e., the energy change in a bulk region due to gravity waves is related to surface processes and hence to $P_p^{\ qra}\pounds_\xi\Gamma^p_{qr}$.

Since the gravitational momentum is intimately connected to Noether current, we can use the above results for variation of gravitational momentum in order to obtain a variation of Noether current as well. From Appendix A.2.1 on imposing on-shell condition, i.e., $E_{ab} = 0$, we get:

$$\sqrt{h}u_a\omega^a(\delta,\pounds_q) + 2\delta\left(\sqrt{h}u_a E^{ab}q_b\right) = \partial_b\left\{\delta\left[\sqrt{h}u_a J^{ab}(q)\right] - 2\sqrt{h}u_a P_p^{\ qr[a}q^{b]}\delta\Gamma^p_{qr}\right\} \qquad (6.29)$$

This expression can be related to the usual Hamiltonian formulation where one relates bulk integral of $\delta(HN + H_\alpha N^\alpha)$ (with H and H^α correspond to constraints in the gravity theory; N and N^α are the standard ADM variables) to a bulk and a surface contribution as:

$$\delta\int_{\mathcal{R}} d^{D-1}x\sqrt{h}(HN + H_\alpha N^\alpha) = \int_{\mathcal{R}} d^{D-1}x\delta B + \int_{\partial\mathcal{R}} d^{D-2}x\delta S \qquad (6.30)$$

The variation present on the left hand side is equivalent to $\delta(2\sqrt{h}u_a E^{ab}\zeta_b)$, where $\zeta^a = Nu^a + N^a$, with $N^a = (0, N^\alpha)$. Then we can identify the bulk term on the right hand side with the symplectic current ω^a. Thus finally the remaining surface contribution turns out to be:

$$\int_{\partial\mathcal{R}} d^{D-2}x\delta S = \int_{\partial\mathcal{R}} d^{D-2}x r_b \left\{\delta\left[\sqrt{h}u_a J^{ab}(\zeta)\right] - 2\sqrt{h}\left(N P_p^{\ qra} + P_p^{\ qr[a} N^{b]}\right)\delta\Gamma^p_{qr}\right\} \tag{6.31}$$

This provides an elegant and simple interpretation of the surface term appearing in the variation of the gravitational action. Thus for Lanczos-Lovelock gravity the Hamiltonian formulation as well can have thermodynamic counterpart.

The above approach of relating surface quantities with bulk energy has also been studied earlier, notably in the context of Virasoro algebra and its associated central charge [18]. In [19] the above approach has been used to derive Wald entropy in Lanczos-Lovelock theories of gravity, which subsequently has been generalized in [20] in order to calculate correction to horizon entropy in higher derivative gravity theories. It has been shown in this context that surface contribution alone is what will lead to the central charge and hence to horizon entropy. Our result strengthens the above feature by showing a connection between total energy and boundary heat energy in all Lanczos-Lovelock theories of gravity.

6.4.2 *Heat Density of the Null Surfaces*

In this final section we will discuss heat density associated with a null surface in Lanczos-Lovelock gravity. Any null surface will be defined using a congruence of null vector ℓ^a, which are tangent as well as normal to the null surface. We will also assume that the null vector ℓ^a satisfies the relation $\ell^2 = 0$, everywhere. The null congruence will be taken to be non-affinely parametrized, such that, $\ell_a = A(x)\nabla_a B(x)$. Then the non-affinity parameter κ can be obtained from the relation:

$$\ell^j \nabla_j \ell^i = \kappa \ell^i; \qquad \kappa = \ell^i \partial_i \ln A \tag{6.32}$$

Since the null vector is both tangent and orthogonal, in order to define a projection we need an auxiliary null vector k^a [21] with the following properties $k^2 = 0$ and $k_a \ell^a = -1$. Then we can introduce a projection tensor $q^a_b = \delta^a_b + \ell^a k_b + \ell_b k^a$ and define an associated covariant derivative D_a.

In the previous sections we have calculated Noether current for ξ^a and its contraction with u_a yielding Noether charge. In the context of null surfaces we will discuss Noether current for ℓ^a and its contraction with ℓ_a itself. This leads to (see Appendix A.2.2)

$$\ell_a J^a(\ell) = \nabla_a\left(\mathcal{K}\ell^a\right) - \kappa\mathcal{K} \tag{6.33}$$

where we have introduced the object $\mathcal{K} = -2P^{abcd}\ell_a k_b \ell_d \nabla_c \ln A$. This in the general relativity limit coincides with κ. Note the formal similarity of the expression on the right hand side, i.e., $\nabla_a (\mathcal{K}\ell^a) - \kappa\mathcal{K}$ for null surface to the expression $D_\alpha \chi^\alpha = \nabla_i \chi^i - a_j \chi^j$ obtained for spacelike surface. Also the expression for χ^a is quiet similar to the expression for $\mathcal{K}$ with ℓ_a and k_a identified with u_a and r_a respectively. Then introducing the covariant derivative D_a on the surface the above expression can be written as (see Appendix A.2.2)

$$\ell_a J^a(\ell) = 2\mathcal{R}_{ab}\ell^a\ell^b + 2\ell_a P_p{}^{qra}\pounds_\ell \Gamma^p_{qr} = D_a\left(\mathcal{K}\ell^a\right) + \frac{d\mathcal{K}}{d\lambda} \tag{6.34}$$

Integrating the above expression over the null surface with integration measure $d\lambda d^{D-2}x\sqrt{q}$ and ignoring the boundary contribution we arrive at:

$$\int d\lambda d^{D-2}x\sqrt{q}\ell_a J^a(\ell) = \int d\lambda d^{D-2}x\sqrt{q}\frac{d\mathcal{K}}{d\lambda} \tag{6.35}$$

This result at the face value shows that for the null surface contraction of the Noether current along ℓ_a is related to *heating* of the boundary surface, with $\mathcal{K}$ being taken as temperature in the Lanczos-Lovelock gravity.

This result is also important from the point of view of variational principle for null surfaces. Such a variational principle based on null surfaces has been carefully investigated in [22]. Following that, given a null surface with null congruence ℓ_a we can construct the functional

$$\mathcal{Q} = \int_{\lambda_1}^{\lambda_2} d\lambda d^{D-2}x\sqrt{q}\left(-2\mathcal{R}_{ab} + T_{ab}\right)\ell^a\ell^b \tag{6.36}$$

Then varying the above action functional with respect to *all* ℓ_a with the constraint $\ell^2 = 0$, we will arrive at: $\mathcal{R}^a_b - (1/2)T^a_b = f(x)\delta^a_b$. This on using Bianchi identity, $\nabla_a E^{ab} = 0 = \nabla_a T^{ab}$ leads to gravitational field equations with an undetermined cosmological constant originating from integration constant.

Then from Eq. (6.34) we can write $\mathcal{R}_{ab}\ell^a\ell^b$ in terms of Lie derivative of the connection and change of $\mathcal{K}$ along the null geodesics such that

$$-2\mathcal{R}_{ab}\ell^a\ell^b = 2\ell_a P_p{}^{qra}\pounds_\ell \Gamma^p_{qr} - \left[D_a\left(\mathcal{K}\ell^a\right) + \frac{d\mathcal{K}}{d\lambda}\right] \tag{6.37}$$

Then substitution of $-2\mathcal{R}_{ab}\ell^a\ell^b$ on the right hand side of Eq. (6.36) leads to the following modified variational principle for Lanczos-Lovelock gravity when boundary terms are neglected as

$$\mathcal{Q} = \int_{\lambda_1}^{\lambda_2} d\lambda d^{D-2}x\sqrt{q}\left[\left(2\ell_a P_p{}^{qra}\pounds_\ell \Gamma^p_{qr} - \frac{d\mathcal{K}}{d\lambda}\right) + T_{ab}\ell^a\ell^b\right] \tag{6.38}$$

Hence varying this Lagrangian with respect to ℓ^a with $\ell^2 = 0$, we can obtain the field equation for gravity with an arbitrary cosmological constant. In the above expression $T_{ab}\ell^a\ell^b$ can be taken as matter heat density Ts, while the rest of the terms represent *heat density* of the null surface itself.

We can always choose the parameter λ such that the null vector ℓ_a is affinely parametrized. In which case $\mathcal{K} = 0$ and the variational principle can be based on the following integral:

$$\bar{\mathcal{Q}} = \int_{\lambda_1}^{\lambda_2} d\lambda d^{D-2}x\sqrt{q}\left[2\ell_a P_p^{\ qra}\pounds_\ell\Gamma^p_{qr} + T_{ab}\ell^a\ell^b\right] \tag{6.39}$$

Again showing explicitly the importance of the Lie derivative term in the derivation of the field equations from an alternative action principle. When there is no matter present the variational principle simplifies considerably leading to

$$\bar{\mathcal{Q}} = \int_{\lambda_1}^{\lambda_2} d\lambda d^{D-2}x\sqrt{q}\left[2\ell_a P_p^{\ qra}\pounds_\ell\Gamma^p_{qr}\right] \tag{6.40}$$

This leads to vacuum field equations when varied over all null surfaces simultaneously. Also integral of this object has an interpretation of heat content over the boundary surface. Thus we observe that at least for affinely parametrized null congruences the variational principle over the null surface acquires a thermodynamic interpretation.

6.5 Discussion

In this chapter our aim was to generalize various results derived in the context of general relativity to all Lanczos-Lovelock theories of gravity. This is a *non-trivial* task since the Lanczos-Lovelock Lagrangian contains higher order terms constructed out of the curvature tensor. Also validity of some result in general relativity does not guarantee its validity in these higher curvature theories, e.g., the expression for entropy in general relativity does not hold in Lanczos-Lovelock gravity. Let us now summarize the key results obtained through this exercise:

- In Sect. 6.2 we have shown that the field equations for Lanczos-Lovelock gravity near an arbitrary null surface is equivalent to a thermodynamic identity. Using a parametrization (known as GNC) for the arbitrary null surface we have shown that the field equations for Lanczos-Lovelock gravity can be used to relate energy momentum tensor with thermodynamic features. This exercise also provides us a definition of energy in an arbitrary spacetime in Lanczos-Lovelock gravity, which in the static case and spherically symmetric case reduces to standard energy definitions.
- Next, in Sect. 6.4.1 we have introduced a gravitational momentum starting from its expression in general relativity. Using this gravitational momentum we have shown that total gravitational plus matter energy in a bulk volume equals the heat density associated with the boundary surface. Also variation of the gravitational

Hamiltonian is directly related to a symplectic structure such that time evolution of this gravitational Hamiltonian in a bulk region can be related to surface effects especially with $P_p^{\ qra}\pounds_\xi \Gamma^p_{qr}$. Using this formalism it is also possible to connect standard Hamiltonian formalism with the thermodynamic features discussed here.
- Finally, in Sect. 6.4.2 we have discussed an alternative variational principle for the null surfaces in Lanczos-Lovelock gravity. It turns out that the variational principle has a nice separation into matter heat density and gravitational heat density associated with the null surface. In this case as well for affine parametrization of null vectors and in vacuum spacetime, i.e., with no matter, the action functional is simply $2\sqrt{-g}P_p^{\ qra}\pounds_\xi \Gamma^p_{qr}$. This provides yet another thermodynamic interpretation for this Lie variation term.

All these results suggest the importance of Noether current in any Lanczos-Lovelock theories of gravity and its relation to the thermodynamic features.

Appendix A

In this appendix, we shall present the supplementary material for this chapter.

A.1 Detailed Expressions Regarding First Law

Let us start with evaluating the following expression in GNC coordinates introduced in the main text, which leads to

$$\begin{aligned}
\frac{1}{2}\left(E^u_u + E^r_r\right) &= E^u_u = E^r_r \\
&= -\frac{1}{2}\frac{1}{16\pi}\frac{1}{2^m}\delta^{ra_1b_1\ldots a_mb_m}_{rc_1d_1\ldots c_md_m} R^{c_1d_1}_{a_1b_1}\ldots R^{c_md_m}_{a_mb_m} \\
&= -\frac{m}{16\pi}\frac{1}{2^{m-1}}\delta^{ruPA_1B_1\ldots A_{m-1}B_{m-1}}_{ruQC_1D_1\ldots C_{m-1}D_{m-1}} R^{uQ}_{uP} R^{C_1D_1}_{A_1B_1}\ldots R^{C_{m-1}D_{m-1}}_{A_{m-1}B_{m-1}} \\
&\quad -\frac{m(m-1)}{16\pi}\frac{1}{2^{m-1}}\delta^{ruPA_1B_1\ldots A_{m-1}B_{m-1}}_{ruQC_1D_1\ldots C_{m-1}D_{m-1}} R^{QC_1}_{uP} R^{uD_1}_{A_1B_1}\ldots R^{C_{m-1}D_{m-1}}_{A_{m-1}B_{m-1}} \\
&\quad -\frac{1}{16\pi}\frac{1}{2^{m+1}}\delta^{rA_1B_1\ldots A_mB_m}_{rC_1D_1\ldots C_mD_m} R^{C_1D_1}_{A_1B_1}\ldots R^{C_mD_m}_{A_mB_m}
\end{aligned} \tag{6.41}$$

Then we obtain:

$$
\begin{aligned}
T^r_r &= 2E^r_r = -\frac{m}{8\pi}\frac{1}{2^{m-1}}\delta^{PA_1B_1\ldots A_{m-1}B_{m-1}}_{QC_1D_1\ldots C_{m-1}D_{m-1}} R^{uQ}_{uP} R^{C_1D_1}_{A_1B_1}\ldots R^{C_{m-1}D_{m-1}}_{A_{m-1}B_{m-1}} \\
&- \frac{m(m-1)}{8\pi}\frac{1}{2^{m-1}}\delta^{PA_1B_1\ldots A_{m-1}B_{m-1}}_{QC_1D_1\ldots C_{m-1}D_{m-1}} R^{QC_1}_{uP} R^{uD_1}_{A_1B_1}\ldots R^{C_{m-1}D_{m-1}}_{A_{m-1}B_{m-1}} \\
&- \frac{1}{16\pi}\frac{1}{2^m}\delta^{A_1B_1\ldots A_mB_m}_{C_1D_1\ldots C_mD_m} R^{C_1D_1}_{A_1B_1}\ldots R^{C_mD_m}_{A_mB_m}
\end{aligned} \tag{6.42}
$$

Now we have the following expression for components of Riemann tensor as:

$$
\begin{aligned}
R^{uQ}_{uP} &= -\frac{1}{2}q^{QE}\partial_u\partial_r q_{PE} - \frac{1}{2}q^{QE}\partial_P\beta_E - \frac{1}{2}\alpha q^{QE}\partial_r q_{PE} - \frac{1}{4}\beta^Q\beta_P \\
&+ \frac{1}{4}\left(q^{QE}\partial_r q_{PF}\right)\left(q^{FL}\partial_u q_{EL}\right) + \frac{1}{2}q^{QE}\beta_A\hat{\Gamma}^A_{EP}
\end{aligned} \tag{6.43a}
$$

$$
R^{AM}_{CD} = \hat{R}^{AM}_{CD} - \frac{1}{4}q^{AE}q^{MB}\Big\{\partial_u q_{CE}\partial_r q_{BD} + \partial_r q_{CE}\partial_u q_{BD} - (C\leftrightarrow D)\Big\} \tag{6.43b}
$$

$$
R^{uN}_{CD} = -\frac{1}{2}q^{MN}\partial_r\partial_C q_{MD} - \frac{1}{4}\beta_C q^{MN}\partial_r q_{MD} - \frac{1}{2}\left(q^{MN}\partial_r q_{CE}\right)\hat{\Gamma}^E_{MD} - (C\leftrightarrow D) \tag{6.43c}
$$

$$
\begin{aligned}
R^{AB}_{uC} &= q^{BD}\Big[\partial_u\hat{\Gamma}^A_{DC} - \frac{1}{2}q^{AE}\partial_C\partial_u q_{DE} - \frac{1}{2}\partial_C q^{AE}\partial_u q_{DE} + \frac{1}{4}\beta^A\partial_u q_{CD} \\
&+ \frac{1}{2}q^{AF}\partial_u q_{EF}\hat{\Gamma}^E_{CD} - \frac{1}{4}\beta_D q^{AE}\partial_u q_{EC} - \frac{1}{2}q^{EF}\partial_u q_{FD}\hat{\Gamma}^A_{CE}\Big]
\end{aligned} \tag{6.43d}
$$

where $\hat{A}$ denotes an object A constructed solely from the transverse metric q_{AB}. Note that for $\partial_u g_{AB} = 0$, we have:

$$
R^{uQ}_{uP} = -\frac{1}{2}q^{QE}\partial_P\beta_E - \frac{1}{2}\alpha q^{QE}\partial_r q_{PE} - \frac{1}{4}\beta^Q\beta_P + \frac{1}{2}q^{QE}\beta_A\hat{\Gamma}^A_{EP} \tag{6.44a}
$$

$$
R^{AM}_{CD} = \hat{R}^{AM}_{CD} \tag{6.44b}
$$

$$
R^{uN}_{CD} = -\frac{1}{2}q^{MN}\partial_r\partial_C q_{MD} - \frac{1}{4}\beta_C q^{MN}\partial_r q_{MD} - \frac{1}{2}\left(q^{MN}\partial_r q_{CE}\right)\hat{\Gamma}^E_{MD} - (C\leftrightarrow D) \tag{6.44c}
$$

$$
R^{AB}_{uC} = 0 \tag{6.44d}
$$

Thus we finally arrive at the following expression:

$$
\begin{aligned}
T^r_r &= -\frac{m}{8\pi}\frac{1}{2^{m-1}}\delta^{PA_1B_1\ldots A_{m-1}B_{m-1}}_{QC_1D_1\ldots C_{m-1}D_{m-1}}\left(-\frac{1}{2}\alpha q^{QE}\partial_r q_{PE}\right) R^{C_1D_1}_{A_1B_1}\ldots R^{C_{m-1}D_{m-1}}_{A_{m-1}B_{m-1}} \\
&- \frac{m}{8\pi}\frac{1}{2^{m-1}}\delta^{PA_1B_1\ldots A_{m-1}B_{m-1}}_{QC_1D_1\ldots C_{m-1}D_{m-1}}\Big[-\frac{1}{2}q^{QE}\partial_u\partial_r q_{PE} - \frac{1}{2}q^{QE}\partial_P\beta_E - \frac{1}{4}\beta^Q\beta_P \\
&+ \frac{1}{4}\left(q^{QE}\partial_r q_{PF}\right)\left(q^{FL}\partial_u q_{EL}\right) + \frac{1}{2}q^{QE}\beta_A\hat{\Gamma}^A_{EP}\Big] R^{C_1D_1}_{A_1B_1}\ldots R^{C_{m-1}D_{m-1}}_{A_{m-1}B_{m-1}}
\end{aligned}
$$

$$
\begin{aligned}
&- \frac{m(m-1)}{8\pi} \frac{1}{2^{m-1}} \delta^{PA_1B_1\ldots A_{m-1}B_{m-1}}_{QC_1D_1\ldots C_{m-1}D_{m-1}} R^{QC_1}_{uP} R^{uD_1}_{A_1B_1} \ldots R^{C_{m-1}D_{m-1}}_{A_{m-1}B_{m-1}} \\
&- \frac{1}{16\pi} \frac{1}{2^m} \delta^{A_1B_1\ldots A_mB_m}_{C_1D_1\ldots C_mD_m} R^{C_1D_1}_{A_1B_1} \ldots R^{C_mD_m}_{A_mB_m}
\end{aligned}
\tag{6.45}
$$

which can be simplified and finally leads to the following expression:

$$
\begin{aligned}
T^r_r &= 2E^r_r = \left(E^u_u + E^r_r\right) \\
&= \frac{m}{8} \frac{1}{2^{m-1}} \left(\frac{\alpha}{2\pi}\right) \left(\delta^{PA_1B_1\ldots A_{m-1}B_{m-1}}_{QC_1D_1\ldots C_{m-1}D_{m-1}} R^{C_1D_1}_{A_1B_1} \ldots R^{C_{m-1}D_{m-1}}_{A_{m-1}B_{m-1}}\right) \left(q^{QE}\partial_r q_{PE}\right) \\
&- \frac{m}{8\pi} \frac{1}{2^{m-1}} \delta^{PA_1B_1\ldots A_{m-1}B_{m-1}}_{QC_1D_1\ldots C_{m-1}D_{m-1}} \Big[-\frac{1}{2} q^{QE}\partial_u\partial_r q_{PE} - \frac{1}{2} q^{QE}\partial_P\beta_E - \frac{1}{4}\beta^Q\beta_P \\
&+ \frac{1}{4}\left(q^{QE}\partial_r q_{PF}\right)\left(q^{FL}\partial_u q_{EL}\right) + \frac{1}{2} q^{QE}\beta_A \hat{\Gamma}^A_{EP} \Big] R^{C_1D_1}_{A_1B_1} \ldots R^{C_{m-1}D_{m-1}}_{A_{m-1}B_{m-1}} \\
&- \frac{m(m-1)}{8\pi} \frac{1}{2^{m-1}} \delta^{PA_1B_1\ldots A_{m-1}B_{m-1}}_{QC_1D_1\ldots C_{m-1}D_{m-1}} R^{QC_1}_{uP} R^{uD_1}_{A_1B_1} \ldots R^{C_{m-1}D_{m-1}}_{A_{m-1}B_{m-1}} \\
&- \frac{1}{16\pi} \frac{1}{2^m} \delta^{A_1B_1\ldots A_mB_m}_{C_1D_1\ldots C_mD_m} R^{C_1D_1}_{A_1B_1} \ldots R^{C_mD_m}_{A_mB_m}
\end{aligned}
\tag{6.46}
$$

This is the expression used in the text. We also have entropy density to be:

$$
\begin{aligned}
s &= 4\pi m\sqrt{q}\mathcal{L}^{(D-2)}_{m-1} \\
&= 4\pi m\sqrt{q}\left(\frac{1}{16\pi} \frac{1}{2^{m-1}} \delta^{A_1B_1\ldots A_{m-1}B_{m-1}}_{C_1D_1\ldots C_{m-1}D_{m-1}} R^{C_1D_1}_{A_1B_1} \ldots R^{C_{m-1}D_{m-1}}_{A_{m-1}B_{m-1}}\right)
\end{aligned}
\tag{6.47}
$$

Then under variation along the radial coordinate i.e. along k^a parametrized by λ we have:

$$
\begin{aligned}
\delta_\lambda s &= 4\pi m\left(\frac{1}{2} q^{AB}\delta_\lambda q_{AB}\right)\sqrt{q}\mathcal{L}^{(D-2)}_{m-1} \\
&- 4\pi m\sqrt{q}\left(\frac{m-1}{16\pi} \frac{1}{2^{m-1}} \delta^{A_1B_1\ldots A_{m-1}B_{m-1}}_{C_1D_1\ldots C_{m-1}D_{m-1}} R^{C_1A}_{A_1B_1} q^{D_1B}\delta_\lambda q_{AB} \ldots R^{C_{m-1}D_{m-1}}_{A_{m-1}B_{m-1}}\right) \\
&= -4\pi m\sqrt{q}\delta_\lambda q_{AB}\Big(-\frac{1}{2} q^{AB}\mathcal{L}^{(D-2)}_{m-1} \\
&+ \frac{m-1}{16\pi} \frac{1}{2^{m-1}} q^{BD_1} \delta^{A_1B_1\ldots A_{m-1}B_{m-1}}_{D_1C_1\ldots C_{m-1}D_{m-1}} R^{AC_1}_{A_1B_1} \ldots R^{C_{m-1}D_{m-1}}_{A_{m-1}B_{m-1}}\Big) \\
&= -4\pi m\sqrt{q}E^{AB}\delta_\lambda q_{AB}
\end{aligned}
\tag{6.48}
$$

where we have:

$$
\begin{aligned}
E^A_B &= -\frac{1}{2}\delta^A_B \mathcal{L}^{(D-2)}_{m-1} + \frac{m-1}{16\pi} \frac{1}{2^{(m-1)}} \delta^{A_1B_1\ldots A_{m-1}B_{m-1}}_{BC_1\ldots C_{m-1}D_{m-1}} R^{AC_1}_{A_1B_1} \ldots R^{C_{m-1}D_{m-1}}_{A_{m-1}B_{m-1}} \\
&= -\frac{1}{2} \frac{1}{16\pi} \frac{1}{2^{m-1}} \delta^{AA_1B_1\ldots A_{m-1}B_{m-1}}_{BC_1D_1\ldots C_{m-1}D_{m-1}} R^{C_1D_1}_{A_1B_1} \ldots R^{C_{m-1}D_{m-1}}_{A_{m-1}B_{m-1}}
\end{aligned}
\tag{6.49}
$$

Hence we obtain:

$$\delta_\lambda s = -4\pi m\sqrt{q}\delta_\lambda q_{AB}\Big(-\frac{1}{2}\frac{1}{16\pi}\frac{1}{2^{m-1}}\delta^{AA_1B_1\ldots A_{m-1}B_{m-1}}_{BC_1D_1\ldots C_{m-1}D_{m-1}}$$
$$\times R^{C_1D_1}_{A_1B_1}\ldots R^{C_{m-1}D_{m-1}}_{A_{m-1}B_{m-1}}\Big)$$
$$= \frac{m}{82^{m-1}}\sqrt{q}\left(\delta^{AA_1B_1\ldots A_{m-1}B_{m-1}}_{BC_1D_1\ldots C_{m-1}D_{m-1}}R^{C_1D_1}_{A_1B_1}\ldots R^{C_{m-1}D_{m-1}}_{A_{m-1}B_{m-1}}\right)q^{BC}\delta_\lambda q_{AC} \tag{6.50}$$

Finally using Eq. (6.50) in Eq. (6.46) we obtain the most general expression for energy as

$$\delta_\lambda E = \delta\lambda\int d\Sigma\Big\{\frac{m}{8\pi}\frac{1}{2^{m-1}}\delta^{PA_1B_1\ldots A_{m-1}B_{m-1}}_{QC_1D_1\ldots C_{m-1}D_{m-1}}\Big[-\frac{1}{2}q^{QE}\partial_u\partial_r q_{PE}-\frac{1}{2}q^{QE}\partial_P\beta_E-\frac{1}{4}\beta^Q\beta_P$$
$$+\frac{1}{4}\left(q^{QE}\partial_r q_{PF}\right)\left(q^{FL}\partial_u q_{EL}\right)+\frac{1}{2}q^{QE}\beta_A\hat{\Gamma}^A_{EP}\Big]R^{C_1D_1}_{A_1B_1}\ldots R^{C_{m-1}D_{m-1}}_{A_{m-1}B_{m-1}}$$
$$+\frac{m(m-1)}{8\pi}\frac{1}{2^{m-1}}\delta^{PA_1B_1\ldots A_{m-1}B_{m-1}}_{QC_1D_1\ldots C_{m-1}D_{m-1}}R^{QC_1}_{uP}R^{uD_1}_{A_1B_1}\ldots R^{C_{m-1}D_{m-1}}_{A_{m-1}B_{m-1}}$$
$$+\frac{1}{16\pi}\frac{1}{2^m}\delta^{A_1B_1\ldots A_mB_m}_{C_1D_1\ldots C_mD_m}R^{C_1D_1}_{A_1B_1}\ldots R^{C_mD_m}_{A_mB_m}\Big\} \tag{6.51}$$

where $d\Sigma = d^{D-2}x\sqrt{q}$ is the integration measure on the null surface.

A.2 Various Identities Used in the Text Regarding Lanczos-Lovelock Gravity

In this subsection we will collect derivation of important identities used in the text while describing the generalization to Lanczos-Lovelock gravity. We will order the derivations as in text.

A.2.1 Gravitational Momentum and related derivations for Einstein-Hilbert Action

In this section we provide derivation to various identities used in Sect. 6.4.1. We start by giving the result for variation of Lanczos-Lovelock Lagrangian:

$$\delta\left(\sqrt{-g}L\right) = \sqrt{-g}E_{ab}\delta g^{ab} - \partial_c\left(2\sqrt{-g}P_p^{\ qrc}\delta\Gamma^p_{qr}\right) \tag{6.52}$$

The Noether current can be written as:

$$J^a(q) = 2\mathcal{R}^a_b q^b + 2P_p^{\ qra}\pounds_q\Gamma^p_{qr}$$
$$= 2E^a_b q^b + 2P_p^{\ qra}\pounds_q\Gamma^p_{qr} + Lq^a$$

$$= 2E^a_b q^b - P^a(q) \tag{6.53}$$

Then multiplying both sides by the four velocity u_a and taking $q_a = \xi_a$ we readily obtain:

$$\begin{aligned} -u_a P^a(\xi) &= u_a J^a(\xi) - 2E^{ab}\xi_a u_b \\ &= D_\alpha\left(2N\chi^\alpha\right) - 2NE^{ab}u_a u_b \end{aligned} \tag{6.54}$$

Let us now consider the variation of the gravitational momentum corresponding to q^a. This has the following expression:

$$\begin{aligned} -\delta\left(\sqrt{-g}P^a\right) &= q^a\delta\left(\sqrt{-g}L\right) + \delta\left(2\sqrt{-g}P_p^{\ qra}\pounds_q\Gamma^p_{qr}\right) \\ &= q^a\left(\sqrt{-g}E_{pq}\delta g^{pq}\right) - q^a\partial_c\left(2\sqrt{-g}P_p^{\ qrc}\delta\Gamma^p_{qr}\right) \\ &+ \pounds_q\left(2\sqrt{-g}P_p^{\ qra}\delta\Gamma^p_{qr}\right) \\ &+ \delta\left(2\sqrt{-g}P_p^{\ qra}\right)\pounds_q\Gamma^p_{qr} - \pounds_q\left(2\sqrt{-g}P_p^{qra}\right)\delta\Gamma^p_{qr} \\ &= \sqrt{-g}E_{pq}\delta g^{pq}q^a + \sqrt{-g}\omega^a + \partial_c\left(2\sqrt{-g}P_p^{\ qr[a}q^{c]}\delta\Gamma^p_{qr}\right) \end{aligned} \tag{6.55}$$

where in arriving at the last line we have used the following relation:

$$\pounds_q Q^a - q^a\partial_c Q^c = \partial_c\left(q^{[c}Q^{a]}\right) \tag{6.56}$$

for the tensor density $Q^c = 2\sqrt{-g}P_p^{\ qrc}\delta\Gamma^p_{qr}$. Also since q^a is a constant vector δ and $\pounds_q$ are assumed to commute. Also the object ω^a is defined as,

$$\sqrt{-g}\omega^a\left(\delta,\pounds_q\right) = \delta\left(2\sqrt{-g}P_p^{\ qra}\right)\pounds_q\Gamma^p_{qr} - \pounds_q\left(2\sqrt{-g}P_p^{\ qra}\right)\delta\Gamma^p_{qr} \tag{6.57}$$

This is the result used in Eq. (6.25). Then we can write the above relation in terms of the Noether current as:

$$\begin{aligned} \delta\left(\sqrt{-g}J^a - 2\sqrt{-g}E^a_b q_b\right) &= \sqrt{-g}E_{pq}\delta g^{pq}q^a + \sqrt{-g}\omega^a \\ &+ \partial_c\left(2\sqrt{-g}P_p^{\ qr[a}q^{c]}\delta\Gamma^p_{qr}\right) \end{aligned} \tag{6.58}$$

The above relation can also be written using the Noether potential as:

$$\begin{aligned} \partial_b\left\{\delta\left(\sqrt{-g}J^{ab}\right) - 2\sqrt{-g}P_p^{\ qr[a}q^{b]}\delta\Gamma^p_{qr}\right\} &= \sqrt{-g}E_{pq}\delta g^{pq}q^a + \sqrt{-g}\omega^a \\ &+ 2\delta\left(\sqrt{-g}E^{ab}q_b\right) \end{aligned} \tag{6.59}$$

While for on-shell (i.e., when $E_{ab} = 0$) we have the following relations:

$$-\delta\left(\sqrt{-g}P^a\right) = \sqrt{-g}\omega^a + \partial_c\left(2\sqrt{-g}P_p^{\ qr[a}q^{c]}\delta\Gamma^p_{qr}\right) \tag{6.60}$$

$$\partial_b \left\{\delta\left(\sqrt{-g}J^{ab}\right) - 2\sqrt{-g}P_p{}^{qr[a}q^{b]}\delta\Gamma^p_{qr}\right\} = \sqrt{-g}\omega^a + 2\delta\left(\sqrt{-g}E^{ab}q_b\right) \quad (6.61)$$

Then integrating the second equation over volume $\mathcal{R}$ with $d^{D-1}x\sqrt{h}$ as integration measure and $q^a = \zeta^a = Nu^a + N^a$ we arrive at:

$$\begin{aligned}
&\delta \int_{\mathcal{R}} d^{D-1}x\sqrt{h}\left(2u_a E^{ab}\zeta_b\right) \\
&= \int_{\mathcal{R}} d^{D-1}x\partial_b\left\{\delta\left[\sqrt{h}u_a J^{ab}(\zeta)\right] - 2\sqrt{h}u_a\left(NP_p{}^{qr[a}u^{b]} + P_p{}^{qr[a}N^{b]}\right)\delta\Gamma^p_{qr}\right\} \\
&- \int_{\mathcal{R}} d^{D-1}x\sqrt{h}u_a\omega^a\left(\delta, \pounds_q\right) \\
&= \int_{\mathcal{R}} d^{D-1}x\partial_b\left\{\delta\left[\sqrt{h}u_a J^{ab}(\zeta)\right] - 2\sqrt{h}\left(Nh^b_a P_p{}^{qra} + P_p{}^{qr[a}N^{b]}\right)\delta\Gamma^p_{qr}\right\} \\
&- \int_{\mathcal{R}} d^{D-1}x\sqrt{h}u_a\omega^a\left(\delta, \pounds_q\right) \qquad (6.62)
\end{aligned}$$

Then we want the variation of the Hamiltonian obtained by contracting the momentum along the four velocity u_a, such that:

$$-\sqrt{h}u_a P^a(\xi) = t_a\sqrt{-g}P^a(\xi) \quad (6.63)$$

where the vector $t_a = -u_a/N$ in the coordinate system under consideration. Then varying the above expression (noting that variation of t_a vanishes) we obtain

$$\begin{aligned}
-\delta\left[\sqrt{h}u_a P^a(\xi)\right] &= t_a\delta\left[\sqrt{-g}P^a(\xi)\right] \\
&= -t_a\left[\sqrt{-g}E_{pq}\delta g^{pq}\xi^a + \sqrt{-g}\omega^a + \partial_c\left(2\sqrt{-g}P_p{}^{qr[a}\xi^{c]}\delta\Gamma^p_{qr}\right)\right] \\
&- \sqrt{h}u_u\omega^a \quad \sqrt{\ g}E_{pq}\delta g^{pq} + \partial_c\left[2\sqrt{-g}u_a P_p{}^{qr[a}u^{c]}\delta\Gamma^p_{qr}\right] \\
&= \sqrt{h}u_a\omega^a - \sqrt{-g}E_{pq}\delta g^{pq} + \partial_c\left[2\sqrt{-g}h^c_a P_p{}^{qra}\delta\Gamma^p_{qr}\right] \quad (6.64)
\end{aligned}$$

where in the second line we have used the standard trick in order to get u_a and in the last line we have used the relation:

$$u_a P_p{}^{qr[a}u^{c]} = h^c_a P_p{}^{qra} \quad (6.65)$$

Defining the gravitational Hamiltonian as:

$$\mathcal{H}_{grav} = -\int d^{D-1}x\sqrt{h}u_a P^a(\xi) \quad (6.66)$$

its variation can be obtained readily from Eq. (6.64) as:

$$\delta\mathcal{H}_{grav} = \int d^{D-1}x\sqrt{h}u_a\omega^a - \int d^{D-1}x\sqrt{-g}E_{pq}\delta g^{pq} + \int d^{D-2}x\, 2r_c\sqrt{q}P_p^{\ qrc}\delta\Gamma^p_{qr} \tag{6.67}$$

where in order to obtain the last term we have used the result that $r_c h^c_a = r_a$ and $\sqrt{-g} = N\sqrt{h}$. The above results are true for arbitrary variations. Applying it to Lie variation along ξ^a we arrive at the following form for Eq. (6.64) as:

$$-\pounds_\xi\left[\sqrt{h}u_a P^a(\xi)\right] = 2\sqrt{-g}E_{pq}\nabla^p\xi^q + \partial_c\left[2\sqrt{-g}h^c_a P_p^{\ qra}\pounds_\xi\Gamma^p_{qr}\right] \tag{6.68}$$

In arriving at the above result we have used the relations: $\pounds_\xi g^{ab} = -(\nabla^a\xi^b + \nabla^b\xi^a)$ and $\omega^a(\pounds_\xi, \pounds_\xi) = 0$. Now using Bianchi identity $\nabla_a E^{ab} = 0$ we arrive at:

$$-\pounds_\xi\left[\sqrt{h}u_a P^a(\xi)\right] = \partial_c\left[2\sqrt{-g}\left(E^{cd}\xi_d + h^c_a P_p^{\ qra}\pounds_\xi\Gamma^p_{qr}\right)\right] \tag{6.69}$$

This is the relation used to arrive at the results in the main text.

A.2.2 Characterizing Null Surfaces

Let us now try to generalize the result for null surfaces to Lovelock gravity with the null vector ℓ_a such that $\ell^2 = 0$ everywhere. For that purpose, we start with the combination:

$$\begin{aligned}
\mathcal{R}_{ab}\ell^a\ell^b &= R_{apqr}P_b^{\ pqr}\ell^a\ell^b \\
&= P^{bpqr}\ell_b\left(R_{apqr}\ell^a\right) \\
&= -P^{bpqr}\ell_b\left(\nabla_q\nabla_r\ell_p - \nabla_r\nabla_q\ell_p\right) \\
&= -2P^{bpqr}\ell_b\nabla_q\nabla_r\ell_p \\
&= \nabla_q\left(-2P^{bpqr}\ell_b\nabla_r\ell_p\right) - \mathcal{S}
\end{aligned} \tag{6.70}$$

where we have defined, the entropy density as:

$$\mathcal{S} = -2P^{bpqr}\nabla_q\ell_b\nabla_r\ell_p \tag{6.71}$$

The Einstein-Hilbert limit can be obtained easily leading to:

$$\begin{aligned}
16\pi\mathcal{S} &= -\left(g^{bq}g^{pr} - g^{br}g^{pq}\right)\nabla_q\ell_b\nabla_r\ell_p \\
&= \nabla_a\ell^b\nabla_b\ell^a - \left(\nabla_i\ell^i\right)^2
\end{aligned} \tag{6.72}$$

as well as,

$$
\begin{aligned}
32\pi P^{bpqr}\ell_b\nabla_r\ell_p &= \left(g^{bq}g^{pr} - g^{br}g^{pq}\right)\ell_b\nabla_r\ell_p \\
&= \ell^q\nabla_r\ell^r - \ell^r\nabla_r\ell^q \\
&= \Theta\ell^q
\end{aligned}
\tag{6.73}
$$

However in Lanczos-Lovelock gravity, the combination, $2P^{bpqr}\ell_b\nabla_r\ell_p$ cannot be written as, $\phi\ell^q$, for arbitrary ϕ, since $2P^{bpqr}\ell_q\ell_b\nabla_r\ell_p \neq 0$. Next let us consider the object, $\ell_a J^a(\ell)$, for which we define, $\ell_a = A\nabla_a B$. Then in Lanczos-Lovelock gravity, we arrive at:

$$
\begin{aligned}
\frac{1}{A}\ell_a J^a(\ell) &= \nabla_b\left[2P^{abcd}\ell_a\ell_d\nabla_c A\frac{1}{A^2}\right] \\
&= \frac{1}{A}\nabla_b\left[2P^{abcd}\ell_a\ell_d\nabla_c \ln A\right] - \frac{1}{A}\left[2P^{abcd}\ell_a\ell_d\nabla_c \ln A\nabla_b \ln A\right]
\end{aligned}
\tag{6.74}
$$

Let us now expand $\nabla_c \ln A$ in canonical null basis, such that we obtain,

$$
\nabla_c \ln A = -\kappa k_c + A\ell_c + B_A e^A_c \tag{6.75}
$$

Note that we obtain, $\ell^c\nabla_c \ln A = \kappa$. In a similar fashion, we can expand, $2P^{abcd}\ell_a\ell_d \nabla_c \ln A$ in the following manner,

$$
2P^{abcd}\ell_a\ell_d\nabla_c \ln A = P\ell^b + Qk^b + R^A e^b_A \tag{6.76}
$$

It is evident that $Q = -2P^{abcd}\ell_a\ell_d\ell_b\nabla_c \ln A = 0$, due to antisymmetry of P^{abcd} in the first two indices. Then it turns out that,

$$
P = -2P^{abcd}\ell_a k_b\ell_d\nabla_c \ln A \equiv \mathcal{K} \tag{6.77}
$$

It is obvious from the Einstein-Hilbert limit, that

$$
\begin{aligned}
16\pi\mathcal{K} &= -\left(g^{ac}g^{bd} - g^{ad}g^{bc}\right)\ell_a k_b\ell_d\nabla_c \ln A \\
&= -\left(g^{ac}g^{bd} - g^{ad}g^{bc}\right)\ell_a k_b\ell_d\left(-\kappa k_c + B_A e^A_c\right) = \kappa
\end{aligned}
\tag{6.78}
$$

Also in the expansion for, $\nabla_c \ln A$, we obtain, B_A is completely arbitrary. Then we can use B_A such that the following relation: $B_Q P^{abcd}\ell_a\ell_d e^A_b e^Q_c = \kappa P^{abcd}\ell_a k_c\ell_d e^A_b$ is satisfied, such that, R^A vanishes. Thus we obtain,

$$
2P^{abcd}\ell_a\ell_d\nabla_c \ln A = \mathcal{K}\ell^b \tag{6.79}
$$

Again, we get,

$$
2P^{abcd}\ell_a\ell_d\nabla_c \ln A\nabla_b \ln A = \mathcal{K}\ell^b\nabla_b \ln A - \kappa\mathcal{K} \tag{6.80}
$$

Thus, finally we arrive at the following result:

$$\ell_a J^a(\ell) = \nabla_a\left(\mathcal{K}\ell^a\right) - \kappa\mathcal{K} \tag{6.81}$$

Again by considering derivative on the null surface, we obtain,

$$\begin{aligned} D_a\left(\mathcal{K}\ell^a\right) &= \left(g^{ab} + \ell^a k^b + \ell^b k^a\right)\nabla_a\left(\mathcal{K}\ell_b\right) \\ &= \nabla_a\left(\mathcal{K}\ell^a\right) + k^b\ell^a\nabla_a\left(\mathcal{K}\ell_b\right) \\ &= \nabla_a\left(\mathcal{K}\ell^a\right) - \kappa\mathcal{K} - \ell^a\nabla_a\mathcal{K} \end{aligned} \tag{6.82}$$

Thus the Noether current contraction can also be written as:

$$\ell_a J^a(\ell) = D_a\left(\mathcal{K}\ell^a\right) + \frac{d\mathcal{K}}{d\lambda} \tag{6.83}$$

References

1. T. Padmanabhan, General relativity from a thermodynamic perspective. Gen. Relativ. Gravit. **46**, 1673 (2014). arXiv:1312.3253 [gr-qc]
2. D. Kothawala, T. Padmanabhan, Thermodynamic structure of Lanczos-Lovelock field equations from near-horizon symmetries. Phys. Rev. D **79**, 104020 (2009). arXiv:0904.0215 [gr-qc]
3. A. Paranjape, S. Sarkar, T. Padmanabhan, Thermodynamic route to field equations in Lancos-Lovelock gravity. Phys. Rev. D **74**, 104015 (2006). arXiv:hep-th/0607240
4. S. Chakraborty, K. Parattu, T. Padmanabhan, Gravitational field equations near an arbitrary null surface expressed as a thermodynamic identity. JHEP **10**, 097 (2015). arXiv:1505.05297 [gr-qc]
5. D. Kothawala, The thermodynamic structure of Einstein tensor. Phys. Rev. D **83**, 024026 (2011). arXiv:1010.2207 [gr-qc]
6. S.A. Hayward, Unified first law of black hole dynamics and relativistic thermodynamics. Class. Quantum Gravity **15**, 3147–3162 (1998). arXiv:gr-qc/9710089 [gr-qc]
7. D. Kastor, S. Ray, J. Traschen, Enthalpy and the mechanics of AdS black holes. Class. Quantum Gravity **26**, 195011 (2009). arXiv:0904.2765 [hep-th]
8. B.P. Dolan, Pressure and volume in the first law of black hole thermodynamics. Class. Quantum Gravity **28**, 235017 (2011). arXiv:1106.6260 [gr-qc]
9. D. Kubiznak, R.B. Mann, P-V criticality of charged AdS black holes. JHEP **07**, 033 (2012). arXiv:1205.0559 [hep-th]
10. A.M. Frassino, D. Kubiznak, R.B. Mann, F. Simovic, Multiple reentrant phase transitions and triple points in lovelock thermodynamics. JHEP **09**, 080 (2014). arXiv:1406.7015 [hep-th]
11. T. Jacobson, R.C. Myers, Black hole entropy and higher curvature interactions. Phys. Rev. Lett. **70**, 3684–3687 (1993). arXiv:hep-th/9305016 [hep-th]
12. T. Clunan, S.F. Ross, D.J. Smith, On Gauss-Bonnet black hole entropy, Class. Quantum Gravity **21**, 3447–3458 (2004). arXiv:gr-qc/0402044 [gr-qc]
13. R.M. Wald, Black hole entropy is the Noether charge. Phys. Rev. D **48**, 3427–3431 (1993). arXiv:gr-qc/9307038 [gr-qc]
14. K. Parattu, B.R. Majhi, T. Padmanabhan, Structure of the gravitational action and its relation with horizon thermodynamics and emergent gravity paradigm, Phys. Rev. D **87**, 124011 (2013). arXiv:gr-qc/1303.1535 [gr-qc], doi:10.1103/PhysRevD.87.124011

15. S. Chakraborty, T. Padmanabhan, Geometrical variables with direct thermodynamic significance in Lanczos-Lovelock gravity. Phys. Rev. D **90**(8), 084021 (2014). arXiv:1408.4791 [gr-qc]
16. B. Julia, S. Silva, Currents and superpotentials in classical gauge invariant theories. 1. Local results with applications to perfect fluids and general relativity. Class. Quantum Gravity **15**, 2173–2215 (1998). arXiv:gr-qc/9804029 [gr-qc]
17. T. Regge, C. Teitelboim, Role of surface integrals in the Hamiltonian formulation of general relativity. Ann. Phys. **88**, 286 (1974)
18. S. Carlip, Entropy from conformal field theory at killing horizons, Class. Quantum Gravity **16**, 3327–3348 (1999). arXiv:gr-qc/9906126 [gr-qc]
19. B.R. Majhi, T. Padmanabhan, Noether current, horizon Virasoro algebra and entropy. Phys. Rev. D **85**, 084040 (2012). arXiv:1111.1809 [gr-qc]
20. B.R. Majhi, S. Chakraborty, Anomalous effective action, Noether current, virasoro algebra and horizon entropy. Eur. Phys. J. C **74**, 2867 (2014). arXiv:1311.1324 [gr-qc]
21. K. Parattu, S. Chakraborty, B.R. Majhi, T. Padmanabhan, Null surfaces: counter-term for the action principle and the characterization of the gravitational degrees of freedom. arXiv:1501.01053 [gr-qc]
22. T. Padmanabhan, A. Paranjape, Entropy of null surfaces and dynamics of spacetime. Phys. Rev. D **75**, 064004 (2007). arXiv:gr-qc/0701003 [gr-qc]

Chapter 7
Null Surface Geometry and Associated Thermodynamics

7.1 Introduction

The development of emergent gravity paradigm has helped us in numerous ways to understand several interesting features of classical gravity itself [1], further bolstering our confidence in the veracity of this approach. In particular, there seems to exist a curious relationship between Einstein's field equations and the structure of null surfaces in the spacetime. Previous works have shown that the above relationship is manifest in *three* different ways:

- There arises an identification between Navier-Stokes equation of fluid dynamics and Einstein's equations. Einstein's equations, when projected on an *arbitrary* null surface, in *any* spacetime, leads to Navier-Stokes equation of fluid dynamics [2, 3].
- The Einstein's equations when projected on to a null surface, one can get [4–8] a thermodynamic identity of the form $T\delta_\lambda S = \delta_\lambda E + P\delta_\lambda V$ in which the symbols have their usual meanings and the variation can be interpreted as changes due to virtual displacement of the null surface along null geodesics parametrized by the affine parameter λ off the surface. Even though initially proved for a few configurations with a high level of symmetry [8, 9], this result has now been generalized for arbitrary null surfaces in both general relativity and Lanczos-Lovelock theories of gravity [10, 11].
- A comparatively more approximate relationship between the null surfaces and Einstein's equations emerged from the early work [12], in which a 'derivation' of Einstein's field equations has been given using the local Rindler horizon as a null surface and the Clausius relation. The above 'derivation' relies heavily on the structure of Raychaudhuri equation as well as the following assumptions: (a) the entropy density is one quarter of the transverse area and, more importantly, (b) the quadratic terms in the Raychaudhuri equation (involving the squares of shear and expansion) can be set to zero. Since neither the Raychaudhuri equation nor the assumption that entropy is proportional to the horizon area hold for theories more

S. Chakraborty, *Classical and Quantum Aspects of Gravity in Relation to the Emergent Paradigm*, Springer Theses, DOI 10.1007/978-3-319-63733-4_7

general than Einstein's gravity, this approach could not naturally be generalized to more general class of theories.

In this chapter, we shall revisit these issues and elaborate further on them using another recent development, motivated by the emergent gravity paradigm. The chapter is organized as follows: In Sect. 7.2 we discuss the Noether current and the associated thermodynamic results for both the $(1+3)$ foliation and for an arbitrary null surface using the Gaussian null coordinates. The thermodynamic interpretation of the reduced gravitational momentum has been given in Sect. 7.3. The final section, i.e., Sect. 7.4 discusses three different projections of the gravitational momentum vis-á-vis an arbitrary null surface and their thermodynamic interpretation. We end the chapter with a short discussion on the results obtained.

7.2 Noether Current and Spacetime Thermodynamics: Null Surfaces

There exists two natural foliations of any spacetime, one based on the $(1+3)$ split and the other adapted to a fiducial null surface. These coordinate charts also come equipped with certain natural vector fields. In the case of the $(1+3)$ split, the time evolution is related to the vector field ξ^a as introduced in [13]. While in the case of the foliation based on a null surface, we again have a natural time evolution field ξ^a given by Eq. (2.51) as well as the null vector ℓ_a which is tangent to the null congruence defining the null surface. It would be interesting to study the Noether current and the gravitational momenta corresponding to the time development vector field associated with null surfaces, which turn out to have direct thermodynamic significance. In this section we shall consider the Noether currents and will take up the properties of gravitational momenta in Sect. 7.4.

Let us consider the thermodynamic interpretation of the Noether current when we use the time development vector field adapted to the null surface in the GNC. To begin with one can consider the following result—Noether charge contained in a bulk region equals the heat content of the boundary, however derived for a subregion of a spacelike hypersurface. It will turn out that a similar result holds for a null surface as well. Given the fact that the Noether current associated with the time development vector for $(1+3)$ split led to a nice thermodynamic interpretation, we will consider the object $\ell_a J^a(\xi)$. (In GNC ξ^a has the components as in Eq. (2.51) and ℓ_a is given by Eq. (2.58a), but of course the results derived here are covariant.). It then turns out that (see Eq. (7.120) of Appendix A.3),

$$16\pi\ell_a J^a(\xi) = J^r(\xi) = \frac{1}{\sqrt{q}}\frac{d}{d\lambda}\left(2\alpha\sqrt{q}\right) \tag{7.1}$$

with λ being the parameter along the null generator ℓ^a, which in GNC is simply u and we have reintroduced the 16π factor. This equation can be integrated over the null surface with an integration measure $\sqrt{q}d^2x d\lambda$ and results into,

$$\int d^2x d\lambda \sqrt{q}\ell_a J^a(\xi) = \int d^2x \left(\frac{\alpha}{2\pi}\right)\left(\frac{\sqrt{q}}{4}\right)\Big|_{\lambda=\lambda_1}^{\lambda=\lambda_2} = Q(\lambda_2) - Q(\lambda_1) \quad (7.2)$$

where

$$Q(\lambda) = \int d^2x \left(\frac{\alpha}{2\pi}\right)\left(\frac{\sqrt{q}}{4}\right) = \int d^2x T s \quad (7.3)$$

is the heat content of the null surface at a given parameter value λ.

This again depicts that the total Noether charge density for the vector field ξ^a, as integrated over the null surface equals the difference between the heat content Q of the two dimensional boundaries located at $\lambda = \lambda_2$ and $\lambda = \lambda_1$. Previously the connection between bulk Noether charge to surface heat density was derived in the context of spacelike hypersurfaces. The result in Eq. (7.2) generalizes the previous connection—between bulk Noether charge and surface heat density in the context of spacelike surfaces—to any arbitrary null surface, by showing that the total Noether charge on a null surface is also expressible as the heat content of the boundary.

At this outset let us mention about two results (see Eqs. (5.1) and (5.50)) of similar nature having relevance with our ongoing discussion. The first one (see Eq. (5.1)) was for a spatial volume $\mathcal{V}$ contained in a space-like hypersurface. In that case, the normal to the surface was u_a and the natural integration measure, for integrating the normal component of a vector field was $u_a\sqrt{h}\,d^3x$. Given the correct measure, we computed the integral $J^a u_a\sqrt{h}\,d^3x$ and found that it is given by a boundary term; we could have also computed the integral in a region contained within two boundary surfaces like, for example, in the shell-like region between two spherical surfaces of radii R_1 and R_2. We would have then found that the Noether charge in the bulk region is the difference between the heat contents associated with the two surfaces.

In the second case, (see Eq. (5.50) in Appendix A) we considered the *flux* of Noether current through a timelike hypersurface with normal $\hat{a}^i$. In this case we calculated the integral of $\hat{a}_p J^p(\xi)$ with the measure being $d^2x dt\sqrt{-g_\perp}$ as fit for a timelike surface and arrived at a similar result. We also mentioned that in this case, the integrand $\hat{a}_p J^p(\xi)$ can be thought of as heat flux per unit time.

In the case of a null surface, our result is similar in spirit to the second one, given by Eq. (5.50). The corresponding integration measure for integrating a vector field over a null surface is given by $\ell_a\sqrt{q}\,d^2x\,d\lambda$ with $\ell^a = dx^a/d\lambda$ being the tangent vector to the null congruence defining the null surface. Therefore, in this case, computation of the Noether charge corresponds to calculating the integral over $J^a\ell_a\sqrt{q}\,d^2x\,d\lambda$. This leads to the difference ΔQ of the heat content at the two boundaries corresponding to two parametrizations $\lambda = \lambda_1$ and $\lambda = \lambda_2$ respectively. In the integrand in this case, which is $J^a\ell_a\sqrt{q}\,d^2x\,d\lambda$, one of the coordinates λ is similar to a time coordinate rather than a spatial coordinate. So one should not think of $J^a\ell_a$ as charge per unit *volume*; instead it represents charge per unit area (flux) per unit time and more appropriately, it is the *rate of production of heat per unit area* of the null surface.

It is possible to proceed further by relating the change in the heat content with the matter energy flux through the null surface. To do this, we use Eq. (2.30) and the Einstein's equation $G_{ab} = 8\pi T_{ab}$ to obtain (on the null surface):

$$16\pi T_{ab}\ell^a\ell^b = 16\pi\ell_a J^a(\xi) - \ell_a g^{bc}\pounds_\xi N^a_{bc} \tag{7.4}$$

The Lie derivative term can be computed directly to yield (see Eq. (7.158) of Appendix A.3):

$$\sqrt{q}\ell_a g^{ij}\pounds_\xi N^a_{ij} = -q_{ab}\pounds_\xi \Pi^{ab} + \frac{d^2\sqrt{q}}{d\lambda^2} \tag{7.5}$$

with $\Pi^{ab} = \sqrt{q}[\Theta^{ab} - q^{ab}(\Theta + \kappa)]$ being the momentum conjugate to q_{ab} and $\Theta^{ab} = q^a_m q^b_n \nabla^m \ell^n$ where Θ is the trace of Θ^{ab}. Integrating the above result over the null surface between two values of the parameter, namely $\lambda = \lambda_1$ and $\lambda = \lambda_2$, and assuming for simplicity that the boundary terms at $\lambda = \lambda_1, \lambda_2$ do not contribute (which assumes $d\mathcal{A}/d\lambda = 0$ at the end points where $\mathcal{A}$ is the area of the null surface), we obtain:

$$-\frac{1}{16\pi}\int d^2x d\lambda\, q_{ab}\pounds_\xi \Pi^{ab} = [Q(\lambda_2) - Q(\lambda_1)] - \int d\lambda d^2x\sqrt{q}T_{ab}\ell^a\ell^b \tag{7.6}$$

This expression explicitly demonstrates that the evolution of the spacetime, which is encoded by the evolution of the momentum Π^{ab} is driven by the difference between (i) heat content Q at the boundary and (ii) the matter heat flux flowing into the null surface. It is possible to rewrite the above in a nicer manner as follows. Defining the surface degrees of freedom as:

$$N_{\rm sur} = A = \int d^2x\sqrt{q} \tag{7.7}$$

and the average temperature over the null surface as:

$$T_{\rm avg} = \frac{1}{A}\int d^2x\,\sqrt{q}\left(\frac{\alpha}{2\pi}\right) \tag{7.8}$$

We further introduce the effective bulk degrees of freedom by:

$$N_{\rm bulk} = \frac{1}{(1/2)T_{\rm avg}}\int d\lambda d^2x\sqrt{q}2\bar{T}_{ab}\ell^a\ell^b \tag{7.9}$$

If the matter heat flux, which is given by the integral in the right hand side, thermalizes at the average temperature of the null surface then, $N_{\rm bulk}$ will represent the effective equipartition degrees of freedom. We now rewrite Eq. (7.6) as,

$$-\frac{1}{8\pi}\int d^2x d\lambda\, q_{ab}\pounds_\xi \Pi^{ab} = \frac{1}{2}\int_{\lambda_2} d^2x \left(\frac{\alpha}{2\pi}\right)\sqrt{q} - \frac{1}{2}\int_{\lambda_1} d^2x \left(\frac{\alpha}{2\pi}\right)\sqrt{q}$$
$$-2\int d\lambda d^2x \sqrt{q} T_{ab}\ell^a \ell^b \tag{7.10}$$

which, on using our definitions introduced above, becomes:

$$-\frac{1}{8\pi}\int d^2x d\lambda\, (q_{ab}\pounds_\xi \Pi^{ab}) = \frac{1}{2}T_{\rm avg}\left[(N_{\rm sur})^{\lambda_2}_{\lambda_1} - N_{\rm bulk}\right] \tag{7.11}$$

This has the interesting interpretation that the evolution of spacetime on a null surface, encoded in the Lie variation of the momentum Π^{ab} along the time development vector, can be thought of as due to the difference between the surface degrees of freedom and the bulk degrees of freedom. This is fully in consonance to the corresponding result in $(1 + 3)$ foliation presented in Eq. (5.2), which was originally obtained in [13].

For the sake of completeness, we would like to clarify the notion of $T_{ab}\ell^a\ell^b$ being the heat flux through the null surface. This concept has been introduced earlier in [12] and arises as follows: Let there be a matter field in the spacetime, with some energy momentum tensor T_{ab}. Around any given spacetime event $\mathcal{P}$, one can always construct local inertial and hence local Rindler frames. We then have an approximate Killing vector field ξ^a, generating boosts and coinciding with the null normal ℓ^a at the null surface. The heat flow vector is defined as the boost energy current, which can be obtained by projecting T_{ab} along ξ^b yielding $T_{ab}\xi^b$. Thus the energy (heat) flux through the null surface have the expression:

$$Q = \int \left(T_{ab}\xi^b\right) d\Sigma^a = \int T_{ab}\xi^b \ell^a \sqrt{q} d^2x d\lambda \tag{7.12}$$

where $\sqrt{q}d^2xd\lambda$ is the integration measure on a null surface generated by the null vector ℓ^a and parametrized by λ. Hence, in the null limit, when $\xi^a \to \ell^a$, $T_{ab}\ell^a\ell^b$ represents the heat flux through the null surface.

The same argument can also be phrased along the following lines. On a null surface one can decompose $T_{ab}\xi^b$ in canonical null basis as, $T_{ab}\xi^b = A\ell_a + Bk_a + C_A e^A_a$. Then the heat flux *through* the null surface is given by the component B along k^a, which is off the surface. This component B is obtained by contracting $T^{ab}\xi_b$ with ℓ_a (since $\ell^2 = 0$ and $\ell_a k^a = -1$). This leads to the heat flux through the null surface to be $T_{ab}\ell^a\ell^b$.

7.3 Reduced Gravitational Momentum and Time Development Vector

In Chap. 2 we have introduced the notion of Noether current as well as the gravitational momentum for general relativity, while the same for Lanczos-Lovelock gravity has been introduced in Chaps. 2 and 6 respectively. In all these cases we notice that the following combinations

$$\mathcal{P}^a = -g^{ij}£_v N^a_{ij}; \qquad \mathcal{P}^a = -2P_p^{\ qra}£_v \Gamma^p_{qr} \tag{7.13}$$

appear quite naturally in them. We shall call this particular combination as *reduced* gravitational momentum $\mathcal{P}^a$. (Later on we will see that $\mathcal{P}^a$ is closely related to the rate of production of heat per unit area on null surfaces.)

The algebraic reason for the occurrence of this combination can be understood as follows. In the thermodynamic interpretation of gravity, the combination $\bar{T}_{ab} = T_{ab} - (1/2)Tg_{ab}$ occurs more naturally than the energy momentum tensor T_{ab}, with the gravitational field equations often arising [13, 14] in the form $2R_{ab} = \bar{T}_{ab}$ rather than $2G_{ab} = T_{ab}$ as it should in the Einstein's gravity. The total momentum of gravity plus matter can be expressed, *on-shell*, as follows:

$$\begin{aligned}\left(P^a + \mathcal{M}^a\right) &= -g^{ij}£_v N^a_{ij} - Rv^a - T^a_b v^b \\ &= -g^{ij}£_v N^a_{ij} + \frac{1}{2}T\delta^a_b v^b - T^a_b v^b \\ &= -g^{ij}£_v N^a_{ij} - \bar{T}^a_b v^b \equiv (\mathcal{P}^a + \bar{\mathcal{M}}^a)\end{aligned} \tag{7.14}$$

where $\bar{\mathcal{M}}^a = -\bar{T}^a_b v^b$ is the matter momentum associated with $\bar{T}^a_b$. This explicitly demonstrates that the vector $\mathcal{P}^a = -g^{ij}£_v N^a_{ij}$ bears the same relation to $\bar{T}^a_b$ as P^a does with T^a_b. Just as $\bar{T}^a_b$ appears more naturally than T^a_b itself in the emergent gravity paradigm, the $\mathcal{P}^a$ will also appear repeatedly in our discussions.

The notion of reduced gravitational momentum can also be generalized to the Lanczos-Lovelock models of gravity as well. For pure m-th order Lanczos-Lovelock gravity we have the following relation $2[m-(D/2)]L = T$. With the help of the same we can write,

$$\begin{aligned}\left(P^a + \mathcal{M}^a\right) &= -2P_p^{\ qra}£_v \Gamma^p_{qr} - Lv^a - T^a_b v^b \\ &= -2P_p^{\ qra}£_v \Gamma^p_{qr} - \frac{1}{2[m-(D/2)]}T\delta^a_b v^b - T^a_b v^b \\ &= -2P_p^{\ qra}£_v \Gamma^p_{qr} - \bar{T}^a_b v^b \equiv \mathcal{P}^a + \bar{\mathcal{M}}^a\end{aligned} \tag{7.15}$$

where the reduced gravitational momentum naturally appears as

$$\mathcal{P}^a = -2P_p^{\ qra}£_v \Gamma^p_{qr} \tag{7.16}$$

In the case of general relativity, the gravitational momentum $\mathcal{P}^a$ as defined in Eq. (7.16) goes over to the $-g^{ij}\pounds_v N^a_{ij}$ term, as it should. We shall now describe some key properties of this reduced gravitational momentum vector in different contexts, while emphasising the results in GNC.

In the case of (1 + 3) foliation one can use spacelike surfaces with normal $u_a = -N\nabla_a t$ (N is the lapse function) and induced metric $h_{ab} = g_{ab} + u_a u_b$, such that the reduced gravitational momentum can be related [13] to the Lie variation of $p^{ab} = \sqrt{h}(Kh^{ab} - K^{ab})$ by:

$$-\sqrt{h}u_a\mathcal{P}^a(\xi) = \sqrt{h}u_a g^{ij}\pounds_\xi N^a_{ij} = -h_{ab}\pounds_\xi p^{ab} \tag{7.17}$$

where, $\xi_a = Nu_a$, is a vector field associated with the (1 + 3) foliation of the spacetime (see e.g., [13]).

In the case of null surfaces as well, one can obtain a similar relation. For a general, non-affine parametrization, i.e., when the null generator ℓ^a satisfies the following relation $\ell^b\nabla_b\ell^a = \kappa\ell^a$, we find that the Lie variation term is given by (see Eq. (7.158) of Appendix A.3):

$$-\sqrt{q}\ell_a\mathcal{P}^a(\xi) = \sqrt{q}\ell_a g^{ij}\pounds_\xi N^a_{ij} = -q_{ab}\pounds_\xi\Pi^{ab} + \frac{d^2\sqrt{q}}{d\lambda^2} \tag{7.18}$$

with $\Pi^{ab} = \sqrt{q}[\Theta^{ab} - q^{ab}(\Theta + \kappa)]$ being the momentum conjugate to q_{ab}. Here, ξ^a is the time evolution vector associated with the null foliations, such that $\xi^a \to \ell^a$ as one approaches the null surface (in particular, see also Eq. (2.51)). Thus in the spirit of spacelike surfaces, for null surfaces as well, the Lie variation of N^c_{ab} is directly related to the Lie variation of the momentum conjugate to the induced metric on the null surface. But in the case of null surfaces there is an extra term contributing only at the boundaries $\lambda = \lambda_1, \lambda_2$.

It follows immediately from straightforward algebra (see Eq. (7.90) of Appendix A.1) that $q_{ab}\pounds_\xi\Pi^{ab}$ is directly related to the object $\mathcal{D} = \Theta_{ab}\Theta^{ab} - \Theta^2$ defined in Eq. (7.97) and can be interpreted as dissipation [3]. This leads to an alternative expression on the null surface for the Lie variation term in the adapted GNC system as (see Eqs. (7.143) and (7.157) in Appendix A.3),

$$\ell_a g^{ij}\pounds_\xi N^a_{ij} = 2\mathcal{D} + \frac{2}{\sqrt{q}}\partial^2_\lambda\left(\sqrt{q}\right) + 2\partial_\lambda\alpha \tag{7.19}$$

Integrating this expression over the null surface located at $r = 0$ with integration measure $d^2x du\,\sqrt{q}$, neglecting total divergence terms and dividing by 16π leads to,

$$\frac{1}{16\pi}\int d^2x d\lambda\,\sqrt{q}\ell_a g^{ij}\pounds_\xi N^a_{ij} = \frac{1}{8\pi}\int d^2x d\lambda\sqrt{q}\mathcal{D} + \int d^2x\left(\frac{\sqrt{q}}{4}\right)d\left(\frac{\alpha}{2\pi}\right) \tag{7.20}$$

The above expression explicitly shows that the Lie variation term when integrated over the null surface, leads to the dissipation and the sdT term (interpreted as $dT = (dT/d\lambda)d\lambda$). Hence the reduced gravitational momentum on the null surface, can be given a natural thermodynamic interpretation.

What is important regarding the above result is that for $\partial_\lambda q_{ab} = 0$ (i.e., the induced metric on the null surface is independent of the parameter along the null generator) the dissipation term identically vanishes and thus Eq. (7.20) can be written as:

$$\frac{1}{16\pi}\int d^2x d\lambda \, \sqrt{q}\ell_a g^{ij}\pounds_\xi N^a_{ij} = Q(\lambda_2) - Q(\lambda_1) \tag{7.21}$$

Hence the Lie variation term in this particular situation is equal to the difference in the heat content allowing us to relate $\mathcal{P}^a$ to the rate of heating of the null surface.

7.4 Projections of Gravitational Momentum on the Null Surface

We shall now take up one further application of the concept of gravitational momentum. Given the thermodynamic properties associated with the null surfaces, we would expect the flow of gravitational momentum vis-á-vis a given null surface to have some importance. To explore this in a detailed fashion, we have to first choose a suitable vector field using which the gravitational momentum can be defined. The most natural choice—as before—is the time evolution vector field ξ^a. Further, in the canonical null basis (ℓ^a, k^a, e^a_A) the gravitational momentum can be decomposed as: $P^a = A\ell^a + Bk^a + C^A e^a_A$. The components of the gravitational momentum A, B and C^A are related to the three projections of P^a by $A = -P^a(\xi)k_a$, $B = -P^a(\xi)\ell_a$ and $C^A = P^a(\xi)e^A_a$. Thus essentially we need to consider the following three components $q^a_b P^b(\xi)$, $k_a P^a(\xi)$ and $\ell_a P^a(\xi)$ respectively to get the complete picture. We will see that each of these terms lead to interesting thermodynamic interpretation. In view of the rather involved calculations, we will first provide a summary of the thermodynamic interpretations of these projections:

- The component $q^b_a P^a(\xi)$ leads to the Navier-Stokes equation for fluid dynamics, using which we can provide yet another justification for the dissipation term introduced in this context. The in depth description is in Sect. 7.4.1.
- The component $k_a P^a(\xi)$, when evaluated on an arbitrary null surface leads to the following result: $TdS = dE + PdV$, i.e., a thermodynamic identity. This helps us to identify a notion of energy associated with an arbitrary null surface. This result is elaborated in Sect. 7.4.2.
- Finally, the component $\ell_a P^a(\xi)$ yields the evolution of null surfaces. This involves both $ds/d\lambda$ and $dT/d\lambda$, where s is the entropy density and T is the temperature associated with the null surface and λ is the parameter along the null generator ℓ_a. This is studied in Sect. 7.4.3.

We shall now explicitly demonstrate how all these three results tie up with the notion of gravitational momentum we have introduced and arise naturally from the three different projections of the gravitational momentum.

7.4.1 Navier-Stokes Equation

The equivalence of field equations for gravity, when projected on a null surface, and the Navier-Stokes equation in fluid dynamics is an important result in the emergent paradigm of gravity. This result generalizes the previous one [15] in the context of black holes, and shows that when Einstein's equations are projected on any null surface and viewed in local inertial frame, they become *identical* to the Navier-Stokes equation [2]. Here we will project the gravitational momentum P^a introduced earlier on an arbitrary null surface, parametrized by GNC and shall demonstrate that the Navier-Stokes equation can be obtained. For this purpose we will use the vector field ξ^a throughout. Even though we will present the results in the context of the adapted coordinates to a null surface (i.e., in GNC), the same result continues to hold in any other parametrization as well.

In order to project the gravitational momenta $-16\pi P^c(\xi) = g^{ab}\pounds_\xi N^c_{ab} + R\xi^c$ on the null surface we need to determine the induced metric q^a_b on the surface, which [16] turns out to be diag(0, 0, 1, 1). Hence it is the angular part of the gravitational momentum which is going to contribute. The vectors ℓ^a and k^a are already given in Eqs. (2.58a) and (2.58b) respectively, which we will use subsequently. In the Navier-Stokes equation there exist two additional vector fields, playing a crucial role. These vectors and their components in the adapted GNC system has the following expressions:

$$\omega_b = \ell^m \nabla_m k_b = \left(\alpha, 0, \frac{1}{2}\beta_A\right); \qquad \omega^a = \left(0, \alpha, \frac{1}{2}\beta^A\right) \tag{7.22}$$

and

$$\Omega_a = \omega_a + \alpha k_a = \left(0, 0, \frac{1}{2}\beta_A\right); \qquad \Omega^a = \left(0, 0, \frac{1}{2}\beta^A\right) \tag{7.23}$$

With the help of [2] it is possible to interpret β_A as the transverse velocity of observers on the null surface. (In particular it can be interpreted as velocity drift of local Rindler observers parallel to the local Rindler horizon.) With this physical motivation in mind, let us now start calculating the projection of P^a on the null surface. Using the coordinates adapted to the given null surface, we arrive at:

$$-16\pi q^a_b P^b(\xi) = q^a_b g^{pq}\pounds_\xi N^b_{pq} = 2\pounds_\xi N^A_{ur} + q^{BC}\pounds_\xi N^A_{BC} \tag{7.24}$$

where we have used the identity $\ell_a q^a_b = 0$. We next need to find out the Lie variation of N^c_{ab}. For that purpose we use the expression $\pounds_v \Gamma^a_{bc} = \nabla_b \nabla_c v^a + R^a{}_{cmb} v^m$, for the Lie variation of Christoffel symbol with respect to an arbitrary vector field v^a and also the result: $N^a_{bc} = Q^{ad}_{be} \Gamma^e_{cd} + Q^{ad}_{ce} \Gamma^e_{bd}$. Thus the Lie variation of N^a_{bc} becomes

$$\begin{aligned} \pounds_v N^a_{bc} &= Q^{ad}_{be} \pounds_v \Gamma^e_{cd} + Q^{ad}_{ce} \pounds_v \Gamma^e_{bd} \\ &= \frac{1}{2}\left(\delta^a_b \nabla_c \nabla_d v^d + \delta^a_c \nabla_b \nabla_d v^d\right) - \frac{1}{2}\left(\nabla_b \nabla_c v^a + \nabla_c \nabla_b v^a\right) \\ &\quad - \frac{1}{2}\left(R^a{}_{bmc} + R^a{}_{cmb}\right) v^m \end{aligned} \tag{7.25}$$

Using the above result we obtain the expressions for Lie variation of N^A_{ur} and N^A_{BC} on the null surface (located at $r = 0$) to be (see Eq. (7.124) of Appendix A.3)

$$\pounds_\xi N^A_{ur} = \frac{1}{2}\partial_u \beta^A = \frac{1}{2} q^{AB} \partial_u \beta_B + \frac{1}{2}\beta_B \partial_u q^{AB} \tag{7.26}$$

$$\pounds_\xi N^A_{BC} = \frac{1}{2}\delta^A_B \partial_C \Theta + \frac{1}{2}\delta^A_C \partial_B \Theta - \partial_u \hat{\Gamma}^A_{BC} \tag{7.27}$$

Substituting these results in Eq. (7.24) we finally arrive at,

$$-16\pi q^p_b P^b = q^p_a \left(g^{bc} \pounds_\xi N^a_{bc}\right) = q^{PB} \partial_u \beta_B + \beta_B \partial_u q^{PB} + q^{PC} \partial_C \Theta - q^{BC} \partial_u \hat{\Gamma}^P_{BC} \tag{7.28}$$

In order to get the Navier-Stokes equation in its familiar form we need to lower the free index in Eq. (7.28) and have to multiply both sides of the equation by $(-1/2)$. Further using Noether current for ξ_a we have from Eq. (2.30)

$$q_{ab}(g^{pq} \pounds_\xi N^b_{pq}) = q_{ab} J^b(\xi) - 2R_{mn} \ell^m q^n_a \tag{7.29}$$

On using Einstein's equations $R_{ab} = 8\pi(T_{ab} - (1/2)T g_{ab})$ and the following result $\ell_a q^a_b = 0$, Eq. (7.28) leads to the desired expression:

$$8\pi T_{mn} \ell^m q^n_a = \frac{1}{2}\beta_A \partial_u \ln\sqrt{q} + \frac{1}{2}\partial_u \beta_A - \partial_A \alpha - \frac{1}{2}\partial_A \partial_u \ln\sqrt{q} + \frac{1}{2} q_{AB} q^{PQ} \partial_u \hat{\Gamma}^B_{PQ} \tag{7.30}$$

which can be simplified along the following lines:

$$
\begin{aligned}
8\pi T_{mn}\ell^m q^n_a &= \frac{1}{2}\beta_A\partial_u \ln\sqrt{q} + \frac{1}{2}\partial_u\beta_A - \partial_A\alpha - \frac{1}{2}\partial_A\partial_u \ln\sqrt{q} \\
&\quad - \frac{1}{2}q_{AB}\hat{\Gamma}^B_{PQ}\partial_u q^{PQ} + \frac{1}{2}\partial_D\left(q^{CD}\partial_u q_{AC}\right) \\
&\quad + \frac{1}{2}\partial_u q^{CF}\partial_C q_{AF} - \frac{1}{2}q_{AB}\partial_u\left(q^{BD}\partial_D \ln\sqrt{q}\right) \\
&= \frac{1}{2}\beta_A\partial_u \ln\sqrt{q} + \frac{1}{2}\partial_u\beta_A - \partial_A\alpha - \partial_A\partial_u \ln\sqrt{q} \\
&\quad - \frac{1}{2}q_{AB}\partial_u q^{BD}\left(\partial_D \ln\sqrt{q}\right) + \frac{1}{2}\partial_D\left(q^{CD}\partial_u q_{AC}\right) \\
&\quad + \frac{1}{2}\partial_u q^{CF}\partial_C q_{AF} - \frac{1}{2}\partial_C q_{AF}\partial_u q^{CF} - \frac{1}{2}q^{CD}\hat{\Gamma}^E_{AC}\partial_u q_{ED}
\end{aligned} \tag{7.31}
$$

In the first line we have used the relation given by Eq. (7.108b) while in the second line Eq. (7.108c) has been used as presented in Appendix A.3. From Eq. (7.107) of Appendix A.3 we obtain:

$$
\begin{aligned}
&\frac{1}{2}\partial_u\beta_A + \partial_B\left(\frac{1}{2}q^{BC}\partial_u q_{AC}\right) + \frac{1}{2}q^{CD}\partial_u q_{AD}\partial_C \ln\sqrt{q} - \frac{1}{2}q^{BD}\partial_u q_{CD}\hat{\Gamma}^C_{AB} \\
&\quad + \partial_u \ln\sqrt{q}\frac{1}{2}\beta_A - \partial_A\partial_u \ln\sqrt{q} - \partial_A\alpha \\
&= q^n_a\pounds_\ell\Omega_n + D_m\sigma^m_a + \Theta\Omega_a - D_a\left(\frac{\Theta}{2} + \alpha\right)
\end{aligned} \tag{7.32}
$$

After some trivial manipulations in Eq. (7.31) and using Eq. (7.32) the following final expression can be immediately obtained:

$$
8\pi T_{mn}\ell^m q^n_a = q^n_a\pounds_\ell\Omega_n + D_m\sigma^m_a + \Theta\Omega_a - D_a\left(\frac{\Theta}{2} + \alpha\right) \tag{7.33}
$$

to be interpreted as the Navier-Stokes equation for fluid dynamics.

The correspondence between Eqs. (7.32) and (7.33) with Navier-Stokes equation for fluid dynamics is based on the following identifications of various geometric quantities associated with the null surface: (i) The momentum density is given by $-\Omega_a/8\pi$. In the coordinates adapted to the null surface, Ω_a has only transverse components to be non-zero, which are given by $(1/2)\beta_A$. This reinforces our earlier interpretation of β_A as the transverse fluid velocity. Moreover, we can identify (ii) the pressure to be $\kappa/8\pi$, (iii) the shear tensor defined as σ_{mn} (see Eq. (7.96) in the Appendix), (iv) the shear viscosity coefficient $\eta = (1/16\pi)$, (v) the bulk viscosity coefficient $\zeta = -1/16\pi$ and finally (vi) an external force given by $F_a = T_{ma}\ell^m$. Thus Eq. (7.32) has the form of a Navier-Stokes equation for a fluid with the convective derivative being replaced by the Lie derivative. Since this equation and its physical interpretation have been extensively discussed in the references cited earlier, we will not repeat them and will confine ourselves to highlighting the dissipation term arising from it.

In order to interpret properly the dissipation term associated with the Navier-Stokes equation we start from the heat density $\mathcal{H} = -4P^{ab}_{cd}\nabla_a\ell^c\nabla_b\ell^d$ (i.e., heat content per unit null surface volume $\sqrt{q}d^2xd\lambda$) used in the variational principle, where $P^{ab}_{cd} = (1/2)(\delta^a_c\delta^b_d - \delta^a_d\delta^b_c)$ and ℓ^a is the null generator of the null surface. To connect up the heat density described above with the dissipation term [3] we consider a virtual displacement of the null surface in which the volume enclosed by the surface changes by $\delta A\delta\lambda$, where δA is a small area element on the two-surface (i.e., $\sqrt{q}d^2x$). Then $\mathcal{H}\delta A d\lambda$ represents the change in heat content of the null surface, which can be obtained by multiplying the heat density $\mathcal{H}$ with the infinitesimal 3-volume element on the null surface. Expanding the expression for the heat density and introducing temperature through local Rindler horizon [3] we finally obtain,

$$\frac{\mathcal{H}\delta A d\lambda}{8\pi} = -\delta A d\lambda\left(2\eta\sigma_{ab}\sigma^{ab} + \zeta\Theta^2\right) + \frac{1}{2}\frac{\kappa}{2\pi}d\delta A \tag{7.34}$$

where the first term stands for the (virtual) dissipation of a viscous fluid during the evolution of the small area element δA of the null surface along the null generator and has the following expression,

$$dE = \delta A d\lambda\left(2\eta\sigma_{ab}\sigma^{ab} + \zeta\Theta^2\right) = \frac{1}{8\pi}\delta A d\lambda\,\mathcal{D} \tag{7.35}$$

So the combination $\mathcal{D} = \left(\Theta_{ab}\Theta^{ab} - \Theta^2\right)$ represents the dissipation as we have indicated earlier. The second term in Eq. (7.34) can be interpreted as $(1/2)kT\,dN$, where $T = (\kappa/2\pi)$ is the temperature and dN is the degrees of freedom on the null surface corresponding to the change in area δA. For affinely parametrized null generator, $\kappa = 0$, which would lead to,

$$\frac{\mathcal{H}\delta A d\lambda}{8\pi} = -\delta A d\lambda\left(2\eta\sigma_{ab}\sigma^{ab} + \zeta\Theta^2\right) = \frac{1}{8\pi}\delta A d\lambda\,\mathcal{D} \tag{7.36}$$

Thus for affinely parametrized null generators (i.e., $\kappa = 0$) the change in the heat content is solely due to the dissipation term $\mathcal{D}$.

Having derived Navier-Stokes equation from the projection of gravitational momentum $P^a(\xi)$ on the null surface, we will now take up the task of projecting it along the other null directions, i.e., k_a and ℓ_a respectively and retrieve the thermodynamic information encoded in them.

7.4.2 A Thermodynamic Identity for the Null Surface

It has earlier been shown that in a wide class of gravity theories, the gravitational field equations near a horizon imply a thermodynamic identity $T\delta_{\bar{\lambda}}S = \delta_{\bar{\lambda}}E + P\delta_{\bar{\lambda}}V$ where the variations are interpreted as being due to virtual displacement of the horizon along the affine parameter $\bar{\lambda}$ of k^a. This result was demonstrated for general relativity and Lanczos-Lovelock theories of gravity when (i) the spacetime admits some symmetry, e.g., staticity or spherical symmetry and (ii) has a horizon. This suggest that this connection—between the field equations of gravity and a thermodynamic identity—is a specific phenomenon that occurs only in solutions containing horizons. But this illusion has been broken in [10] for general relativity and in [11] for Lanczos-Lovelock theories of gravity. There it was shown that gravitational field equations near any generic null surface in both general relativity and Lanczos-Lovelock theories of gravity lead to a thermodynamic relation: $T\delta_{\bar{\lambda}}S = \delta_{\bar{\lambda}}E + P\delta_{\bar{\lambda}}V$.

Here we will show that this thermodynamic identity is essentially contained in the gravitational momentum for ξ^a and can be retrieved from its component along ℓ^a which amounts to compute the projection $P^a(\xi)k_a$ (this picks out the component along ℓ^a because $k^2 = 0$ and $\ell^a k_a = -1$). Since all the details are similar to the one presented in [10], we will be brief and just indicate the manner in which the result can be obtained. (For detailed derivation see Appendix C of [10]). From Eq. (2.32) we obtain,

$$-k_a P^a(\xi) = k_a J^a(\xi) - 2G^a_b \xi^b k_a \tag{7.37}$$

In the second term we can use the Einstein's equations, i.e., $2G_{ab} = T_{ab}$ leading to the combination $T_{ab}\xi^a k^b$, which is the work function (effective pressure) for the matter (see e.g., [17, 18]). The work function when integrated over the two-surface, will yield the average force $\bar{F}$ due to matter flux on the null surface. The first term, viz., the projection of the Noether current $k_a J^a(\xi)$ has been evaluated explicitly in the earlier work [10]. Taking a cue from the result of [10] for the projection of the Noether current we arrive at:

$$\bar{F}\delta\bar{\lambda} = T\delta_{\bar{\lambda}}S - \delta_{\bar{\lambda}}E\ , \tag{7.38}$$

In the above expression, $\bar{F}$ stands for the average force on the null surface, T corresponds to the null surface temperature obtained using local Rindler observers, $S = (A/4)$ is the entropy and E represents the energy given by:

$$E = \frac{1}{2}\int d\bar{\lambda}\left(\frac{\chi}{2}\right) - \frac{1}{8\pi}\int d^2x\partial_u\sqrt{q} - \frac{1}{16\pi}\int d\bar{\lambda}\int d^2x\sqrt{q}\left\{\frac{1}{2}\beta_A\beta^A\right\}\ . \tag{7.39}$$

where χ is the Euler characteristics of the two-surface. (For a detailed discussion see [10]). A much simpler expression covering most of the interesting cases can be obtained by setting (a) $\beta_A|_{r=0} = 0$ and (ii) $\partial_A\alpha = 0$ on the null surface [10]. On imposing these conditions we arrive at the following simpler expression for the energy as,

$$E = \frac{1}{2}\int d\bar{\lambda}\left(\frac{\chi}{2}\right) - \frac{1}{8\pi}\int d^2x\partial_u\sqrt{q} = \frac{1}{2}\int d\bar{\lambda}\left(\frac{\chi}{2}\right) - \frac{1}{8\pi}\partial_u A \tag{7.40}$$

Hence the projection of the gravitational momentum along ℓ_a is equivalent to the thermodynamic identity $TdS = dE + PdV$.

7.4.3 Evolution of the Null Surface

Finally, we consider the remaining component of the gravitational momentum $P^a(\xi)$, which is along k^a. If we expand any vector field in the null basis (ℓ^a, k^a, e^a_A) as $v^a = A\ell^a + Bk^a + C^A e^a_A$, the component along k^a can be obtained by the contraction $\ell_a v^a$, since $\ell^2 = 0$ and $\ell_a k^a = -1$ on the null surface. For a given null surface we will use our adapted coordinates, i.e., GNC and will demonstrate that this component is intimately connected with evolution of the spacetime.

In this adapted coordinate system the scalar $-\ell_a P^a(\xi)$ has the following expression,

$$-16\pi\ell_a P^a(\xi) = \ell_a g^{bc}\pounds_\xi N^a_{bc} = 2\pounds_\xi N^r_{ur} + q^{AB}\pounds_\xi N^r_{AB} \tag{7.41}$$

where the first line follows from the fact that on the null surface $\ell^2 = 0$. Both the Lie variation terms have been explicitly calculated in Appendix A.3. From the expressions obtained there, the projection of gravitational momentum along k^a turns out to be (see Eqs. (7.143) and (7.157) of Appendix A.3),

$$-16\pi\ell_a P^a(\xi) = \ell_a g^{bc}\pounds_\xi N^a_{bc} = 2\partial_\lambda\alpha + 2\mathcal{D} + \frac{2}{\sqrt{q}}\partial^2_\lambda\sqrt{q} \tag{7.42}$$

Here $\mathcal{D}$ represents the dissipation term obtained earlier through the Navier-Stokes equation and has the definition $\mathcal{D} = (\Theta_{ab}\Theta^{ab} - \Theta^2)$.

To get a physical interpretation for the above result we integrate this expression over the null surface with the volume measure $\sqrt{q}d^2xd\lambda$ (for the null vector ℓ^a in GNC the parameter λ is just u) and then divide by proper factors of π, and ignore the surface term (which arises from the third term) to obtain:

$$\frac{1}{16\pi}\int d\lambda d^2x\sqrt{q}\ell_a g^{ij}\pounds_\xi N^a_{ij} = \frac{1}{8\pi}\int d\lambda d^2x\sqrt{q}\mathcal{D} + \int d^2x\ sdT \tag{7.43}$$

Here $s = \sqrt{q}/4$ is the entropy density and $dT \equiv (dT/d\lambda)d\lambda$. This leads to the interpretation that $\ell_a P^a(\xi)$ and $\ell_a g^{bc} £_\xi N^a_{bc}$ can be interpreted as the heating rate of the null surface per unit area. When integrated over the affine parameter λ and the proper area $\sqrt{q}\, d^2x$, it leads to the heating of the spacetime due to the dissipation (given by the first term in the right hand side) and the integral of sdT which is the second term.

Incidentally, one can provide a similar interpretation for the matter energy flux which crosses the null surface as well. Using a corresponding expression for $R_{ab}\ell^a\xi^b$ and using the Einstein's equations, we can easily show that (see Eq. (7.144) of Appendix A.3)

$$T_{ab}\ell^a\ell^b = \frac{1}{\sqrt{q}}\left(\frac{\alpha}{2\pi}\right)\frac{\partial}{\partial u}\left(\frac{\sqrt{q}}{4}\right) - \frac{1}{8\pi}\mathcal{D} - \frac{1}{8\pi}\frac{1}{\sqrt{q}}\frac{\partial^2\sqrt{q}}{\partial u^2} \tag{7.44}$$

Integrating both sides of the expression over the null surface and ignoring boundary contributions, we get

$$\int d^2x d\lambda\, \sqrt{q} T_{ab}\ell^a\ell^b + \frac{1}{8\pi}\int d^2x d\lambda \sqrt{q}\mathcal{D} = \int d^2x T ds \tag{7.45}$$

This tells us that the heating of the null surface due to incoming matter flux plus the heat generated by the dissipation is equal to the integral of $T\partial_u s$ over the null surface. This reconfirms our earlier interpretation of the projection of the gravitational momentum contributing to the heating of the null surface.

These results can further be used to re-express the heat content of the null surface which was earlier used in the thermodynamic variational principle. Two equivalent forms of the variational principle can be constructed, differing from each other by a total divergence term and based on $R_{ab}\ell^a\ell^b$ and the Lie variation term. These two variational principles (neglecting possible surface contributions) have the following expressions

$$\begin{aligned} Q_1 &\equiv \int d\lambda d^2x \sqrt{q}\left(-\frac{1}{8\pi}R_{ab}\ell^a\ell^b + T_{ab}\ell^a\ell^b\right) \\ &= \int d\lambda d^2x \sqrt{q}\Big[\frac{1}{8\pi}\mathcal{D} + T_{ab}\ell^a\ell^b\Big] - \int d^2x\, T ds \end{aligned} \tag{7.46}$$

and,

$$\begin{aligned} Q_2 &\equiv \int d\lambda d^2x \sqrt{q}\left[\frac{1}{16\pi}\ell_a g^{ij} £_\xi N^a_{ij} + T_{ab}\ell^a\ell^b\right] \\ &= \int d\lambda d^2x \sqrt{q}\Big[\frac{1}{8\pi}\mathcal{D} + T_{ab}\ell^a\ell^b\Big] + \int d^2x\, s dT \end{aligned} \tag{7.47}$$

Note that both these variational principles have the dissipation term $\mathcal{D}$ and matter energy flux through the null surface $T_{ab}\ell^a\ell^b$ in common as emphasized earlier. However Q_1 is connected to TdS while, the other one is connected to SdT. Thus both the variational principles have natural thermodynamic interpretation.

7.5 Conclusions

Since we have already described the physical consequences of the results in the various sections themselves, we shall limit ourselves to summarising the key conclusions in this section.

- Given a Vector field v^a, it is possible to construct three currents: (a) the Noether current $J^a(v)$, (b) the gravitational momentum $P^a(v)$ and (c) the reduced gravitational momentum $\mathcal{P}^a(v)$. Interestingly enough, one can attribute *thermodynamic* meaning to all these quantities which are usually considered to be purely *geometrical* in origin. For example, the conserved current J^a, associated with the time-development vector field ξ^a of the spacetime, leads to a conserved charge (i.e., integral of $u_a J^a(\xi)$ defined either on a spacelike surface or on a null surface) that can be related to the boundary heat density Ts, where T corresponds to the Unruh-Davies temperature and s stands for the entropy density.
- The gravitational field equations can also be derived from a thermodynamic variational principle, which essentially extremises the total heat density of all the null surfaces in the spacetime. This variational principle can also be expressed directly in terms of the total gravitational momentum, thereby providing it with a simple physical interpretation.
- One can associate with any null surface two null vector fields ℓ_a and k_a with the following properties: $\ell_a k^a = -1$, ℓ_a being the tangent vector to the congruence defining the null surface, as well as the 2-metric $q_{ab} = g_{ab} + \ell_a k_b + \ell_b k_a$. These structures define three natural projections of the gravitational momentum (namely, $P^a\ell_a$, $P^a k_a$, $P^a q_{ab}$), all of which have thermodynamic significance. The first one leads to the description of time evolution of the null surface in terms of suitably defined bulk and surface degrees of freedom associated with it; the second leads to a thermodynamic identity which can be written in the simple form $TdS = dE + PdV$; the third leads to a Navier-Stokes equation for the transverse degrees of freedom on the null surface which can also be interpreted as a drift velocity.

These results again demonstrate that the emergent gravity paradigm enriches our understanding of the spacetime dynamics and specially the structure of null surfaces, by allowing a rich variety of thermodynamic backdrops for the geometrical variables.

Appendix A

In this appendix, we shall present the supplementary material for this chapter.

A.1 General Analysis Regarding Null Surfaces

We will start with a null vector $\ell_a = A\nabla_a B$, which satisfies the condition $\ell^2 = 0$ only over a single surface, which is the null surface under our consideration. Then we obtain

$$\ell^a \nabla_a \ell^b = \kappa \ell^b \tag{7.48}$$

where we have the following expression for κ

$$\kappa = \ell^a \partial_a \ln A + \tilde{\kappa}; \qquad \tilde{\kappa} = -\frac{1}{2} k^a \partial_a \ell^2 \tag{7.49}$$

The last relation defining $\tilde{\kappa}$ can also be written as $\nabla_a \ell^2 = 2\tilde{\kappa}\ell_a$. The derivation of the result goes as follows, let us expand $\nabla_b \ell^2$ in canonical null basis, i.e., $\nabla_a \ell^2 = C\ell_a + Dk_a + E_A e^A_a$. Then both $E_A = e^a_A \nabla_a \ell^2$ and $D = -\ell^a \nabla_a \ell^2$ vanishes, since variation of ℓ^2 along the null surface vanishes. This shows that the only non-zero component of $\nabla_a \ell^2$ is along ℓ_a. Then it turns out that [16]

$$\nabla_i \ell^i = \Theta + \kappa + \tilde{\kappa} \tag{7.50}$$

where, $\Theta = q^{ma} q_{mb} \nabla_a \ell^b$. Note that the term $\tilde{\kappa}$ enters the picture as $\ell^2 = 0$ only on the null surface. With this setup let us now find out $R_{ab}\ell^a \ell^b$ in detail, which leads to,

$$\begin{aligned} R_{ab}\ell^a \ell^b &= \ell^j \left(\nabla_i \nabla_j \ell^i - \nabla_j \nabla_i \ell^i\right) \\ &\quad - \nabla_i \left(\ell^i \nabla_j \ell^i\right) - \nabla_j \left(\ell^i \nabla_i \ell^i\right) - \nabla_i \ell^j \nabla_j \ell^i + \left(\nabla_i \ell^i\right)^2 \\ &= \nabla_i \left(\ell^j \nabla_j \ell^i - \ell^i \nabla_j \ell^j\right) - \left(\nabla_i \ell^j \nabla_j \ell^i - \left(\nabla_i \ell^i\right)^2\right) \end{aligned} \tag{7.51}$$

However in general, $\ell^j \nabla_j \ell^i = \kappa \ell^i$ is not true, it only holds on the null surface (when $\ell^2 = 0$ everywhere this relation is also true everywhere). Since we were doing the calculation for the most general case, $\ell^2 \neq 0$ in the above expression we cannot substitute $\ell^j \nabla_j \ell^i = \kappa \ell^i$, since it appears inside the derivative. Thus for the special case when $\ell^2 = 0$ everywhere, we will arrive at the following result

$$R_{ab}\ell^a \ell^b = -\nabla_i \left(\Theta \ell^i\right) - \left(\nabla_i \ell^j \nabla_j \ell^i - \left(\nabla_i \ell^i\right)^2\right) \tag{7.52}$$

In order to simplify things quiet a bit we will compute the last term $\left(\nabla_i \ell^j \nabla_j \ell^i - (\nabla_i \ell^i)^2\right)$ which we designate by $\mathcal{S}$. Then we start by calculating the following object on the null surface

$$\begin{aligned}
\Theta^{ab} &= q^a_m q^b_n \nabla^m \ell^n \\
&= \left(\delta^a_m + \ell^a k_m + k^a \ell_m\right)\left(\delta^b_n + \ell^b k_n + k^b \ell_n\right) \nabla^m \ell^n \\
&= \nabla^a \ell^b + \ell^a k_m \nabla^m \ell^b + \kappa k^a \ell^b + \ell^b k_n \nabla^a \ell^n + \ell^a \ell^b k_m k_n \nabla^m \ell^n \\
&- \kappa k^a \ell^b + \tilde{\kappa} \ell^a k^b - \tilde{\kappa} \ell^a k^b + \kappa k^a k^b \ell^2 \\
&= \nabla^a \ell^b + \ell^a k_m \nabla^m \ell^b + \ell^b k_n \nabla^a \ell^n + \ell^a \ell^b k_m k_n \nabla^m \ell^n
\end{aligned} \tag{7.53}$$

In arriving at the third line we have used the following results: $\ell^a \nabla_a \ell^i = \kappa \ell^i$ and $\ell_a \nabla_b \ell^a = \tilde{\kappa} \ell_b$. Then we can reverse the above equation leading to

$$\nabla_a \ell_b = \Theta_{ab} - \ell_a k^m \nabla_m \ell_b - \ell_b k_n \nabla_a \ell^n - \ell_a \ell_b k_m k_n \nabla^m \ell^n \tag{7.54}$$

From the above equation we can derive two very important identity:

$$\begin{aligned}
\left(\nabla_a \ell_b\right)\left(\nabla^a \ell^b\right) &= \Theta_{ab}\Theta^{ab} + \ell^a \ell_b k^m \nabla_m \ell^b k^n \nabla_a \ell_n + \ell_a \ell^b k_n \nabla^a \ell^n k^m \nabla_m \ell_b \\
&= \Theta_{ab}\Theta^{ab} + 2\left(\tilde{\kappa} k^m \ell_m\right)\left(\kappa \ell_n k^n\right) \\
&= \Theta_{ab}\Theta^{ab} + 2\kappa\tilde{\kappa}
\end{aligned} \tag{7.55}$$

In the same spirit we will arrive at

$$\begin{aligned}
\left(\nabla_a \ell_b\right)\left(\nabla^b \ell^a\right) &= \Theta_{ab}\Theta^{ab} + \ell^a \ell_b k_n \nabla_a \ell^n k^m \nabla^b \ell_m + \ell_a \ell^b k^m \nabla_m \ell^a k^n \nabla_n \ell_b \\
&= \Theta_{ab}\Theta^{ab} + \left(\tilde{\kappa} k^m \ell_m\right)\left(\tilde{\kappa} k^n \ell_n\right) + \left(\kappa \ell_n k^n\right)\left(\kappa \ell_m k^m\right) \\
&= \Theta_{ab}\Theta^{ab} + \kappa^2 + \tilde{\kappa}^2
\end{aligned} \tag{7.56}$$

The extrinsic curvature for null surfaces, i.e., Θ_{ab} can be given a very natural interpretation. This essentially follows from [16]. There the expression for Θ_{ab} in terms of Lie variation of q_{ab} along the null generator ℓ_a was obtained as,

$$\Theta_{ab} = \frac{1}{2} q^m_a q^n_b \pounds_\ell q_{mn} \tag{7.57}$$

Now expanding out the Lie derivative term we obtain,

$$\pounds_\ell q_{mn} = \ell^i \partial_i q_{mn} + q_{ma} \partial_n \ell^a + q_{an} \partial_m \ell^a \tag{7.58}$$

Which on being substituted in Eq. (7.57) immediately leads to,

$$\Theta_{ab} = \frac{1}{2} q^m_a q^n_b \ell^i \partial_i q_{mn} + \frac{1}{2} q_{ai} q^n_b \partial_n \ell^i + \frac{1}{2} q_{bi} q^m_a \partial_m \ell^i \tag{7.59}$$

Now on the null surface $q_{ab} = q_{AB}$ as the only non-zero component. Hence the above equation can be written as,

$$\Theta_{ab} = \Theta_{AB} = \frac{1}{2}\frac{d}{d\lambda}q_{AB} + \frac{1}{2}q_{AC}\partial_B \ell^C + \frac{1}{2}q_{BC}\partial_A \ell^C \tag{7.60}$$

On the null surface $q^a_b \ell^b = 0$, which in this coordinate system leads to $\ell^A = 0$ on the null surface. Since $\partial_A \ell^2$ represent derivatives on the null surface it also vanishes. If $\ell^2 = 0$ everywhere, then also ℓ^A is identically zero everywhere. Hence we have

$$\Theta_{ab} = \frac{1}{2}\frac{d}{d\lambda}q_{AB} \tag{7.61}$$

There is another way to get this result. If e^a_A are the basis vectors on the null surface and if ℓ_a, e^a_A forms coordinate basis vectors, then $q_{AB} = q_{ab}e^a_A e^b_B$ is a scalar under 4-dimensional coordinate transformation. This immediately leads to the previous expression. For more discussions along identical lines see [16].

Now the expression for the quantity $\mathcal{S}$ can be obtained as

$$\begin{aligned}
\mathcal{S} &= \nabla_i \ell^j \nabla_j \ell^i - \left(\nabla_i \ell^i\right)^2 \\
&= \Theta_{ab}\Theta^{ab} + \kappa^2 + \tilde{\kappa}^2 - (\Theta + \kappa + \tilde{\kappa})^2 \\
&= \left(\Theta_{ab}\Theta^{ab} - \Theta^2\right) - 2\Theta\left(\kappa + \tilde{\kappa}\right) - 2\kappa\tilde{\kappa}
\end{aligned} \tag{7.62}$$

Using the general expression for $R_{ab}\ell^a\ell^b$ we obtain the following form:

$$\begin{aligned}
R_{ab}\ell^a\ell^b &= \nabla_i\left(\ell^j\nabla_j\ell^i - \ell^i\nabla_j\ell^j\right) - \mathcal{S} \\
&= \nabla_i\left(\ell^j\nabla_j\ell^i - \ell^i\nabla_j\ell^j\right) - \left(\Theta_{ab}\Theta^{ab} - \Theta^2\right) + 2\Theta\left(\kappa + \tilde{\kappa}\right) + 2\kappa\tilde{\kappa}
\end{aligned} \tag{7.63}$$

For the situation where, $\ell^2 = 0$ everywhere we finally arrive at the following simplified expression

$$\begin{aligned}
R_{ab}\ell^a\ell^b &= -\left(\Theta_{ab}\Theta^{ab} - \Theta^2\right) + 2\Theta\kappa + \nabla_i\left(\kappa\ell^i - [\Theta + \kappa]\,\ell^i\right) \\
&= -\left(\Theta_{ab}\Theta^{ab} - \Theta^2\right) + \Theta\kappa - \frac{1}{\sqrt{q}}\frac{d}{d\lambda}\left(\sqrt{q}\Theta\right)
\end{aligned} \tag{7.64}$$

Let us now try to derive the Raychaudhuri equation starting from the basic properties of null surfaces. We start with the following result

$$\begin{aligned}
\ell^a\nabla_a\left(\nabla_c\ell_d\right) &= \ell^a\nabla_a\nabla_c\ell_d \\
&= R_{dbac}\ell^a\ell^b + \ell^a\nabla_c\nabla_a\ell_d \\
&= \nabla_c\left(\ell^a\nabla_a\ell_d\right) - \nabla_a\ell_d\nabla_c\ell^a - R_{bdac}\ell^b\ell^a
\end{aligned} \tag{7.65}$$

Then contraction of the indices c, d leads to the following result

$$\ell^a \nabla_a \left(\nabla_c \ell^c\right) = \nabla_c \left(\ell^a \nabla_a \ell^c\right) - \nabla_a \ell_b \nabla^b \ell^a - R_{ab} \ell^a \ell^b \tag{7.66}$$

Otherwise we can rewrite it in a different manner which exactly coincide with Eq. (7.51). On using Eq. (7.56) and the decomposition: $\Theta_{ab} = (1/2)\Theta q_{ab} + \sigma_{ab} + \omega_{ab}$ we arrive at

$$\begin{aligned}
\ell^a \nabla_a \left(\nabla_c \ell^c\right) - \nabla_c \left(\ell^a \nabla_a \ell^c\right) &= -\Theta_{ab}\Theta^{ab} - \kappa^2 - \tilde{\kappa}^2 - R_{ab}\ell^a \ell^b \\
&= -\frac{1}{2}\Theta^2 - \sigma^{ab}\sigma_{ab} + \omega_{ab}\omega^{ab} - \kappa^2 - \tilde{\kappa}^2 - R_{ab}\ell^a \ell^b
\end{aligned} \tag{7.67}$$

For the situation where, $\ell^2 = 0$ the left hand side is just: $d\left(\Theta + \kappa\right)/d\lambda$ and the first term on the right hand side is $d\kappa/d\lambda + \kappa(\Theta + \kappa)$ the above equation leads to

$$\begin{aligned}
\frac{d\Theta}{d\lambda} &= \kappa\Theta + \kappa^2 - \nabla_a \ell_b \nabla^b \ell^a - R_{ab}\ell^a \ell^b \\
&= \kappa\Theta - \Theta_{ab}\Theta^{ab} - R_{ab}\ell^a \ell^b \\
&= \kappa\Theta - \frac{1}{2}\Theta^2 - \sigma^{ab}\sigma_{ab} + \omega_{ab}\omega^{ab} - R_{ab}\ell^a \ell^b
\end{aligned} \tag{7.68}$$

where to arrive at the last line we have used the following decomposition: $\Theta_{ab} = (1/2)\Theta q_{ab} + \sigma_{ab} + \omega_{ab}$. This is precisely the one obtained in [19] though in a completely different manner.

The next object to consider is the quantity $\ell_a J^a(\ell)$. This can be obtained by using the identity for Noether current leading to,

$$\begin{aligned}
\frac{1}{A}\ell_a J^a(\ell) &= \nabla_b \left(\left\{\ell^a \ell^b - \ell^2 g^{ab}\right\} \frac{\nabla_a A}{A^2}\right) \\
&= \nabla_b \left[\frac{1}{A}\ell^b \left(\kappa - \tilde{\kappa}\right)\right] - \nabla_b \left(\ell^2 \frac{\nabla^b A}{A^2}\right) \\
&= \frac{1}{A}\nabla_i \left[\left(\kappa - \tilde{\kappa}\right)\ell^i\right] - \frac{1}{A}\left(\kappa - \tilde{\kappa}\right)^2 - \frac{\nabla^b A}{A^2}\nabla_b \ell^2 \\
&= \frac{1}{A}\nabla_i \left[\left(\kappa - \tilde{\kappa}\right)\ell^i\right] - \frac{1}{A}\left(\kappa - \tilde{\kappa}\right)^2 - \frac{2}{A}\tilde{\kappa}\left(\kappa - \tilde{\kappa}\right)
\end{aligned} \tag{7.69}$$

This can be written in a slightly modified manner as,

$$\begin{aligned}
\ell_a J^a(\ell) &= \nabla_i \left[\left(\kappa - \tilde{\kappa}\right)\ell^i\right] - \left(\kappa^2 - \tilde{\kappa}^2\right) \\
&= \ell^i \nabla_i \left(\kappa - \tilde{\kappa}\right) + \left(\kappa - \tilde{\kappa}\right)\left(\Theta + \kappa + \tilde{\kappa}\right) - \left(\kappa^2 - \tilde{\kappa}^2\right) \\
&= \frac{d}{d\lambda}\left(\kappa - \tilde{\kappa}\right) + \Theta\left(\kappa - \tilde{\kappa}\right)
\end{aligned} \tag{7.70}$$

The above expression can be simplified significantly by noting that $\Theta = d(\ln\sqrt{q})/d\lambda$, which leads to

$$\ell_a J^a(\ell) = \frac{1}{\sqrt{q}} \frac{d}{d\lambda} \left[(\kappa - \tilde{\kappa}) \sqrt{q} \right] \tag{7.71}$$

Again, we have

$$\begin{aligned}
D_a \left[(\kappa - \tilde{\kappa}) \ell^a \right] &= \left(g^{ab} + \ell^a k^b + \ell^b k^a \right) \nabla_a \left[(\kappa - \tilde{\kappa}) \ell_b \right] \\
&= \nabla_a \left[(\kappa - \tilde{\kappa}) \ell^a \right] + \left(\ell^a k^b + \ell^b k^a \right) \left[(\kappa - \tilde{\kappa}) \nabla_a \ell_b + \ell_b \nabla_a (\kappa - \tilde{\kappa}) \right] \\
&= \nabla_a \left[(\kappa - \tilde{\kappa}) \ell^a \right] - \kappa (\kappa - \tilde{\kappa}) - \tilde{\kappa} (\kappa - \tilde{\kappa}) - \frac{d}{d\lambda} (\kappa - \tilde{\kappa}) \qquad (7.72) \\
&= (\kappa - \tilde{\kappa}) (\Theta + \kappa + \tilde{\kappa}) - \left(\kappa^2 - \tilde{\kappa}^2 \right) \\
&= (\kappa - \tilde{\kappa}) \frac{d \ln \sqrt{q}}{d\lambda} \qquad (7.73)
\end{aligned}$$

Thus we arrive at

$$\begin{aligned}
\ell_a J^a(\ell) &= D_a \left[(\kappa - \tilde{\kappa}) \ell^a \right] + \frac{d}{d\lambda} (\kappa - \tilde{\kappa}) + (\kappa - \tilde{\kappa}) (\kappa + \tilde{\kappa}) - \left(\kappa^2 - \tilde{\kappa}^2 \right) \\
&= D_a \left[(\kappa - \tilde{\kappa}) \ell^a \right] + \frac{d}{d\lambda} (\kappa - \tilde{\kappa}) \qquad (7.74)
\end{aligned}$$

From the expression of the Noether current we get,

$$\ell_a J^a(\ell) = 2 R_{ab} \ell^a \ell^b + \ell_a g^{ij} \pounds_\ell N^a_{ij} \tag{7.75}$$

The above equation can be used to write, $g^{ab} \pounds_\ell N^c_{ab}$ in terms of κ and R_{ab}. For that purpose we use Eq. (2.30) and insert Eq. (7.70) leading to

$$\begin{aligned}
\ell_a g^{ij} \pounds_\ell N^a_{ij} &= \ell_a J^a(\ell) - 2 R_{ab} \ell^a \ell^b \\
&= \frac{d}{d\lambda} (\kappa - \tilde{\kappa}) + \Theta (\kappa - \tilde{\kappa}) - 2 R_{ab} \ell^a \ell^b \\
&= \frac{1}{\sqrt{q}} \frac{d}{d\lambda} \left[\sqrt{q} (\kappa - \tilde{\kappa}) \right] - 2 R_{ab} \ell^a \ell^b \\
&= \frac{2}{\sqrt{q}} \frac{d}{d\lambda} \left(\sqrt{q} \kappa \right) - 2 R_{ab} \ell^a \ell^b - \frac{1}{\sqrt{q}} \frac{d}{d\lambda} \left[\sqrt{q} (\kappa + \tilde{\kappa}) \right] \qquad (7.76)
\end{aligned}$$

The above expression when integrated over the null surface with integration measure, $\sqrt{q} d^2 x d\lambda$ and then being divided by 16π leads to

$$\frac{1}{16\pi}\int d^2x d\lambda\sqrt{q}\left\{\ell_a g^{ij}\pounds_\ell N^a_{ij}+\frac{1}{\sqrt{q}}\frac{d}{d\lambda}\left[\sqrt{q}\left(\kappa+\tilde{\kappa}\right)\right]\right\}$$
$$=\frac{1}{8\pi}\int d^2x\sqrt{q}\kappa|_1^2-\frac{1}{8\pi}\int d^2x d\lambda\sqrt{q}R_{ab}\ell^a\ell^b \qquad (7.77)$$

Then on using the field equation: $R_{ab}=8\pi\left[T_{ab}-(1/2)g_{ab}T\right]=8\pi\bar{T}_{ab}$ and then being substituted in Eq. (7.77) we arrive at the following expression

$$\frac{1}{4}\int d^2x\sqrt{q}\left(\frac{\kappa}{2\pi}\right)|_1^2-\int d^2x d\lambda\sqrt{q}\;T_{ab}\ell^a\ell^b$$
$$=\frac{1}{16\pi}\int d^2x d\lambda\sqrt{q}\left\{\ell_a g^{ij}\pounds_\ell N^a_{ij}+\frac{1}{\sqrt{q}}\frac{d}{d\lambda}\left[\sqrt{q}\left(\kappa+\tilde{\kappa}\right)\right]\right\} \qquad (7.78)$$

where the last equality follows from the fact that $\ell^2=0$ on the null surface. The above equation can be written in a more abstract form as

$$\frac{1}{16\pi}\int d^2x d\lambda\sqrt{q}\ell_a g^{ij}\pounds_\ell N^a_{ij}=\left[\frac{1}{4}\int_{\lambda_2}d^2x\sqrt{q}\left(\frac{\kappa}{2\pi}\right)-\frac{1}{4}\int_{\lambda_1}d^2x\sqrt{q}\left(\frac{\kappa}{2\pi}\right)\right]$$
$$-\int d^2x d\lambda\sqrt{q}\;T_{ab}\ell^a\ell^b-\frac{1}{16\pi}\int d^2x d\lambda\frac{d}{d\lambda}\left[\sqrt{q}\left(\kappa+\tilde{\kappa}\right)\right] \qquad (7.79)$$

As an illustration when $\ell^2=0$ everywhere, we have $\tilde{\kappa}=0$, then Eq. (7.79) leads to,

$$\frac{1}{16\pi}\int d^2x d\lambda\sqrt{q}\ell_a g^{ij}\pounds_\ell N^a_{ij}=\frac{1}{2}\left[\int_{\lambda_2}d^2x\frac{\sqrt{q}}{4}\left(\frac{\kappa}{2\pi}\right)-\int_{\lambda_1}d^2x\frac{\sqrt{q}}{4}\left(\frac{\kappa}{2\pi}\right)\right]$$
$$-\int d^2x d\lambda\sqrt{q}\;T_{ab}\ell^a\ell^b \qquad (7.80)$$

We will now try to obtain an expression for the quantity $\ell_a g^{ij}\pounds_\ell N^a_{ij}$ independently. For that we start with the symmetric and anti-symmetric part of the derivative $\nabla_a\ell_b$ such that:

$$S^{ab}=\nabla^a\ell^b+\nabla^b\ell^a,\qquad J^{ab}=\nabla^a\ell^b-\nabla^b\ell^a \qquad (7.81)$$

Then we have the following result: $\nabla^a\ell^b=(1/2)(S^{ab}+J^{ab})$, which on being substituted in the identity:

$$\nabla_b\left(\nabla^a\ell^b\right)-\nabla^a\left(\nabla_b\ell^b\right)=R^a_b\ell^b \qquad (7.82)$$

leads to the following identification:

$$g^{ab}\pounds_\ell N^c_{ab} = -\nabla_b\left(S^{bc} - g^{bc}S\right) \tag{7.83}$$

Hence we arrive at the following relation:

$$\begin{aligned}
\ell_a g^{bc}\pounds_\ell N^a_{bc} &= -\ell_a\nabla_b\left[\nabla^a\ell^b + \nabla^b\ell^a - 2g^{ab}\nabla_c\ell^c\right] \\
&= -\nabla_b\left[\ell^a\nabla_a\ell^b + \ell_a\nabla^b\ell^a - 2\ell^b\left(\nabla_c\ell^c\right)\right] \\
&+ \nabla_a\ell_b\nabla^b\ell^a + \nabla_a\ell_b\nabla^a\ell^b - 2\left(\nabla_c\ell^c\right)^2 \\
&= -\nabla_b\left[\ell^a\nabla_a\ell^b + \ell_a\nabla^b\ell^a - 2\ell^b\left(\nabla_c\ell^c\right)\right] \\
&+ 2\Theta_{ab}\Theta^{ab} + (\kappa+\tilde{\kappa})^2 - 2\left(\Theta+\kappa+\tilde{\kappa}\right)^2 \\
&= -\nabla_b\left[\ell^a\nabla_a\ell^b + \ell_a\nabla^b\ell^a - 2\ell^b\left(\nabla_c\ell^c\right)\right] \\
&+ 2\left(\Theta_{ab}\Theta^{ab} - \Theta^2\right) - 4\Theta\left(\kappa+\tilde{\kappa}\right) - (\kappa+\tilde{\kappa})^2
\end{aligned} \tag{7.84}$$

For the case $\ell^2 = 0$ the term within bracket can be written in a simplified manner such that Lie derivative term gets simplified leading to:

$$\begin{aligned}
\ell_a g^{bc}\pounds_\ell N^a_{bc} &= 2\left(\Theta_{ab}\Theta^{ab} - \Theta^2\right) - 4\Theta\kappa - \kappa^2 + \nabla_b\left[(2\Theta+\kappa)\,\ell^b\right] \\
&= 2\left(\Theta_{ab}\Theta^{ab} - \Theta^2\right) - 4\Theta\kappa - \kappa^2 + (2\Theta+\kappa)\,(\Theta+\kappa) + \frac{d}{d\lambda}\,(2\Theta+\kappa) \\
&= 2\left(\Theta_{ab}\Theta^{ab} - \Theta^2\right) - \Theta\kappa + \frac{d}{d\lambda}\kappa + \frac{2}{\sqrt{q}}\frac{d}{d\lambda}\left(\sqrt{q}\Theta\right)
\end{aligned} \tag{7.85}$$

If the null generator is affinely parametrized then $\kappa = 0$ and Eq. (7.85) reduces to:

$$\ell_a g^{bc}\pounds_\ell N^a_{bc} = 2\left(\Theta_{ab}\Theta^{ab} - \Theta^2\right) + \frac{2}{\sqrt{q}}\frac{d}{d\lambda}\left(\sqrt{q}\Theta\right) \tag{7.86}$$

While for the null generator ℓ^a in GNC we have (see Appendix A.3 Eq. (7.133)):

$$\ell_a g^{bc}\pounds_\ell N^a_{bc} = 2\left(\Theta_{ab}\Theta^{ab} - \Theta^2\right) + \frac{2}{\sqrt{q}}\frac{d^2\sqrt{q}}{d\lambda^2} + 2\frac{d\kappa}{d\lambda} - \frac{2}{\sqrt{q}}\frac{d}{d\lambda}\left(\sqrt{q}\kappa\right) \tag{7.87}$$

Through the above analysis we have obtained expressions for $\ell_a J^a(\ell)$, $R_{ab}\ell^a\ell^b$ and $\ell_a g^{bc}\pounds_\ell N^a_{bc}$.

It turns out from the above analysis that $\Theta_{ab}\Theta^{ab} - \Theta^2$ can be given a more physical meaning by considering Lie variation of gravitational momentum. This can be obtained by considering variation of the gravitational momentum first:

$$\begin{aligned} q_{ab}\delta\Pi^{ab} &= q_{ab}\delta\left[\sqrt{q}\left(\Theta^{ab}-\Theta q^{ab}\right)\right] \\ &= q_{ab}\sqrt{q}\delta\Theta^{ab}-2\sqrt{q}\delta\Theta-\sqrt{q}\Theta q_{ab}\delta q^{ab}-\Theta\delta\sqrt{q} \\ &= \sqrt{q}q_{ab}\delta\Theta^{ab}-2\sqrt{q}\delta\Theta+\Theta\delta\sqrt{q} \end{aligned} \tag{7.88}$$

Now specializing to Lie variation we arrive at:

$$\begin{aligned} -q_{ab}£_\ell\Pi^{ab} &= -\Theta£_\ell\sqrt{q}-\sqrt{q}q_{ab}£_\ell\Theta^{ab}+2\sqrt{q}£_\ell\Theta \\ &= -\sqrt{q}£_\ell\Theta+\sqrt{q}\Theta^{ab}£_\ell q_{ab}-\Theta£_\ell\sqrt{q}+2\sqrt{q}£_\ell\Theta \\ &= 2\sqrt{q}\left(\Theta^{ab}\Theta_{ab}-\Theta^2\right)+£_\ell\left(\sqrt{q}\Theta\right) \\ &= \sqrt{q}\ell_a g^{bc}£_\ell N^a_{bc}-\frac{d^2\sqrt{q}}{d\lambda^2} \end{aligned} \tag{7.89}$$

where in the last line we have used Eq. (7.86). Here the quantities Θ_{ab} and Θ can be defined as: $\Theta_{ab} = (1/2)£_\ell q_{ab}$ and $\Theta = £_\ell \ln\sqrt{q}$. In GNC parametrization: $(1/2)£_\ell q_{ab} = (1/2)\partial_u q_{ab}$ and $£_\ell \ln\sqrt{q} = \partial_u \ln\sqrt{q}$. Thus the Lie variation of gravitational momentum for affine parametrization is directly related to $\mathcal{D}$, i.e. to $(\Theta_{ab}\Theta^{ab}-\Theta^2)$.

For non-affine parametrization the gravitational momentum associated with null surfaces can be taken as $\Pi^{ab} = \sqrt{q}(\Theta^{ab}-(\Theta+\kappa)q^{ab})$. Then we readily arrive at the Lie variation expression:

$$-q_{ab}£_\ell\Pi^{ab} = 2\sqrt{q}\left(\Theta^{ab}\Theta_{ab}-\Theta^2\right)+£_\ell\left(\sqrt{q}\Theta\right)+2\sqrt{q}£_\ell\kappa \tag{7.90}$$

Since κ is a scalar Lie variation term can be written as: $£_\ell\kappa = d\kappa/d\lambda$, where λ is the parameter along ℓ^a. If we consider the null generator ℓ^a from GNC then we arrive at (see Eq. (7.87)):

$$-q_{ab}£_\ell\Pi^{ab} = \sqrt{q}\ell_a g^{bc}£_\ell N^a_{bc}-\frac{d^2\sqrt{q}}{d\lambda^2}+\frac{2}{\sqrt{q}}\frac{d}{d\lambda}\left(\sqrt{q}\kappa\right) \tag{7.91}$$

while if we have defined the conjugate momenta Π^{ab} as $\Pi^{ab} = \sqrt{q}(\Theta^{ab}-\Theta q^{ab})$, then the above relation could have been written as,

$$-q_{ab}£_\ell\Pi^{ab} = \sqrt{q}\ell_a g^{bc}£_\ell N^a_{bc}-\frac{d^2\sqrt{q}}{d\lambda^2}+2\kappa\frac{d}{d\lambda}\left(\sqrt{q}\right) \tag{7.92}$$

A.2 Some Useful Results Associated with Null Foliation

There are a few more geometrical quantities associated with the null vectors associated with GNC, which we will introduce for ready reference later on. The first one corresponds to the induced metric q_{ab} on the null surface, defined as,

$$q_{ab} \equiv g_{ab} + \ell_a k_b + \ell_b k_a; \qquad q^a_b \equiv \delta^a_b + \ell^a k_b + \ell_b k^a \tag{7.93}$$

Note that both $\ell_a q^a_b$ and $k_a q^a_b$ identically vanishes on the null surface, thanks to the relation $\ell_a k^a = -1$; we can think of q^a_b as a projector on to the $r = 0$ surface, which is two-dimensional. Using this projector, we can define extrinsic curvature on a null surface:

$$\Theta_{ab} \equiv q^m_a q^n_b \nabla_m \ell_n = \frac{1}{2} q^m_a q^n_b \pounds_\ell q_{mn} \tag{7.94}$$

If λ is the parameter along the null generator ℓ_a on the null surface, the only nonzero components of Θ_{ab} are (see Eq. (7.61) of Appendix A.1),

$$\Theta_{AB} = \frac{1}{2}\frac{d}{d\lambda} q_{AB} \tag{7.95}$$

The trace of Θ_{ab}, is $\Theta = q_{ab}\Theta^{ab}$ and it is useful to define the trace free *shear tensor* σ_{ab} as,

$$\sigma_{ab} \equiv \Theta_{ab} - \frac{1}{2} q_{ab}\Theta \tag{7.96}$$

Then, as described in Ref. [3] we can introduce shear viscosity coefficient, $\eta = (1/16\pi)$ and the bulk viscosity coefficient $\zeta = -(1/16\pi)$ as well as the dissipation term by:

$$\mathcal{D} \equiv 8\pi \left(2\eta\sigma_{ab}\sigma^{ab} + \zeta\Theta^2\right) = \Theta_{ab}\Theta^{ab} - \Theta^2 \tag{7.97}$$

The importance of σ_{ab} and $\mathcal{D}$—which will occur repeatedly in our discussion—arises from the following fact: It turns out that Einstein's equations, when projected on to any null surface in any spacetime, takes the form of a Navier-Stokes equations [2] with σ_{ab} acting as a viscous tensor and η, ζ acting as bulk and shear viscosity coefficients, In that case, the apparent viscous dissipation is given by $\mathcal{D}$. (The conceptual issues related to this 'dissipation without dissipation' since there can be no real, irreversible, drain of energy are clarified in [2]). In the GNC, we have on the null surface,

$$\mathcal{D} = \frac{1}{4} q^{ac} q^{bd} \partial_u q_{ab} \partial_u q_{cd} - \left(\partial_u \ln \sqrt{q}\right)^2 \tag{7.98}$$

which vanishes when $\partial_u q_{ab} = 0$ on the null surface. These results have been used in the main text.

A.3 Derivation of Various Expressions Used in Text

This appendix will contain derivations of most of the results that we have used in the main text. The derivations will be arranged in the same order as that in the main text. First we will present derivations related to the Navier-Stokes equation and then we will present the requisite derivations of subsequent sections.

A.3.1 Derivation Regarding Navier-Stokes Equation

The first thing to compute is the Lie derivative of the object N^c_{ab}. This can be obtained starting from the first principle, i.e., using expression for N^c_{ab} in terms of Γ^a_{bc} and then using Lie variation of the connection. This immediately leads to:

$$\begin{aligned}
\pounds_v N^a_{bc} &= Q^{ad}_{be}\pounds_v \Gamma^e_{cd} + Q^{ad}_{ce}\pounds_v \Gamma^e_{bd} \\
&= Q^{ad}_{be}\left(\nabla_c\nabla_d v^e + R^e_{\ dmc}v^m\right) + Q^{ad}_{ce}\left(\nabla_b\nabla_d v^e + R^e_{\ dmb}v^m\right) \\
&= \frac{1}{2}\left(\delta^a_b\nabla_c\nabla_d v^d + \delta^a_c\nabla_b\nabla_d v^d\right) - \frac{1}{2}\left(\nabla_b\nabla_c v^a + \nabla_c\nabla_b v^a\right) \\
&- \frac{1}{2}\left(R^a_{\ bmc} + R^a_{\ cmb}\right)v^m
\end{aligned} \tag{7.99}$$

In the above expression the second term in the last line can be written as:

$$\begin{aligned}
\left(\nabla_b\nabla_c v^a + \nabla_c\nabla_b v^a\right) &= 2\partial_b\partial_c v^a + 2\Gamma^a_{bd}\partial_c v^d + 2\Gamma^a_{cd}\partial_b v^d - 2\Gamma^d_{bc}\partial_d v^a \\
&+ v^d\left(\partial_b\Gamma^a_{cd} + \partial_c\Gamma^a_{bd}\right) - 2\Gamma^d_{bc}\Gamma^a_{de}v^e + \Gamma^a_{bd}\Gamma^d_{ce}v^e + \Gamma^a_{cd}\Gamma^d_{be}v^e
\end{aligned} \tag{7.100}$$

In order to compute the Lie variation of N^c_{ab} along the transverse direction we need the two objects $\pounds_\ell N^A_{ur}$ and $\pounds_\ell N^A_{BC}$. For the evaluation of $\pounds_\ell N^A_{ur}$ the following identities will be useful:

$$\begin{aligned}
\left(\nabla_b\nabla_c\ell^a + \nabla_c\nabla_b\ell^a\right)^A_{ur} &= 2\partial_u\partial_r\ell^A + 2\Gamma^A_{ud}\partial_r\ell^d + 2\Gamma^A_{rd}\partial_u\ell^d - 2\Gamma^d_{ur}\partial_d\ell^A \\
&+ v^d\left(\partial_u\Gamma^A_{rd} + \partial_r\Gamma^A_{ud}\right) - 2\Gamma^d_{ur}\Gamma^A_{de}\ell^e + \Gamma^A_{ud}\Gamma^d_{re}v^e + \Gamma^A_{rd}\Gamma^d_{ue}\ell^e \\
&= 2\partial_u\beta^A + 4\alpha\Gamma^A_{ur} + 2\beta^B\Gamma^A_{uB} - 2\beta^A\Gamma^r_{ur} + \partial_u\Gamma^A_{ru} + \partial_r\Gamma^A_{uu} \\
&- 2\Gamma^d_{ur}\Gamma^A_{du} + \Gamma^A_{ud}\Gamma^d_{ur} + \Gamma^A_{rd}\Gamma^d_{uu}
\end{aligned} \tag{7.101}$$

and

$$\begin{aligned}
&\left(\nabla_b\nabla_c\ell^a+\nabla_c\nabla_b\ell^a\right)^A_{ur}+\left[\left(R^a{}_{bmc}+R^a{}_{cmb}\right)\ell^m\right]^A_{ur}\\
&\quad=2\partial_u\beta_A+4\alpha\Gamma^A_{ur}+2\beta^B\Gamma^A_{uB}-2\beta^A\Gamma^r_{ur}\\
&\quad+\partial_u\Gamma^A_{ru}+\partial_r\Gamma^A_{uu}-2\Gamma^d_{ur}\Gamma^A_{du}+\Gamma^A_{ud}\Gamma^d_{ur}+\Gamma^A_{rd}\Gamma^d_{uu}+R^A{}_{uur}\\
&\quad=2\partial_u\beta^A+2\beta^B\Gamma^A_{uB}+2\partial_u\Gamma^A_{ur}\\
&\quad=\partial_u\beta^A+\beta^Bq^{AC}\partial_uq_{BC} \qquad (7.102)
\end{aligned}$$

While for $\pounds_\ell N^A_{BC}$ we have:

$$\begin{aligned}
&\left(\nabla_b\nabla_c\ell^a+\nabla_c\nabla_b\ell^a\right)^A_{BC}+\left[\left(R^a{}_{bmc}+R^a{}_{cmb}\right)\ell^m\right]^A_{BC}\\
&\quad=-2\beta^A\Gamma^r_{BC}+\partial_B\Gamma^A_{uC}+\partial_C\Gamma^A_{uB}-2\Gamma^d_{BC}\Gamma^A_{ud}\\
&\quad+\Gamma^A_{Bd}\Gamma^d_{uC}+\Gamma^A_{Cd}\Gamma^d_{uB}+\partial_u\hat\Gamma^A_{BC}-\partial_C\Gamma^A_{Bu}+\Gamma^A_{ud}\Gamma^d_{BC}\\
&\quad-\Gamma^A_{Cd}\Gamma^d_{uB}+\partial_u\hat\Gamma^A_{BC}-\partial_B\Gamma^A_{Cu}+\Gamma^A_{ud}\Gamma^d_{BC}-\Gamma^A_{Bd}\Gamma^d_{uC}\\
&\quad=-2\beta^A\Gamma^r_{BC}+2\partial_u\hat\Gamma^A_{BC} \qquad (7.103)
\end{aligned}$$

These are the expressions used to get expressions in Sect. 7.4.1. From the vector Ω_a given in Eq. (7.23) we can calculate the Lie variation along ℓ^a leading to,

$$\begin{aligned}
\pounds_\ell\Omega_n&=\ell^m\partial_m\Omega_n+\Omega_m\partial_n\ell^m\\
&=\left(0,\frac{1}{2}\beta_A\beta^A,\frac{1}{2}\partial_u\beta_A\right) \qquad (7.104)
\end{aligned}$$

and equivalently,

$$\begin{aligned}
\pounds_\ell\Omega^n&=\ell^m\partial_m\Omega^n-\Omega^m\partial_m\ell^n\\
&=\left(0,0,\frac{1}{2}\partial_u\beta^A\right) \qquad (7.105)
\end{aligned}$$

Also,

$$\begin{aligned}
D_m\Theta^m_a&=\partial_B\Theta^B_A+\partial_C\ln\sqrt{q}\,\Theta^C_A-\hat\Gamma^C_{AB}\Theta^B_C\\
&=\frac{1}{2}\partial_B\left(q^{BC}\partial_uq_{AC}\right)+\frac{1}{2}q^{CD}\partial_C\ln\sqrt{q}\,\partial_uq_{AD}-\frac{1}{2}q^{BD}\partial_uq_{CD}\hat\Gamma^C_{AB} \qquad (7.106)
\end{aligned}$$

Using these results we finally obtain,

$$\begin{aligned}
&q^n_a\pounds_\ell\Omega_n+D_m\Theta^m_a+\Theta\Omega_a-D_a\left(\Theta+\alpha\right)\\
&\quad=\frac{1}{2}\partial_u\beta_A+\partial_B\left(\frac{1}{2}q^{BC}\partial_uq_{AC}\right)+\frac{1}{2}q^{CD}\partial_uq_{AD}\partial_C\ln\sqrt{q}\\
&\quad-\frac{1}{2}q^{BD}\partial_uq_{CD}\hat\Gamma^C_{AB}+\partial_u\ln\sqrt{q}\frac{1}{2}\beta_A-\partial_A\partial_u\ln\sqrt{q}-\partial_A\alpha \qquad (7.107)
\end{aligned}$$

In raising the free index of the above equation the following identities can be useful

$$-q^{CD}q^{AB}\partial_u q_{BC}\partial_D \ln\sqrt{q} = \partial_u\left(q^{AD}\partial_D \ln\sqrt{q}\right) - q^{AD}\partial_u\partial_D \ln\sqrt{q} \tag{7.108a}$$

$$\begin{aligned}\partial_u\left(q^{AD}\partial_D \ln\sqrt{q}\right) - q^{AB}\partial_u q^{CF}\partial_C q_{BF} - q^{AB}\partial_D\left(q^{CD}\partial_u q_{BC}\right)&\\ = -\partial_u\left(q^{BC}\hat{\Gamma}^A_{BC}\right) + q^{CF}\partial_u q^{AB}\partial_C q_{BF} - q^{AB}\partial_D q^{CD}\partial_u q_{BC}&\\ = -\partial_u\left(q^{BC}\hat{\Gamma}^A_{BC}\right)&\end{aligned} \tag{7.108b}$$

$$q^{AB}q^{CD}\partial_u q_{ED}\hat{\Gamma}^E_{BC} = \hat{\Gamma}^A_{FC}\partial_u q^{CF} - q^{AB}\partial_C q_{BF}\partial_u q^{CF} \tag{7.108c}$$

Moreover we also have:

$$\begin{aligned}R_{ab}\ell^a q^b_c &= R_{uA} = G_{uA}\\ &= \frac{1}{2}\partial_u\beta_A - \partial_A\alpha + \frac{1}{2}q^{BC}\partial_u\partial_B q_{CA} + \frac{1}{2}\partial_B q^{BC}\partial_u q_{AC} - \partial_u\partial_A \ln\sqrt{q}\\ &+ \frac{1}{2}\beta_A\partial_u \ln\sqrt{q} + \frac{1}{2}q^{BC}\partial_u q_{AC}\partial_B \ln\sqrt{q} - \frac{1}{2}q^{BD}\partial_u q_{CD}\hat{\Gamma}^C{}_{AB}\end{aligned} \tag{7.109}$$

It can be checked that this expression coincides exactly with Eq. (7.32) in Sect. 7.4.1 as it should.

To bring out the physics associated with Noether current and its various projections we compute the Noether potential and hence the Noether current completely in GNC for the vector ξ^a. To start with we provide all the components of the tensor $\nabla_a\xi_b$, which are:

$$(\nabla_a\xi_b)_{uu} = -r\partial_u\alpha;\quad (\nabla_a\xi_b)_{ur} = \alpha + r\partial_r\alpha;\quad (\nabla_a\xi_b)_{ru} = -\alpha - r\partial_r\alpha \tag{7.110}$$

$$(\nabla_a\xi_b)_{uA} = r\partial_A\alpha - r\partial_u\beta_A;\quad (\nabla_a\xi_b)_{Au} = -r\partial_A\alpha; \tag{7.111}$$

$$\begin{aligned}(\nabla_a\xi_b)_{rA} = -\frac{1}{2}\beta_A - \frac{1}{2}r\partial_r\beta_A;\quad (\nabla_a\xi_b)_{rr} = 0;\quad (\nabla_a\xi_b)_{Ar} = \frac{1}{2}\beta_A + \frac{1}{2}r\partial_r\beta_A;&\\ (\nabla_a\xi_b)_{AB} = \frac{1}{2}\partial_u q_{AB} + \frac{1}{2}r\left(\partial_A\beta_B - \partial_B\beta_A\right);&\\ (\nabla_a\xi_b)_{BA} = \frac{1}{2}\partial_u q_{AB} - \frac{1}{2}r\left(\partial_A\beta_B - \partial_B\beta_A\right)&\end{aligned} \tag{7.112}$$

Then components of Noether potential $J_{ab} = \nabla_a\xi_b - \nabla_b\xi_a$ have the following expression:

$$J_{uu} = 0;\qquad J_{ur} = 2\alpha + 2r\partial_r\alpha;\qquad J_{uA} = 2r\partial_A\alpha - r\partial_u\beta_A \tag{7.113}$$

$$J_{rA} = -\beta_A - r\partial_r\beta_A;\qquad J_{AB} = r\left(\partial_A\beta_B - \partial_B\beta_A\right) \tag{7.114}$$

The upper components of Noether potential can be obtained as

$$J^{uu}=0; \qquad J^{ur}=-2\alpha-2r\partial_r\alpha-r\beta_A\beta^A-r^2\beta^A\partial_r\beta_A \tag{7.115}$$

$$J^{uA}=-\beta^A-rq^{AB}\partial_r\beta_B; \qquad J^{rr}=r^3\beta^A\beta^B\left(\partial_A\beta_B-\partial_B\beta_A\right) \tag{7.116}$$

$$\begin{aligned}J^{rA}&=2rq^{AB}\partial_B\alpha-rq^{AB}\partial_u\beta_B-2r^2\alpha q^{AB}\partial_r\beta_B-r^3\beta^2q^{AB}\partial_r\beta_B\\&-r^2q^{AB}\beta^C\left(\partial_B\beta_C-\partial_C\beta_B\right)+2r^2\beta^A\partial_r\alpha-r^3\beta^A\beta^B\partial_r\beta_B\end{aligned} \tag{7.117}$$

$$\begin{aligned}J^{AB}&=-r\beta^Aq^{BC}\left(\beta_C+r\partial_r\beta_C\right)\\&+r\beta^Bq^{AC}\left(\beta_C+r\partial_r\beta_C\right)+rq^{AC}q^{BD}\left(\partial_C\beta_D-\partial_D\beta_C\right)\end{aligned} \tag{7.118}$$

Using the above components of Noether potential the components of Noether current can be obtained as

$$J^u(\xi)=-4\partial_r\alpha-\beta^2-2\alpha\partial_r\ln\sqrt{q}-\frac{1}{\sqrt{q}}\partial_A\left(\sqrt{q}\beta^A\right) \tag{7.119}$$

$$J^r(\xi)=2\alpha\partial_u\ln\sqrt{q}+2\partial_u\alpha \tag{7.120}$$

$$J^A(\xi)=\frac{1}{\sqrt{q}}\partial_u\left(\sqrt{q}\beta^A\right)+q^{AB}\partial_u\beta_B-2q^{AB}\partial_B\alpha \tag{7.121}$$

Note that $k_aJ^a(\xi)=-J^u(\xi)$, $q^a_bJ^b(\xi)=J^A(\xi)$ and finally $\ell_aJ^a(\xi)=J^r(\xi)$. As we will see all of them matches with our desired expressions. Also in the stationary limit we have $\partial_u\alpha=\partial_u\beta_A=\partial_uq_{AB}=0$, which in particular tells us that $J^r=0$. Hence in the static limit Noether current is on the null surface since its component along k^a (which is $-\ell_aJ^a(\xi)$) vanishes. Also in this case we have:

$$\begin{aligned}&\left(\nabla_b\nabla_c\xi^a+\nabla_c\nabla_b\xi^a\right)^A_{ur}+\left[\left(R^a{}_{bmc}+R^a{}_{cmb}\right)\xi^m\right]^A_{ur}\\&\qquad=\partial_u\Gamma^A_{ru}+\partial_r\Gamma^A_{uu}-\Gamma^d_{ur}\Gamma^A_{du}+\Gamma^A_{rd}\Gamma^d_{uu}+R^A{}_{uur}\\&\qquad=2\partial_u\Gamma^A_{ur}=-\partial_u\beta^A\end{aligned} \tag{7.122}$$

as well as,

$$\begin{aligned}&\left(\nabla_b\nabla_c\xi^a+\nabla_c\nabla_b\xi^a\right)^A_{BC}+\left[\left(R^a{}_{bmc}+R^a{}_{cmb}\right)\xi^m\right]^A_{BC}\\&\qquad=\partial_B\Gamma^A_{uC}+\partial_C\Gamma^A_{uB}-2\Gamma^d_{BC}\Gamma^A_{ud}\\&\qquad+\Gamma^A_{Bd}\Gamma^d_{uC}+\Gamma^A_{Cd}\Gamma^d_{uB}+\partial_u\hat{\Gamma}^A_{BC}-\partial_C\Gamma^A_{Bu}+\Gamma^A_{ud}\Gamma^d_{BC}\\&\qquad-\Gamma^A_{Cd}\Gamma^d_{uB}+\partial_u\hat{\Gamma}^A_{BC}-\partial_B\Gamma^A_{Cu}+\Gamma^A_{ud}\Gamma^d_{BC}-\Gamma^A_{Bd}\Gamma^d_{uC}\\&\qquad=2\partial_u\hat{\Gamma}^A_{BC}\end{aligned} \tag{7.123}$$

Using these two results we arrive at:

$$£_\xi N^A_{ur} = \frac{1}{2}\partial_u \beta^A \qquad (7.124)$$

$$£_\xi N^A_{BC} = \frac{1}{2}\delta^A_B \partial_C \Theta + \frac{1}{2}\delta^A_C \partial_B \Theta - \partial_u \hat{\Gamma}^A_{BC} \qquad (7.125)$$

These are the expressions used in Sect. 7.4.1.

A.3.2 Derivation Regarding Spacetime Evolution

We need to consider the object $\ell_a g^{ij} £_\ell N^a_{ij}$ in GNC. This in turn requires us to obtain expressions for $£_\ell N^r_{ur}$ and $£_\ell N^r_{AB}$. Then using the identity for Lie variation of N^c_{ab} we can obtain both the Lie variations. For that purpose we have:

$$\frac{1}{2}\left(\delta^a_b \nabla_c \nabla_d \ell^d + \delta^a_c \nabla_b \nabla_d \ell^d\right)^r_{ur} = \frac{1}{2}\partial_u \Theta + \partial_u \alpha \qquad (7.126)$$

$$\left(\nabla_b \nabla_c \ell^a + \nabla_c \nabla_b \ell^a\right)^r_{ur} = 2\partial_u \alpha \qquad (7.127)$$

$$\left[-\frac{1}{2}\left(R^a{}_{bmc} + R^a{}_{cmb}\right)\ell^m\right]^r_{ur} = 0 \qquad (7.128)$$

$$\left(\nabla_b \nabla_c \ell^a + \nabla_c \nabla_b \ell^a\right)^r_{AB} = \alpha \partial_u q_{AB} - \frac{1}{2} q^{CD} \partial_u q_{AC} \partial_u q_{BD} \qquad (7.129)$$

$$\left[-\frac{1}{2}\left(R^a{}_{bmc} + R^a{}_{cmb}\right)\ell^m\right]^r_{AB} = -\frac{1}{2}\alpha \partial_u q_{AB} + \frac{1}{2}\partial_u^2 q_{AB} - \frac{1}{4} q^{CD} \partial_u q_{AC} \partial_u q_{BD} \qquad (7.130)$$

This immediately leads to

$$£_\ell N^r_{ur} = \frac{1}{2}\partial_u^2 \ln \sqrt{q} \qquad (7.131)$$

$$£_\ell N^r_{AB} = -\alpha \partial_u q_{AB} + \frac{1}{2}\partial_u^2 q_{AB} \qquad (7.132)$$

Combining all the pieces and using the results $\Theta = \partial_u \ln \sqrt{q}$ and $\Theta_{AB} = (1/2)\partial_u q_{AB}$, which is the only non-zero component of Θ_{ab}, [16] we finally obtain

$$\begin{aligned} \ell_a g^{ij} £_\ell N^a_{ij} &= 2£_\ell N^r_{ur} + q^{AB} £_\ell N^r_{AB} \\ &= -2\alpha \partial_u \ln \sqrt{q} + 2\partial_u^2 \ln \sqrt{q} - \frac{1}{2}\partial_u q_{AB} \partial_u q^{AB} \\ &= 2\partial_u \alpha + 2\left(\Theta_{ab}\Theta^{ab} - \Theta^2\right) + \frac{2}{\sqrt{q}}\frac{d^2 \sqrt{q}}{du^2} - \frac{2}{\sqrt{q}}\frac{d}{du}\left(\sqrt{q}\alpha\right) \end{aligned} \qquad (7.133)$$

which can also be obtained from a completely different viewpoint. For sake of completeness we will illustrate the alternative methods as well. For the null vector ℓ^a in the adapted GNC system we have:

$$\left(\ell^c\nabla_c\ell^a\right)^u = \alpha + r\beta^2 + \mathcal{O}(r^2); \qquad \left(\ell^c\nabla_c\ell^a\right)^r = r\partial_u\alpha + 2r\alpha^2 + \mathcal{O}(r^2)$$
$$\left(\ell^c\nabla_c\ell^a\right)^A = r\alpha\beta^A + rq^{CA}\partial_C\alpha + \mathcal{O}(r^2) \tag{7.134}$$

Hence on $r = 0$ surface, we have $\kappa = \alpha$, as well as, $\tilde{\kappa} = -(1/2)k^a\nabla_a\ell^2 = \alpha$. Now we will use the Raychaudhuri equation to get $R_{ab}\ell^a\ell^b$ and hence the Lie variation term. In this case we have, $du = d\lambda$, thus Raychaudhuri equation reduces to the following form (see Eq. (7.66))

$$\ell^a\nabla_a\left(\Theta + 2\alpha\right) = \nabla_c\left(\ell^a\nabla_a\ell^c\right) - \nabla_a\ell_b\nabla^b\ell^a - R_{ab}\ell^a\ell^b \tag{7.135}$$

where, the $\Theta + 2\alpha$ term comes from $\nabla_i\ell^i$. Then we have,

$$\begin{aligned}\nabla_c\left(\ell^a\nabla_a\ell^c\right) &= \partial_c\left(\ell^a\nabla_a\ell^c\right) + \ell^a\nabla_a\ell^c\partial_c\ln\sqrt{q}\\ &= 2\alpha^2 + \alpha\partial_u\ln\sqrt{q} + 2\partial_u\alpha\end{aligned} \tag{7.136}$$

Thus non zero components of $B_{ab} = \nabla_a\ell_b$ are as follows:

$$B_{ur} = \alpha; \qquad B_{rA} = \frac{1}{2}\beta_A; \qquad B_{AC} = \frac{1}{2}\partial_u q_{AC} \tag{7.137}$$

From which it can be easily derived that, $B_{ab}B^{ba} = 2\alpha^2 - (1/4)\partial_u q_{AB}\partial_u q^{AB}$. Thus we obtain

$$\begin{aligned}R_{ab}\ell^a\ell^b &= -\partial_u\Theta + 2\alpha^2 + \Theta\alpha - B_{ab}B^{ba}\\ &= \alpha\Theta - \frac{1}{2}q^{AB}\partial_u^2 q_{AB} - \frac{1}{4}\partial_u q_{AB}\partial_u q^{AB}\\ &= \alpha\Theta - \frac{1}{\sqrt{q}}\partial_u^2\sqrt{q} + \left(\partial_u\ln\sqrt{q}\right)^2 - \Theta_{ab}\Theta^{ab}\end{aligned} \tag{7.138}$$

where Θ_{ab} has the only non-zero component to be, $\Theta_{AB} = (1/2)\partial_u q_{AB}$. For the GNC null normal ℓ_a, the Noether current vanishes, such that Lie variation of N^a_{bc} turns out to have the following expression

$$\begin{aligned}\ell_a g^{ij}\pounds_\ell N^a_{ij} &= -2R_{ab}\ell^a\ell^b\\ &= 2\partial_u\alpha + 2\left(\Theta_{ab}\Theta^{ab} - \Theta^2\right) + \frac{2}{\sqrt{q}}\frac{d^2\sqrt{q}}{du^2} - \frac{2}{\sqrt{q}}\frac{d}{du}\left(\sqrt{q}\alpha\right)\end{aligned} \tag{7.139}$$

The components of $S_{ab} = \nabla_a\ell_b + \nabla_b\ell_a$ In GNC are as follows:

$$
\begin{aligned}
S_{uu} &= 2r\partial_u\alpha - 4r\alpha^2 + \mathcal{O}(r^2); \qquad S_{ur} = 2\alpha + 2r\partial_r\alpha + r\beta^2 + \mathcal{O}(r^2) \\
S_{uA} &= -r\beta^B\partial_u q_{AB} + 2r\partial_A\alpha - 2r\alpha\beta_A + \mathcal{O}(r^2); \qquad S_{rr} = 0 \\
S_{rA} &= \beta_A + r\partial_r\beta_A - r\beta^C\partial_r q_{CA} + \mathcal{O}(r^2) \\
S_{AB} &= \partial_u q_{AB} + 2r\alpha\partial_r q_{AB} + r\left(D_A\beta_B + D_B\beta_A\right) + \mathcal{O}(r^2) \qquad (7.140)
\end{aligned}
$$

Thus the trace at $r = 0$ leads to: $S = 4\alpha + 2\partial_u \ln\sqrt{q}$. Thus we arrive at the following expression (see Eq. (7.83) of Appendix A.1)

$$
\ell_a g^{ij}\pounds_\ell N^a_{ij} = 2\partial_u\left(\Theta + 2\alpha\right) - \partial_b S^{rb} - \Gamma^r_{bc}S^{bc} - S^{rc}\partial_c\ln\sqrt{q} \qquad (7.141)
$$

Then the upper components of S_{ab} necessary for the above computation are the followings:

$$
\begin{aligned}
S^{ur} &= S_{ur} + r\beta^A S_{rA} = 2\alpha + 2r\partial_r\alpha + 2r\beta^2 + \mathcal{O}(r^2) \\
S^{rr} &= 2r\partial_u\alpha + 4r\alpha^2 + \mathcal{O}(r^2) \\
S^{rA} &= 4\alpha r\beta^A + 2rq^{AB}\partial_A\alpha - 2r\alpha\beta^A + \mathcal{O}(r^2) \qquad (7.142)
\end{aligned}
$$

The mixed components leads to nothing new so we have not presented them. From Eq. (7.141) the expression for Lie derivative can be obtained as:

$$
\begin{aligned}
\ell_a g^{ij}\pounds_\ell N^a_{ij} &= 2\partial_u\left(\Theta + 2\alpha\right) - \partial_u S^{ru} - \partial_r S^{rr} - \partial_A S^{rA} + 4\alpha^2 + 2\Theta_{ab}\Theta^{ab} - 2\alpha\Theta \\
&= 2\left(\Theta_{ab}\Theta^{ab} - \Theta^2\right) + \frac{2}{\sqrt{q}}\frac{d}{d\lambda}\left(\sqrt{q}\Theta\right) - 2\alpha\Theta + 4\partial_u\alpha + 4\alpha^2 - 4\partial_u\alpha - 4\alpha^2 \\
&= 2\left(\Theta_{ab}\Theta^{ab} - \Theta^2\right) + \frac{2}{\sqrt{q}}\frac{d}{d\lambda}\left(\sqrt{q}\Theta\right) - 2\alpha\Theta \qquad (7.143)
\end{aligned}
$$

which exactly matches with Eq. (7.139). The same can be ascertained for Eq. (7.138) by computing $R_{ab}\ell^a\ell^b$ on the null surface i.e. in the $r \to 0$ limit, directly leading to:

$$
\begin{aligned}
R_{ab}\ell^a\ell^b &= R_{uu} = \partial_a\Gamma^a_{uu} - \partial_u\Gamma^a_{ua} + \Gamma^a_{uu}\Gamma^b_{ab} - \Gamma^a_{ub}\Gamma^b_{ua} \\
&= \partial_u\Gamma^u_{uu} + \partial_r\Gamma^r_{uu} + \partial_A\Gamma^A_{uu} - \partial_u^2\ln\sqrt{q} + \Gamma^u_{uu}\partial_u\ln\sqrt{q} - \Gamma^a_{ub}\Gamma^b_{ua} \\
&= 2\alpha^2 - \partial_u^2\ln\sqrt{q} + \alpha\partial_u\ln\sqrt{q} - \Gamma^u_{ub}\Gamma^b_{uu} - \Gamma^r_{ub}\Gamma^b_{ur} - \Gamma^A_{ub}\Gamma^b_{uA} \\
&= -\partial_u^2\ln\sqrt{q} + \alpha\partial_u\ln\sqrt{q} - \Theta_{ab}\Theta^{ab} \qquad (7.144)
\end{aligned}
$$

which under some manipulations will match exactly with Eq. (7.138). Then in GNC we obtain in identical fashion, the following expression for heat density,

$$
\begin{aligned}
\mathcal{S} &= \nabla_i\ell_j\nabla^j\ell^i - \left(\nabla_i\ell^i\right)^2 \\
&= \left(2\alpha^2 - (1/4)\partial_u q_{AB}\partial_u q^{AB}\right) - \left(\Theta + 2\alpha\right)^2 \\
&= -2\alpha^2 - 4\alpha\Theta - \Theta^2 + \Theta_{ab}\Theta^{ab} \qquad (7.145)
\end{aligned}
$$

This on integration over the null surface leads to,

$$\frac{1}{8\pi}\int dud^2x\sqrt{q}\mathcal{S} = \frac{1}{8\pi}\int dud^2x\sqrt{q}\left(\Theta_{ab}\Theta^{ab} - \Theta^2\right) - \frac{1}{4\pi}\int dud^2x\sqrt{q}\alpha^2 - 4\int d^2xTds \quad (7.146)$$

Let us now write the integral form of $R_{ab}\ell^a\ell^b$, for that we note the integration measure to be $dud^2x\sqrt{q}$. Thus on integration with proper measure and $(1/8\pi)$ factor leads to

$$\frac{1}{8\pi}\int dud^2x\sqrt{q}R_{ab}\ell^a\ell^b = -\frac{1}{8\pi}\int dud^2x\sqrt{q}\mathcal{D} - \frac{1}{8\pi}\frac{d\mathcal{A}_\perp}{d\lambda}\Big|_1^2 + \int d^2xTs|_1^2 - \int d^2xsdT \quad (7.147)$$

which can be written in a slightly modified manner as:

$$\frac{1}{8\pi}\int dud^2x\sqrt{q}R_{ab}\ell^a\ell^b = -\frac{1}{8\pi}\int dud^2x\sqrt{q}\mathcal{D} - \frac{1}{8\pi}\frac{d\mathcal{A}_\perp}{d\lambda}\Big|_1^2 + \int d^2xTds \quad (7.148)$$

Also the Lie variation term (with all the surface contributions kept) on being integrated over the null surface we obtain

$$\begin{aligned}\frac{1}{16\pi}&\int dud^2x\sqrt{q}\times \ell_a g^{ij}\pounds_\ell N^a_{ij}\\ &= \frac{1}{8\pi}\int dud^2x\sqrt{q}\mathcal{D} + \frac{1}{8\pi}\frac{d\mathcal{A}_\perp}{d\lambda}\Big|_1^2 - \int d^2x\left(\frac{\alpha}{2\pi}\right)d\left(\frac{\sqrt{q}}{4}\right)\\ &= -\int d^2x\; Tds + \frac{1}{8\pi}\int dud^2x\sqrt{q}\mathcal{D} + \frac{1}{8\pi}\frac{d\mathcal{A}_\perp}{d\lambda}\Big|_1^2 \end{aligned}\quad (7.149)$$

To calculate Lie variation for ξ^a we need to calculate $\nabla_a\xi_b + \nabla_b\xi_a - S_{ab}$. This tensor has the following components:

$$\begin{aligned} S_{uu} &= -2r\partial_u\alpha, &\quad S_{ur} &= 0\\ S_{uA} &= -r\partial_u\beta_A, &\quad S_{rr} &= 0\\ S_{rA} &= 0. &\quad S_{AB} &= \partial_u q_{AB} \end{aligned}\quad (7.150)$$

Thus in the null limit obtained from the relation: $r \to 0$ we arrive at the result that all the components of S_{ab} vanishes except for the S_{AB} components. If we want to satisfy the Killing condition for ξ^a on the null surface we would require $\partial_u q_{AB} = 0$. From the above relations it is clear that $\nabla_a\xi^a = \Theta$. Moreover we also have,

$$\kappa = -k_b \xi^a \nabla_a \xi^b = -\Gamma^b_{ac} k_b \xi^a \xi^c = \Gamma^u_{uu} = \alpha \tag{7.151a}$$

$$\tilde{\kappa} = -\frac{1}{2} k_b \nabla^b \xi^2 = \frac{1}{2} \partial_r \left(-2r\alpha\right) = -\alpha \tag{7.151b}$$

which shows that for ξ^a, $\kappa = \tilde{\kappa}$. Thus even without the condition $\partial_u q_{AB} = 0$, we arrive at the relation $\kappa = -\tilde{\kappa} = \alpha$. Moreover Lie variation of N^a_{bc} along ξ^a can be obtained by computing the following objects:

$$\frac{1}{2}\left(\delta^a_b \nabla_c \nabla_d \xi^d + \delta^a_c \nabla_b \nabla_d \xi^d\right)^r_{ur} = \frac{1}{2}\partial_u \Theta \tag{7.152}$$

$$\left(\nabla_b \nabla_c \xi^a + \nabla_c \nabla_b \xi^a\right)^r_{ur} = -2\partial_u \alpha \tag{7.153}$$

$$\left[-\frac{1}{2}\left(R^a_{\ bmc} + R^a_{\ cmb}\right)\xi^m\right]^r_{ur} = 0 \tag{7.154}$$

$$\left(\nabla_b \nabla_c \xi^a + \nabla_c \nabla_b \xi^a\right)^r_{AB} = -\alpha \partial_u q_{AB} - \frac{1}{2} q^{CD} \partial_u q_{AC} \partial_u q_{BD} \tag{7.155}$$

$$\left[-\frac{1}{2}\left(R^a_{\ bmc} + R^a_{\ cmb}\right)\xi^m\right]^r_{AB} = -\frac{1}{2}\alpha \partial_u q_{AB} + \frac{1}{2}\partial_u^2 q_{AB} - \frac{1}{4} q^{CD} \partial_u q_{AC} \partial_u q_{BD} \tag{7.156}$$

which can be used to obtain the Lie variation term associated with ξ^a as,

$$\begin{aligned} \ell_a g^{ij} \pounds_\xi N^a_{ij} &= 2\partial_u \alpha + 2\left(\Theta_{ab}\Theta^{ab} - \Theta^2\right) + \frac{2}{\sqrt{q}}\partial_u^2 \sqrt{q} \\ &= \frac{2}{\sqrt{q}}\partial_u \left(\alpha\sqrt{q}\right) + \ell_a g^{ij} \pounds_\ell N^a_{ij} \end{aligned} \tag{7.157}$$

Then using the momentum $\Pi^{ab} = \sqrt{q}[\Theta^{ab} - q^{ab}(\Theta + \kappa)]$ conjugate to the induced metric q_{ab} from Eq. (7.90) we immediately arrive at,

$$-q_{ab} \pounds_\xi \Pi^{ab} = \sqrt{q}\,\ell_a g^{ij} \pounds_\xi N^a_{ij} - \frac{d^2 \sqrt{q}}{d\lambda^2} \tag{7.158}$$

These expressions are used to obtain Eq. (7.19). Also the variational principles in this context are:

$$\begin{aligned} Q_1 &= \int d\lambda d^2x \sqrt{q}\left(-\frac{1}{8\pi} R_{ab}\ell^a \ell^b + T_{ab}\ell^a \ell^b\right) \\ &= \int d\lambda d^2x \sqrt{q}\left[\frac{1}{8\pi}\mathcal{D} + T_{ab}\ell^a \ell^b\right] - \int d^2x\, T ds + \frac{1}{8\pi}\frac{d\mathcal{A}_\perp}{d\lambda}\bigg|_1^2 \end{aligned} \tag{7.159a}$$

$$\begin{aligned} Q_2 &= \int d\lambda d^2x \sqrt{q}\left[\frac{1}{16\pi}\ell_a g^{ij} \pounds_\xi N^a_{ij} + T_{ab}\ell^a \ell^b\right] \\ &= \int d\lambda d^2x \sqrt{q}\left[\frac{1}{8\pi}\mathcal{D} + T_{ab}\ell^a \ell^b\right] + \int d^2x\, s dT + \frac{1}{8\pi}\frac{d\mathcal{A}_\perp}{d\lambda}\bigg|_1^2 \end{aligned} \tag{7.159b}$$

These are the expressions used in Sect. 7.4.3.

References

1. T. Padmanabhan, Lessons from Classical Gravity about the Quantum Structure of Spacetime. arXiv:1012.4476 [gr-qc]
2. T. Padmanabhan, Entropy density of spacetime and the Navier-Stokes fluid dynamics of null surfaces. Phys. Rev. D **83**, 044048 (2011). arXiv:1012.0119 [gr-qc]
3. S. Kolekar, T. Padmanabhan, Action principle for the fluid-gravity correspondence and emergent gravity. Phys. Rev. D **85**, 024004 (2012). arXiv:1109.5353 [gr-qc]
4. T. Padmanabhan, Classical and quantum thermodynamics of horizons in spherically symmetric space-times. Class. Quantum Gravity **19**, 5387–5408 (2002). arXiv:gr-qc/0204019 [gr-qc]
5. R.-G. Cai, S.P. Kim, First law of thermodynamics and Friedmann equations of Friedmann-Robertson-Walker universe. JHEP **0502**, 050 (2005). arXiv:hep-th/0501055 [hep-th]
6. M. Akbar, R.-G. Cai, Friedmann equations of FRW universe in scalar-tensor gravity, f(R) gravity and first law of thermodynamics. Phys. Lett. B **635**, 7–10 (2006). arXiv:hep-th/0602156 [hep-th]
7. D. Kothawala, S. Sarkar, T. Padmanabhan, Einstein's equations as a thermodynamic identity: the cases of stationary axisymmetric horizons and evolving spherically symmetric horizons. Phys. Lett. B **652**, 338–342 (2007). arXiv:gr-qc/0701002 [gr-qc]
8. A. Paranjape, S. Sarkar, T. Padmanabhan, Thermodynamic route to field equations in Lancos-Lovelock gravity. Phys. Rev. D **74**, 104015 (2006). arXiv:hep-th/0607240 [hep-th]
9. D. Kothawala, T. Padmanabhan, Thermodynamic structure of Lanczos-Lovelock field equations from near-horizon symmetries. Phys. Rev. D **79**, 104020 (2009). arXiv:0904.0215 [gr-qc]
10. S. Chakraborty, K. Parattu, T. Padmanabhan, Gravitational field equations near an arbitrary null surface expressed as a thermodynamic identity. JHEP **10**, 097 (2015). arXiv:1505.05297 [gr-qc]
11. S. Chakraborty, Lanczos-Lovelock gravity from a thermodynamic perspective. JHEP **08**, 029 (2015). arXiv:1505.07272 [gr-qc]
12. T. Jacobson, Thermodynamics of space-time: the Einstein equation of state. Phys. Rev. Lett. **75**, 1260–1263 (1995). arXiv:gr-qc/9504004 [gr-qc]
13. T. Padmanabhan, General relativity from a thermodynamic perspective. Gen. Relativ. Gravit. **46**, 1673 (2014). arXiv:1312.3253 [gr-qc]
14. S. Chakraborty, T. Padmanabhan, Evolution of spacetime arises due to the departure from holographic equipartition in all Lanczos-Lovelock theories of gravity. Phys. Rev. D **90**(12), 124017 (2014). arXiv:1408.4679 [gr-qc]
15. T. Damour, Surface effects in black hole physics, in *Proceedings of the Second Marcel Grossmann Meeting on General Relativity* (1982)
16. K. Parattu, S. Chakraborty, B.R. Majhi, T. Padmanabhan, Null Surfaces: Counter-term for the Action Principle and the Characterization of the Gravitational Degrees of Freedom. arXiv:1501.01053 [gr-qc]
17. S.A. Hayward, Unified first law of black hole dynamics and relativistic thermodynamics. Class. Quantum Gravity **15**, 3147–3162 (1998). arXiv:gr-qc/9710089 [gr-qc]
18. D. Kothawala, The thermodynamic structure of Einstein tensor. Phys. Rev. D **83**, 024026 (2011). arXiv:1010.2207 [gr-qc]
19. T. Padmanabhan, *Gravitation: Foundations and Frontiers* (Cambridge University Press, Cambridge, UK, 2010)

Chapter 8
Entropy of a Generic Null Surface from Its Associated Virasoro Algebra

8.1 Introduction

It is well known that one can associate thermodynamic variables like entropy (S) and temperature (T) with any null surfaces. These surfaces are perceived to be one-way membranes, by the class of observers who do not cross them (see e.g., Chap. 8 of [1]). For example, observers at constant spatial coordinates, located at $r > 2M$ in the Schwarzschild spacetime will associate a temperature $T = (1/8\pi M)$ with the black hole horizon [2–5] situated at $r = 2M$. In a similar fashion, an observer at constant spatial coordinates ($x > 0$) in a Rindler spacetime (with the metric $ds^2 = -g^2x^2dt^2 + d\mathbf{x}^2$) will associate a temperature $T = (g/2\pi)$ with the Rindler horizon [6, 7] located at $x = 0$. Both these observers will also associate an entropy density (i.e., entropy per unit area) of $(1/4)$ with the respective horizons. (In fact, the near horizon metric of the Schwarzschild black hole can always be reduced to the Rindler metric; this itself suggests that, as long as physical phenomena are reasonably local, we should get similar results in both these cases.) The freely falling observers, on the other hand, in either spacetime will not perceive the relevant null surface ($r = 2M$ in the black hole case and $x = 0$ in the Rindler case) as endowed with thermodynamic properties, because these observers will eventually cross these null surfaces; therefore, these null surfaces do not act as one-way membranes for these observers (see [8, 9] and also references therein).

For a broad class of, rather generic, null surfaces in an arbitrary spacetime, it is possible to introduce a set of observers who do not cross these null surfaces and perceive them as one-way membranes. It is, therefore, important to investigate in the light of previous discussion whether, in such a general context, these observers will associate thermodynamic variables with such a generic null surface. Several earlier works have shown that there exist a multitude of deep connections between the properties of null surfaces and gravitational dynamics which suggest that this could be true [8, 9]. If so, then we can obtain a *grand unified* picture of the connection between thermodynamics and null surfaces, *with all the previously known cases being reduced to just special cases of this general result.*

S. Chakraborty, *Classical and Quantum Aspects of Gravity in Relation to the Emergent Paradigm*, Springer Theses, DOI 10.1007/978-3-319-63733-4_8

In this chapter, we will explicitly demonstrate that this is indeed the case. In particular, we will show that one can associate a very natural Virasoro algebra (e.g., [10]) with a general class of null surfaces. This Virasoro algebra does have a central charge which, through Cardy's formula [11, 12], leads to the entropy of the associated null surface. The explicit calculation shows that the entropy per unit area is still one quarter, i.e., (1/4) which is consistent with the standard results for black hole horizons, cosmological horizons etc. in the context of general relativity.

This chapter is based broadly on the approach pioneered by Carlip [13, 14] who first proved similar results in the specific case of a black hole horizon. To obtain the general result applicable for generic null surfaces derived in this chapter, we will add certain new ingredients and shall modify the earlier approaches suitably.

8.2 The Formalism

It is known that the Bekenstein entropy can also be derived by exploiting the near horizon conformal symmetries [13, 15–19]. This procedure imposes some suitable fall-off conditions on the metric near the horizon and subsequently one investigates the symmetries that preserve those fall-off conditions. The algebra of the conserved charges associated with these symmetries turns out to be a Virasoro algebra with a central extension. At this stage one can use Cardy's formula [11, 12] and compute the entropy associated with the horizon from the central extension. The result agrees with the standard expression for entropy, viz., one-quarter per unit area. (For a review of this approach and references, see, e.g., [14].) Thus the entropy of a black hole spacetime can also be determined purely from the local symmetries at its event horizon. We shall make these ideas more explicit and concrete in the sequel.

A new ingredient was added to this particular approach to black hole entropy couple of years back, which further emphasized the local nature of this procedure. In this approach [20–23], one uses the diffeomorphism invariance of the Gibbons-Hawking-York surface term in the gravitational action, as well as the near horizon's symmetries, in order to obtain the Virasoro algebra and the entropy. We will be exclusively using this formalism in what follows.

It is well known that [1]: (i) the Einstein-Hilbert action does not possess a valid functional derivative with respect to the metric, or in other words the variational problem for gravity is not well posed and (ii) this situation can be remedied by adding a suitable boundary term to the action. This boundary term contains normal derivatives of the metric and corresponds to the integral of the extrinsic curvature $K = -\nabla_a N^a$ over the boundary, which has an induced metric h_{ab} (here N_a is the unit normal to the boundary surface). It is possible to convert the boundary integral to a volume integral via the Gauss theorem (with any suitable extension of N^a away from the boundary):

$$\int_{\text{Boundary}} \sqrt{-h}\, d^3x\, K = \int \sqrt{-g}\, d^4x\, \nabla_a (K N^a(x^\mu)). \tag{8.1}$$

The invariance of this boundary term under diffeomorphisms leads to a conserved Noether charge. In particular, this Noether charge associated with coordinate transformations (or, diffeomorphisms), related to the local symmetries near the horizon, allows one to obtain a natural Virasoro algebra on the event horizon. The central extension of this Virasoro algebra leads to the correct expression for entropy through Cardy's formula. This formalism was later on successfully extended to cosmological black holes [24] and also to black hole spacetimes endowed with a cosmological event horizon [25].

The previous works cited above, essentially deals with stationary black hole spacetimes having a Killing vector field which becomes null on the Killing horizon. Our aim in this chapter is to extend the above mentioned procedure to a generic null surface which, of course, will not be a Killing horizon. We shall therefore consider a general null surface and describe it in terms of some suitable null coordinates, called the Gaussian null coordinates in this situation (see e.g., [26, 27]), defined locally in the vicinity of the surface, and then shall use the spacetime geometry near that surface. Such null surfaces, as we shall see below, are *much more general* than the Killing horizons. We shall then use the surface term formalism [20–23] mentioned above in order to find out the conserved charges associated with the local symmetry of the null surface geometry. We will further show that, for a wide class of null surfaces, satisfying rather mild physical requirements, it is possible to define a natural Virasoro algebra, whose central extension will lead to the entropy.

The rest of this chapter is organized as follows: in the next section we will derive the Virasoro algebra and subsequently the expression for the entropy in Sect. 8.3. Finally we will end with a summary of our result.

8.3 Entropy Associated with an Arbitrary Null Surface

The geometry in the local neighborhood of a generic null surface can be described using the Gaussian Null Coordinates (elaborated in Sect. 2.4 and henceforth called as GNC). We will now impose the following condition on the GNC metric: *On the null surface*, located at $r = 0$, the functions α and q_{AB} are assumed to be independent of u and one of the coordinates, say, x^1. We stress that we impose no conditions on the spacetime away from the null surface and no conditions on β_A anywhere. (If one thinks of x^1 as analogous to an angular coordinate and u as analogous to the time coordinate, our condition has a superficial similarity to stationarity and axi-symmetry on the null surface at $r = 0$. It can be shown, however, that the null surface with these conditions is not a Killing horizon and that our geometry is much more general.

We shall briefly outline the derivation of the Virasoro algebra and the entropy, omitting algebraic details, which are similar to those in the previous works and can possibly be found in, e.g., [20]. If ζ^a is a vector field that generates the diffeomorphisms, the invariance of the surface term of the Einstein-Hilbert action, yields a conserved ($\nabla_a J^a = 0$) Noether current $J^a = \nabla_b J^{ab}$ with:

$$J^{ab}[\zeta] = \frac{K}{8\pi}(\zeta^a N^b - \zeta^b N^a) \tag{8.2}$$

The corresponding Noether charge is given by:

$$Q[\zeta] = \frac{1}{2}\int d\Sigma_a J^a = \frac{1}{2}\int dq_{ab} J^{ab} \tag{8.3}$$

where Σ is some suitable hypersurface and the last integral is over the 2-surface on the boundary. For our purpose, this would be the spatial 2-surface with metric q_{ab}. The area element on that surface is given by $dq_{ab} = \sqrt{q}\,(N_a M_b - N_b M_a) d^2x^A$, where N^a and M^a are respectively the unit spacelike and timelike normals to the two-surface. We take them to have the following form (see e.g., Eqs. (2.52) and (2.58)):

$$\begin{aligned} N^a &= \left(\frac{1}{\sqrt{2r\alpha}}, \sqrt{2r\alpha}, \frac{r\beta^A}{\sqrt{2r\alpha}}\right) \\ M^a &= \left(\frac{1}{\sqrt{2r\alpha}}, 0, 0, 0\right) \end{aligned} \tag{8.4}$$

so that:

$$K = -\nabla_a N^a = -\sqrt{\alpha/2r} + \mathcal{O}(r) \tag{8.5}$$

The Lie bracket algebra of the Noether charge can be obtained, starting from the following relation:

$$[Q[\zeta_m], Q[\zeta_n]] := \frac{1}{2}\int dq_{ab}(\zeta_m^a J^b[\zeta_n] - \zeta_n^a J^b[\zeta_m]) \tag{8.6}$$

We now have to choose appropriate diffeomorphism generators $\{\zeta_m\}$, such that the near null surface geometry of Eq. (2.46) is preserved. Since the essential feature of the null surface we want to capture, viz., the blocking of the information is solely determined by the (u, r) sector of the metric, i.e., by g_{uu}, g_{ur} and g_{rr}, we need to pick up only those vector fields which acts as an isometries on these metric components at $r = 0$. Setting $\pounds_\zeta g_{ab} = \zeta^c \partial_c g_{ab} + g_{c(a}\partial_{b)}\zeta^c = 0$ for these components, we find that

$$\zeta^a \equiv \{F(u, x^A), -rF'(u, x^A), 0, 0\} \tag{8.7}$$

where a prime denotes differentiation with respect to u. We have further set all the transverse components of ζ^a to zero, since all such components yield vanishing contributions, when substituted into Eqs. (8.3) and (8.6).

Following the standard procedure, at the next step we will try to find out a set of vector fields which obey an infinite dimensional discrete Lie algebra on a circle on or around $r = 0$. In order to do this, we expand the fields as $\zeta = \Sigma A_m \zeta_m$, where m are integers and ζ_m's are regarded as the mode (or, eigen) functions. The desired Lie bracket algebra then turns out to be,

$$[\zeta_m, \zeta_n] = -i(m-n)\zeta_{m+n} \tag{8.8}$$

An obvious choice for the mode functions being, $\zeta_m \equiv F(u, x^A) = \ell^{-1} e^{im(\ell u + p_i x^A)}$, where ℓ is an arbitrary constant. Substituting this expression for mode functions into Eqs. (8.3) and (8.6) along with our assumption that neither α nor $\sqrt{q}$ depend upon the u, x^1 on the null surface, gives us the conserved charges

$$Q_m = \frac{1}{8\pi\ell} \int d^2x \sqrt{q} \alpha \delta_{m,0} \tag{8.9}$$

as well as their associated algebra (see Appendix for a derivation)

$$\begin{aligned}[Q_m, Q_n] = &-\frac{i}{8\pi\ell} \int d^2x \sqrt{q} \alpha (m-n) \delta_{m+n,0} \\ &-\frac{i\ell m^3}{16\pi} \int d^2x \frac{\sqrt{q}}{\alpha} \delta_{m+n,0}\end{aligned} \tag{8.10}$$

This is clearly a Virasoro algebra with a central extension, and looks very similar to the one we get for the Killing horizons [20] obtained earlier in the literature. However, here α is *not* a constant, so it cannot be pulled out of the integration unlike in the case when the null surface is actually a Killing horizon. However we can also work with the associated *densities* and hence easily obtain an expression for the entropy density. The Q_0 and the central charge C for Eqs. (8.3) and (8.6) are given by

$$Q_0 = \frac{1}{8\pi\ell} \int d^2x \sqrt{q} \alpha, \quad C = \frac{3\ell}{4\pi} \int d^2x \frac{\sqrt{q}}{\alpha} \tag{8.11}$$

We now wish to apply Cardy's formula [11, 12] in order to obtain the entropy density associated with our generic null surface. To do this one may note that a small area element $\Delta A \equiv d^2x$ will contribute $\Delta Q_0 \equiv \mathcal{Q}_0 \Delta A$ (to Q_0) and an amount $\Delta C \equiv \mathcal{C} \Delta A$ to the central charge where $\mathcal{Q}_0 = \sqrt{q}\alpha/(8\pi\ell)$ and $\mathcal{C} = 3\ell\sqrt{q}/(4\pi\alpha)$ are the integrands in (8.11). Using Cardy's formula, we finally associate with this area element ΔA the following entropy

$$\Delta S = 2\pi \sqrt{\frac{\mathcal{C}\,\Delta A \mathcal{Q}_0\,\Delta A}{6}} = \frac{\sqrt{q}}{4}\,\Delta A \tag{8.12}$$

The crucial square root in Cardy's formula allows us to interpret the resulting expression in terms of an entropy density as:

$$s \equiv \frac{\Delta S}{\Delta A} = \sqrt{\frac{C\mathcal{Q}_0}{6}} = \frac{\sqrt{q}}{4} \tag{8.13}$$

Since we are considering a very general situation, we can not have a result equivalent to the constancy of α, (which is the analogue of surface gravity for the Killing case) on the null surface. Nevertheless, it is interesting that if we apply the Cardy's formula to the contribution from each area element, we can essentially work with the local densities and hence arrive at the expected result. Thus we find that even though the temperature associated with the null surface is not a constant, one can still obtain an appropriate entropy density.

8.4 Summary

We strongly believe that this result provides a key "missing link" in the study of null surfaces vis-á-vis gravitational thermodynamics. This is mainly due to the facts elaborated below:

- The derivation presented in this chapter is remarkably local. This seems reasonable, since all physics, including thermodynamics of horizons, should have a proper local description. This is because, operationally, all the relevant measurements must be local. The locality in our derivation is mainly based on three facts: (i) We consider diffeomorphisms near the null surface and use only the structural features of the metric near and on the horizon. (ii) We use the behaviour of the surface term in the Einstein-Hilbert action under diffeomorphism with the boundary being the relevant null surface. Again, no bulk construction is required or become necessary. (iii) We further demonstrate that a local version of the Cardy's formula indeed gives the correct answer.
- We have shown that when a null surface (need not be a Killing horizon) is perceived to be a one-way membrane by a particular congruence of observers, they will certainly associate an entropy with it. This directly links the inaccessibility of information with entropy associated with null surfaces acting as one way membranes, which is gratifying. Moreover, the result holds for a very wide class of null surfaces and can also be naturally generalized to arbitrary spacetime dimensions, in a straightforward manner. We only needed to impose certain minimal conditions on the metric, that too only *on* the null surface. Otherwise. *the spacetime is completely arbitrary, away from the null surface.*
- All the previous results known in the literature in the context of black holes, cosmology, non-inertial frames etc. just become special cases of this very general result. Such a unified perspective will definitely be useful in further investigations.

Appendix

For the sake of completeness, we derive the bracket for the Noether charge associated with the vectors as given in Eq. (8.4).

The normalized normal to r = constant surface corresponds to,

$$N_a = \left(0, \frac{1}{\sqrt{2r\alpha + r^2\beta^2}}, 0, 0\right) \tag{8.14}$$

$$N^a = \left(\frac{1}{\sqrt{2r\alpha + r^2\beta^2}}, \sqrt{2r\alpha + r^2\beta^2}, \frac{r\beta^A}{\sqrt{2r\alpha + r^2\beta^2}}\right) \tag{8.15}$$

The extrinsic curvature of the $r = 0$ surface can be computed easily using the above normalized normal N_a as,

$$\begin{aligned} K &= -\nabla_a N^a = -\partial_a N^a - N^a \partial_a \ln\sqrt{-g} \\ &= -\partial_u\left(\frac{1}{\sqrt{2r\alpha + r^2\beta^2}}\right) - \partial_r\sqrt{2r\alpha + r^2\beta^2} \\ &= -\frac{\alpha}{\sqrt{2r\alpha + r^2\beta^2}} \end{aligned} \tag{8.16}$$

The corresponding non-zero component of the Noether potential associated with the Noether charge on $r = 0$ surface yield,

$$\begin{aligned} J^{ur} &= K\xi^u n^r - K\xi^r n^u \\ &= -\left(\alpha - \frac{\partial_u \alpha}{2\alpha}\right)\left(F + \frac{\partial_u F}{2\alpha}\right) \end{aligned} \tag{8.17}$$

Finally the relevant component of the Noether current for calculation of the bracket between Noether charges becomes

$$\begin{aligned} J^r &= \partial_u J^{ru} + \partial_A J^{rA} + J^{rA}\partial_A \ln\sqrt{-g} \\ &= \partial_u\left[\alpha\left(F + \frac{\partial_u F}{2\alpha}\right)\right] \\ &= \alpha\partial_u F + \frac{1}{2}\partial_u^2 F \end{aligned} \tag{8.18}$$

The other normalized vector (the time evolution vector) corresponds to,

$$M^a = \left(\frac{1}{\sqrt{2r\alpha}}, 0, 0, 0\right) \tag{8.19}$$

$$M_a = \left(-\sqrt{2r\alpha}, \frac{1}{\sqrt{2r\alpha}}, -\frac{r\beta^A}{\sqrt{2r\alpha}}\right) \tag{8.20}$$

Hence one arrives at on the $r = 0$ surface,

$$\begin{aligned} d\Sigma_{ab}J^{ab} &= -d^2x\sqrt{q}\,(n_a M_b - n_b M_a)\,J^{ab} \\ &= 2d^2x\sqrt{q}\alpha\left(F + \frac{\partial_u F}{2\alpha}\right) \end{aligned} \tag{8.21}$$

Thus the Noether charge becomes,

$$Q[\xi] = \frac{1}{2}\int d\Sigma_{ab}J^{ab} = \int d^2x\sqrt{q}\alpha\left(F + \frac{\partial_u F}{2\alpha}\right) \tag{8.22}$$

The commutator reads,

$$\begin{aligned} [Q_1, Q_2] &= \int d^2x\sqrt{q}\left\{\left(F_1 + \frac{\partial_u F_1}{2\alpha}\right)\left(\alpha\partial_u F_2 \frac{1}{2}\partial_u^2 F\right) - (1 \leftrightarrow 2)\right\} \\ &= \int d^2x\sqrt{q}\Big[\alpha\,(F_1\partial_u F_2 - F_2\partial_u F_1) + \frac{1}{2}\left(F_1\partial_u^2 F_2 - F_2\partial_u^2 F_1\right) \\ &+ \frac{1}{4\alpha}\left(\partial_u F_1\partial_u^2 F_2 - \partial_u F_2\partial_u^2 F_1\right)\Big] \end{aligned} \tag{8.23}$$

Then with the ansatz for $F(u)$ as in Eq. (8.8) will lead to the Fourier space commutator bracket Eq. (8.10).

References

1. T. Padmanabhan, *Gravitation: Foundations and Frontiers* (Cambridge University Press, Cambridge, UK, 2010)
2. J.D. Bekenstein, Black holes and entropy. Phys. Rev. D **7**, 2333–2346 (1973)
3. J. D. Bekenstein, Generalized second law of thermodynamics in black hole physics. Phys. Rev. D **9**, 3292–3300 (1974)
4. J.D. Bekenstein, Statistical black hole thermodynamics. Phys. Rev. D **12**, 3077–3085 (1975)
5. S. Hawking, Particle creation by black holes. Commun. Math. Phys. **43**, 199–220 (1975)
6. P. Davies, Scalar particle production in Schwarzschild and Rindler metrics. J. Phys. A **8**, 609–616 (1975)
7. W. Unruh, Notes on black hole evaporation. Phys. Rev. D **14**, 870 (1976)
8. T. Padmanabhan, Thermodynamical aspects of gravity: new insights. Rept. Prog. Phys. **73**, 046901 (2010). arXiv:0911.5004 [gr-qc]
9. T. Padmanabhan, General relativity from a thermodynamic perspective. Gen. Rel. Grav. **46**, 1673 (2014). arXiv:1312.3253 [gr-qc]

10. B. Zwiebach, *A first course in string theory*, (Cambridge University Press, 2006), http://www.cambridge.org/uk/catalogue/catalogue.asp?isbn=0521831431
11. J.L. Cardy, Operator Content of Two-Dimensional Conformally Invariant Theories. Nucl. Phys. B **270**, 186–204 (1986)
12. H.W.J. Bloete, J.L. Cardy, M.P. Nightingale, Conformal ivariance, the central charge, and universal finite size amplitudes at criticality. Phys. Rev. Lett. **56**, 742–745 (1986)
13. S. Carlip, Entropy from conformal field theory at killing horizons. Class. Quant. Grav. **16**, 3327–3348 (1999). arXiv:gr-qc/9906126 [gr-qc]
14. S. Carlip, Black hole thermodynamics. Int. J. Mod. Phys. D **23**, 1430023 (2014). arXiv:1410.1486 [gr-qc]
15. O. Dreyer, A. Ghosh, J. Wisniewski, Black hole entropy calculations based on symmetries. Class. Quant. Grav. **18**, 1929–1938 (2001). arXiv:hep-th/0101117 [hep-th]
16. S. Silva, Black hole entropy and thermodynamics from symmetries. Class. Quant. Grav. **19**, 3947–3962 (2002). arXiv:hep-th/0204179 [hep-th]
17. S. Carlip, Symmetries, horizons, and black hole entropy. Gen. Rel. Grav. **39**, 1519–1523 (2007). arXiv:0705.3024 [gr-qc]. [Int. J. Mod. Phys.D17,659(2008)]
18. O. Dreyer, A. Ghosh, A. Ghosh, Entropy from near-horizon geometries of Killing horizons. Phys. Rev. D **89**(2), 024035 (2014). arXiv:1306.5063 [gr-qc]
19. G. Kang, J.-i. Koga, M.-I. Park, Near horizon conformal symmetry and black hole entropy in any dimension. Phys. Rev. D **70**, 024005 (2004). arXiv:hep-th/0402113 [hep-th]
20. B.R. Majhi, T. Padmanabhan, Noether current, horizon virasoro algebra and entropy. Phys. Rev. D **85**, 084040 (2012). arXiv:1111.1809 [gr-qc]
21. B.R. Majhi, T. Padmanabhan, Noether current from the surface term of gravitational action, virasoro algebra and horizon entropy. Phys. Rev. D **86**, 101501 (2012). arXiv:1204.1422 [gr-qc]
22. B.R. Majhi, T. Padmanabhan, Thermality and heat content of horizons from infinitesimal coordinate transformations. arXiv:1302.1206 [gr-qc]
23. B.R. Majhi, Conformal transformation, near horizon symmetry, virasoro algebra and entropy. Phys. Rev. D **90**(4), 044020 (2014). arXiv:1404.6930 [gr-qc]
24. B.R. Majhi, Thermodynamics of Sultana-Dyer black hole. JCAP **1405**, 014 (2014). arXiv:1403.4058 [gr-qc]
25. S. Bhattacharya, A note on entropy of de Sitter black holes. Eur. Phys. J. C **76**(3), 112 (2016). arXiv:1506.07809 [gr-qc]
26. S. Chakraborty, T. Padmanabhan, Thermodynamical interpretation of the geometrical variables associated with null surfaces. Phys. Rev. D **92**(10), 104011 (2015). arXiv:1508.04060 [gr-qc]
27. K. Parattu, S. Chakraborty, B.R. Majhi, T. Padmanabhan, Null surfaces: counter-term for the action principle and the characterization of the gravitational degrees of freedom. arXiv:1501.01053 [gr-qc]

Part IV
Classical Gravity, Quantum Matter

Chapter 9
A Quantum Peek Inside the Black Hole Event Horizon

9.1 Introduction, Motivation and Summary of Results

It is well known that, in the presence of a gravitationally collapsing structure forming a black hole and a quantum field in the Unruh vacuum state, an observer far away from the black hole will see a flux of thermal radiation at late times [1, 2]. This result, which arises from the study of quantum fields in the curved spacetime, has led to several fascinating developments (see e.g., the textbooks and the reviews, [3–12]) in general relativity. While probing this result from different perspectives, quantum field theory in the region *outside* the black hole event horizon has been studied extensively in the literature. However, somewhat surprisingly, there has been much less emphasis in the study of quantum field theory *inside* the event horizon (for some earlier work, similar in spirit, see e.g., [13–18]). The purpose of this chapter is to investigate quantum field theory inside the horizon in the context of a collapsing dust sphere in $(1 + 1)$ spacetime. As we shall see, such a study leads to several curious and interesting results which have their counterparts in $(3 + 1)$ dimensions.

In order to elaborate what is involved, let us consider the Penrose diagram in Fig. 9.1 (top left) describing a gravitationally collapsing body. It is clear from the figure that there are four distinct spacetime regions—marked A, B, C and D. Of these, region D—which is inside the collapsing body and outside the event horizon—is the least interesting one for our purposes. Even though the time dependence of the metric will lead to particle production in this region, we do not expect any universal behaviour here; the results will depend on the details of the collapse. Let us next consider region C which is outside both the collapsing body and the event horizon. This region is of primary importance—and has been extensively investigated in the literature—in connection with the black hole radiation. This is schematically illustrated in Fig. 9.1 (top right) by an outgoing null ray that straddles just outside the horizon and escapes to future null infinity. The thermal nature of the black hole radiation arises essentially due to the exponential redshift suffered by this null ray as it travels from just outside the collapsing matter to future null infinity. While this ray is inside the collapsing matter during part of its travel, the details of the collapse

S. Chakraborty, *Classical and Quantum Aspects of Gravity in Relation to the Emergent Paradigm*, Springer Theses, DOI 10.1007/978-3-319-63733-4_9

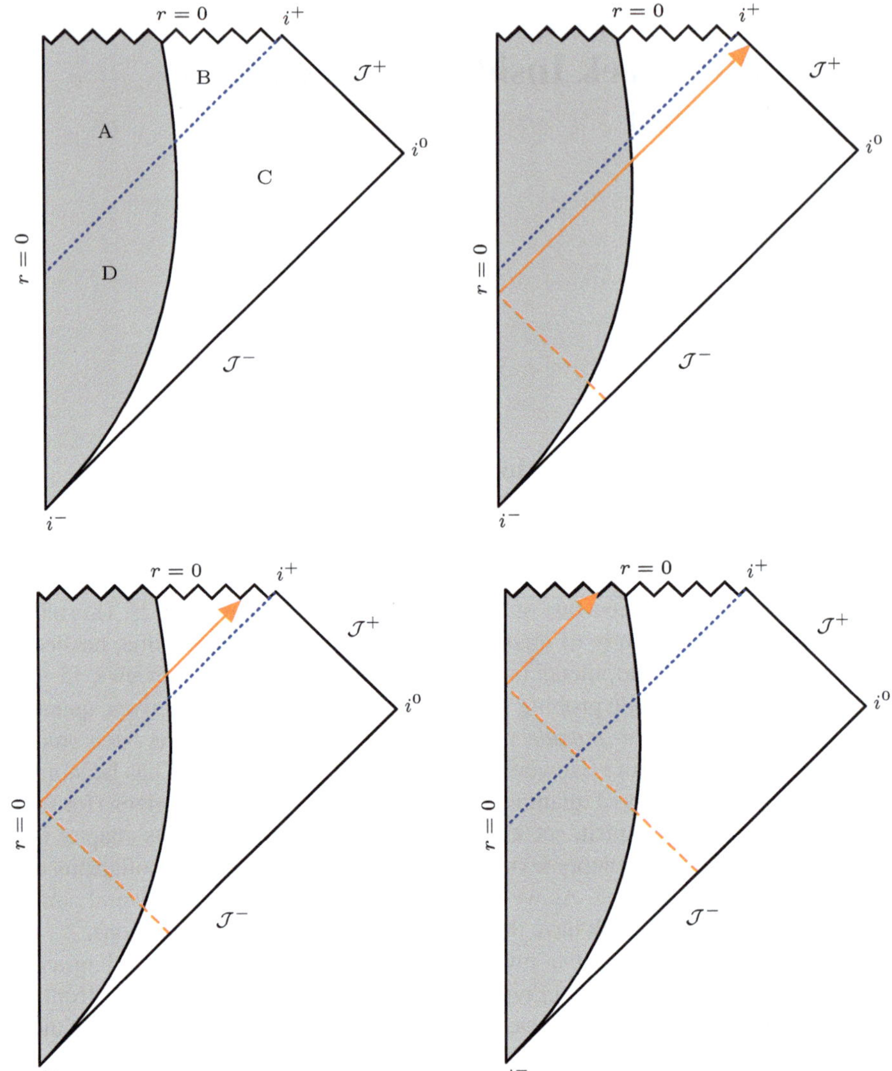

Fig. 9.1 The Penrose diagram in the *top left* figure illustrates the four spacetime regions A, B, C, D we are interested in. The other three figures describe the three null rays relevant to this chapter. See text for detailed discussion

are sub-dominant to the effect of the exponential redshift at late times. We can investigate the black hole evaporation scenario vis-á-vis different kinds of observers in this region: like, e.g., asymptotic and non-asymptotic static observers, radial and inspiraling free-fallers, observers moving in circular orbits etc.; all these cases indeed have been studied in the literature (see, for some recent works, Refs. [19–21] which contain references to earlier papers). In this chapter too, we will briefly discuss the physics in this region since recovering the standard results provides a 'calibration test' for our approach and calculations.

But what we will concentrate on are the regions B and A which are *inside* the event horizon. (We have not found any extensive and systematic investigation of these regions in published literature which was one of the key motivations.) The examination of the Penrose diagram in Fig. 9.1 reveals the following facts about these two regions.

- Region B is inside the event horizon but *outside* the collapsing body. Being a vacuum region in a spherically symmetric geometry, this region is indeed described by a Schwarzschild metric. But, if we use the standard (t, r, θ, ϕ) Schwarzschild coordinates, then r is like a *time* coordinate in this region due to the flip of signs in the metric coefficients at $r < 2M$. Naively speaking, this makes the geometry "time dependent" (due to the dependence of geometry on r) in this region. Alternatively, one can describe this region using a coordinate system which is non-singular at the event horizon like, for e.g., the Kruskal coordinates, in which the line element takes the following form:

$$ds^2 = \frac{32M^3}{r} e^{-r/2M} \left(-dT^2 + dX^2\right) + dL_{\perp}^2 \tag{9.1}$$

where, r is given as an implicit function of X and T via the relation: $(r/2M - 1)e^{r/2M} = X^2 - T^2$ and the transverse line element is $dL_{\perp}^2 = r^2 d\theta^2 + r^2 \sin^2\theta d\phi^2$. Once again, the metric will be time dependent because of its dependence in the Kruskal time coordinate T. All these suggest that we expect a non-trivial dynamics for the quantum field in the region B.

A typical null ray in this region, shown in Fig. 9.1 (bottom left), is trapped and hits the singularity *in the region outside the collapsing matter* in a finite time. (More precisely, it gets *arbitrarily* close to the singularity in the course of time.) It is instructive to compare the null ray in this region with the null ray in the region C mentioned above. (That is, we compare the figures at top right and bottom left of Fig. 9.1.) The two relevant null rays propagate straddling the event horizon just outside and just inside. While the outside ray reaches future null infinity (and plays a key role in the description of the black hole evaporation), it is not clear how the physical parameters vary along the ray which travels just inside the event horizon but gets *arbitrarily* close to the singularity at late times. This is important because the accumulation of energy density due to these modes of the vacuum *which are trapped inside the event horizon* can have important back reaction effects. One key aim of this chapter is to study the physical properties of the quantum field at the events along the null ray like the one shown in the bottom left figure of Fig. 9.1.

- Region A is inside the collapsing body as well as inside the event horizon. Here too we can study the behaviour of the quantum field along the events in the path of the null ray shown in bottom right of the Fig. 9.1. This null ray also hits the singularity but inside the collapsing matter. The key difference between the events along the rays shown in bottom right and those in bottom left, is the following: In the former case, one could compare the energy density of the quantum field with that of the collapsing matter; We expect both to diverge as we approach the singularity inside the collapsing matter. The key question is to determine which one diverges faster in order to ascertain the effects of back reaction. But in the latter case (corresponding to the situation in bottom left figure), we have no matter energy density to compare with the energy density of quantum fields close to the singularity. Therefore, any divergence in the energy density of the quantum field will potentially have important consequences for the back reaction.

We shall focus mostly on the regions A and B in this chapter. We will use C mainly for "calibration"; that is, to arrange the quantum state in such a way that an asymptotic observer sees the standard Hawking flux at late times. After choosing such a state, we will study the dynamical evolution of the quantum field and its regularized energy momentum tensor in the regions A and B. The region D is of comparatively less interest but we mention the results regarding this region in the appendix for the sake of completeness.

Methodology. The Klein-Gordon equation for a massless scalar field can be solved exactly in $(1 + 1)$ dimension, by exploiting the conformal invariance of the theory, which is a fairly standard procedure (see for a review [22]). The conformal invariance of the massless scalar field in $(1 + 1)$ dimension (and the fact that any $(1+1)$ dimensional metric can be written in conformally flat form) makes the relevant mode functions just plane waves. Therefore, the dynamics of a massless scalar field can be reduced essentially to ray tracing. As argued usually in the literature [7, 22], we expect the main conclusions to carry forward to $(3 + 1)$ dimensions, because the dominant contribution to the Hawking effect comes from s-waves even in $(3 + 1)$ dimensions.

Once the solution $\phi(x)$ to the Klein-Gordon equation is known (with suitable boundary conditions ensuring that one obtains the standard black hole evaporation in region C of the spacetime), our next task is to construct physically meaningful observables. There are two standard approaches which have been pursued in the literature in this context. One possibility is to introduce the particle detectors [23–26] in the spacetime moving on various trajectories. The particle content determined by the detector is then given essentially by the Fourier transform of the two-point correlation function of the field in the relevant quantum state. This leads to a Planckian spectrum for an asymptotic detector at rest in the Schwarzschild spacetime which agrees with the standard interpretation of black hole evaporation. Unfortunately, the response of the detector essentially measures the nature of vacuum fluctuations and is sensitive to the history of the trajectory because it is defined using an integral over the proper time. We cannot use the particle content determined by such a detector for estimating the back reaction effects of the quantum field on the spacetime. This is easily seen from the fact that a uniformly accelerated detector in flat spacetime will detect a thermal spectrum of particles but these "particles" do not back react on the spacetime in a generally covariant manner.

Since our primary interest is to study the effect of quantum fields on the background geometry, we need to use a more covariant diagnostic. Such a diagnostic is provided by the (regularised) stress-energy tensor of the quantum field in the given quantum state. It is generally believed that, at least in the semi-classical regime, this regularized expectation value $\langle T_{ab}\rangle_{\rm reg}$ can be used as a source in Einstein's equations to study the effects of back reaction. In particular, when $\langle T_{ab}\rangle_{\rm reg}$ is comparable to the classical source of geometry T_{ab}, we will expect back reaction effects to be significant.

It should be noted that these two diagnostics for describing the quantum field—viz., the detector response or $\langle T_{ab}\rangle_{\rm reg}$—will, in general, give different results. Again, in the flat spacetime Minkowski vacuum, an accelerating detector will see a thermal spectrum but the regularized expectation value of stress-energy tensor will remain zero. For our purpose, it is clearly more meaningful to study $\langle T_{ab}\rangle_{\rm reg}$ rather than the detector response and we will concentrate on this study in this chapter. However, in the last section of this chapter, we will give the relevant results for the detector response, for the sake of comparison and completeness.

The components of the stress-energy tensor depends on the choice of the coordinate system and hence could inherit the pathological characteristics of the coordinate system. It is therefore better to use physically well defined *scalar* quantities that have an invariant meaning. The most natural scalar quantities at any event in spacetime can be constructed in the freely falling frame at that event which eliminates any acceleration effects. If we fill the spacetime by a suitable congruence of radially freely falling observers with four-velocity u^a, then we can construct [27–33] two useful scalar quantities, at any event in spacetime, by the following definition:

$$\mathcal{U} = \langle \hat{T}_{ab}\rangle u^a u^b; \qquad \mathcal{F} = -\langle \hat{T}_{ab}\rangle u^a n^b \tag{9.2}$$

Here n^a is the normal in the radial direction (such that $u_a n^a = 0$), $\mathcal{U}$ is the energy density and $\mathcal{F}$ is the flux at that event as measured by a freely falling observer.

As we said before, the most natural choice for u^a is the four-velocity of the freely falling observers which are free of the acceleration effects. Further, we can fill the entire spacetime with the freely falling observers, which allows a fairly uniform description of *all* the regions of the spacetime. But, in principle, one can also define the scalars in Eq. (9.2) corresponding to *any* observer with a given four-velocity u^a and corresponding n^a. This is relevant in regions C and D of the spacetime where one can also introduce static observers and compute the energy density and flux as measured by them.

Before proceeding further one needs to clarify couple of issues related to this approach. First, note that the $r = 0$ singularity is a mathematically ill-defined event. Even though in the Kruskal coordinates in Fig. 9.2 (both in (1+1) dimension and

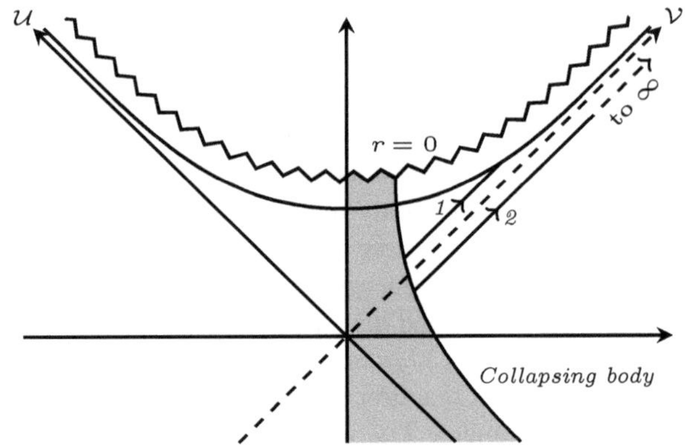

Fig. 9.2 The collapse scenario in the Kruskal coordinates indicating the $r = 0$ and $r = \epsilon$ surfaces. Of the two null rays which are marked, the ray 1 reaches arbitrarily close to the singularity in finite time while the ray 2 propagates to asymptotic infinity at late times. We will be interested in the energy density of the quantum field along the events in the path of ray 1, while ray 2 will be used for calibration. See text for detailed discussion

in (3 + 1) dimension) it is drawn as a hyperbola, with distinct (T, X) coordinates along it, related by $T^2 - X^2 = 1$ one needs to be careful while dealing with the mathematically ill-defined nature of this event. (For example, consider the two events on which the two null rays in bottom-left and bottom-right figures of Fig. 9.1 hit the singularity. One could worry whether these two events are physically distinct.) This is particularly true in the (3 + 1) dimension in which the angular part vanishes at $r = 0$. To keep our discussion physically unambiguous, we will work throughout this chapter with a surface infinitesimally close to $r = 0$ but non-singular. Since $r \neq 0$ on this surface (see Fig. 9.2) any two separate points on it *are indeed distinct*. Thus by working on a $r = \epsilon$ surface, with arbitrary small ϵ these issues can be avoided.

Second, the divergences in the *densities* in the (3 + 1) dimension can sometimes be spurious because the volume factor can become arbitrarily small. So one can have a situation in which $\mathcal{U}$ diverges but $\mathcal{U}\sqrt{h}d^3x$ remains finite just because of geometrical considerations. So before we declare the energy density to be divergent in (3 + 1) dimension we need to ensure that the total energy contained inside a small volume itself diverges, rather than just the energy density. This requires us to define a 3-volume element inside the event horizon. (One cannot, of course, use $4\pi r^2 dr$ because inside the event horizon, $r =$ constant surfaces are spacelike.) When we use a freely-falling observer with the four-velocity u^a to define $\mathcal{U}$, the appropriate 3-volume measure to use is the one corresponding to the same observer in the synchronous coordinates. Using this measure and integrating $\mathcal{U}\sqrt{h}d^3x$ over a small volume around the singularity, we can determine the proper nature of divergence near $r = 0$. We will show later that the energy *density* $\mathcal{U}$ diverges in both regions A and B in (1+1) spacetime dimensions. However in (3 + 1) dimension the divergence in region A is compensated by the shrinking volume measure (thereby making the energy inside a small region around the singularity finite), while the divergence in region B persists, even though it is milder (being only $\epsilon^{-4/3}$ compared to ϵ^{-2} in the (1+1) dimensions).

We will work with the Oppenheimer-Snyder model [34–40], which corresponds to collapsing dust (matter with zero pressure) that forms a black hole. In this case, the spacetime inside the collapsing matter is described by the closed Friedmann metric. To compute the (regularised) stress-energy tensor of the field in a particular quantum state, we use the tools of the standard conformal field theory techniques. (We expect the results to map to the s-wave sector of the scalar field in the collapsing background in $(3+1)$ case which is the usual assumption made in literature; see e.g., [7, 22]) This essentially implies that at any point of the spacetime, we have ingoing and outgoing (spherical) waves and hence the whole manifold can be coordinatized using these null rays. The incoming ray comes directly from $\mathcal{J}^-$ and the outgoing ray comes from $\mathcal{J}^-$ after a reflection from the vertical line in the Penrose diagrams, representing $r = 0$. As for the vacuum state, we shall work with the natural in-vacuum that is uniquely defined on $\mathcal{J}^-$. The relation of this state with the Unruh vacuum and its effects in the region C for the case of thin null shell collapse have been discussed in the refs. [19–21] and will not be repeated here.

The plan of this chapter is as follows. We briefly review the matching of the interior and exterior parts of the spacetime for a collapsing dust ball in Sect. 9.2 and then introduce the double null coordinate spanning all of the spacetime. In Sect. 9.3, we consider the scalar field on this background and compute the components of the regularised stress-energy tensor for same. Subsequently, we compute the energy density and fluxes for various cases in Sect. 9.4. Introducing the radially infalling observers, we study the behaviour of the invariant observables $\mathcal{U}$ and $\mathcal{F}$ along the three null rays mentioned earlier in Sect. 9.5. Lastly, for the sake of completeness, we discuss the results regarding the detector response in terms of the effective temperature in Sect. 9.6. The last section contains the concluding remarks.

9.2 The Gravitational Collapse Geometry

In this section, we briefly review the junction conditions for matching the interior Friedmann universe to the exterior Schwarzschild spacetime.We will then introduce a useful global coordinate system (which we call the double null coordinates) and express both the interior and exterior coordinates in terms of this double null coordinates in a conformally flat form, in the $(1+1)$ sector of the spacetime. This allows us to determine the conformal factor of the metric from which we can calculate the regularised stress-energy tensor.

9.2.1 *Junction Conditions*

We consider the collapse of a spherical region filled with pressure-free dust in the $(3+1)$ spacetime dimensions. The inside region of the dust sphere is homogeneous and isotropic and the metric interior to the dust will be a closed ($k = 1$) Friedmann model with the line element

$$ds^2_{\rm int} = -d\tau^2 + a^2(\tau)d\chi^2 + a^2(\tau)dL^2_\perp \tag{9.3}$$

with τ denoting proper time of the dust particles, comoving with x^μ = constant and $dL^2_\perp = \sin^2\chi d\Omega^2$. A more convenient form for the line element can be obtained by transforming the proper time to conformal time η via the relation:

$$\eta = \int \frac{d\tau}{a(\tau)} \tag{9.4}$$

in terms of which the interior metric reduces to the following form:

$$ds^2_{\rm int} = a^2(\eta)\left(-d\eta^2 + d\chi^2 + dL^2_\perp\right). \tag{9.5}$$

The Einstein's equations for the interior Friedmann universe filled with dust can be solved in a parametric form leading to the results:

$$a(\eta) = \frac{1}{2}a_{\rm max}\,(1 + \cos\eta) \tag{9.6}$$

$$\tau(\eta) = \frac{1}{2}a_{\rm max}\,(\eta + \sin\eta) \tag{9.7}$$

In these conformal coordinates the surface of the dust sphere is taken to be located at some value $\chi = \chi_0$ and the collapse starts at $\eta = \tau = 0$ and ends at $\eta = \pi, \tau = (\pi/2)a_{\rm max}$. The total energy contained within the dust sphere is constant and can be determined in terms of the quantity $a_{\rm max}$ as:

$$\rho a^3 = \text{constant} = \frac{3}{8\pi}a_{\rm max} \tag{9.8}$$

The exterior region, which is spherically symmetric and empty, is described by Schwarzschild metric but in a different set of coordinates. The spherical symmetry both inside and outside suggests that the angular coordinates for both the metric can be taken to be identical. Hence the outside line element has the following expression (we shall use the units with $2M = 1$ henceforth):

$$ds^2_{\rm ext} = -\left(1 - \frac{1}{r}\right)dt^2 + \left(1 - \frac{1}{r}\right)^{-1}dr^2 + dL^2_\perp \tag{9.9}$$

where we have $dL^2_\perp = r^2 d\Omega^2$. The above line element can also be written in terms of outgoing Eddington-Finkelstein coordinates $v = t + r^*$, with $r^* = r + \ln(r - 1)$, such that the line element reduces to

$$ds_{\text{ext}}^2 = -\left(1 - \frac{1}{r}\right) dv^2 + 2dvdr + dL_{\perp}^2 \tag{9.10}$$

In the outside coordinates the surface of the collapsing matter is characterized by $r = r(\tau)$ and $t = t(\tau)$, where, r and t are the Schwarzschild coordinates with τ being the internal time coordinate. The above surface is also determined by $\chi = \chi_0$. Therefore the connecting equation for radial coordinate is [10, 38]:

$$r(\eta) = a(\eta) \sin \chi_0 = \frac{1}{2} a_{\max} \sin \chi_0 \, (1 + \cos \eta) \tag{9.11}$$

Next we need to solve for the time coordinate, which is obtained by solving the following differential equation:

$$\left(1 - \frac{1}{r}\right) \frac{dt}{d\tau} = \text{constant} \tag{9.12}$$

Matching the extrinsic curvature on both sides of the surface of the dust sphere we can fix the constant to be $\cos \chi_0$. This leads to the following differential equation for Schwarzschild time as:

$$\frac{dt}{d\eta} = a_{\max} \cos \chi_0 \cos^2 \frac{\eta}{2} + \frac{a_{\max} \cos \chi_0 \cos^2 \frac{\eta}{2}}{a_{\max} \sin \chi_0 \cos^2 \frac{\eta}{2} - 1} \tag{9.13}$$

Thus the time coordinate can be obtained from the above equation as:

$$\begin{aligned} t(\eta) = &\left\{ \left(1 + \frac{a_{\max} \sin \chi_0}{2}\right) \cot \chi_0 \right\} \eta + a_{\max} \cos \chi_0 \sin \frac{\eta}{2} \cos \frac{\eta}{2} \\ &+ \ln \left| \frac{\sin^2 \frac{\eta}{2} - \cos^2 \frac{\eta}{2} + a_{\max} \sin \chi_0 \cos^2 \frac{\eta}{2} + 2 \cot \chi_0 \sin \frac{\eta}{2} \cos \frac{\eta}{2}}{\left(a_{\max} \sin \chi_0 \cos^2 \frac{\eta}{2} - 1\right)} \right| \end{aligned} \tag{9.14}$$

The time corresponding to horizon crossing of the collapsing surface can be obtained by setting $r = 1$, which leads to, $\eta_H = \pi - 2\chi_0$. Note that as $\eta \to \eta_H$, $t(\eta)$ diverges; thus, for an outside observer, the surface of the imploding matter takes infinite time to reach the event horizon. The a_{max} and χ_0 are not independent, since mass of the imploding dust ball is:

$$M = \frac{4\pi}{3} \rho r^3 = \frac{4\pi}{3} \rho a^3 \sin^3 \chi_0 = \frac{1}{2} a_{\max} \sin^3 \chi_0 \tag{9.15}$$

Thus with $2M = 1$ we are left with the condition $a_{max} \sin^3 \chi_0 = 1$. With these conditions we find the connection between the Eddington-Finkelstein coordinate v in the exterior region to that of interior region to be:

$$v(\eta) = \left\{\cot\chi_0\left(1 + \frac{a_{\max}\sin\chi_0}{2}\right)\right\}\eta + a_{\max}\sin\chi_0\left[\cos\frac{\eta}{2} + \cot\chi_0\sin\frac{\eta}{2}\right]\cos\frac{\eta}{2} + 2\ln\left|\sin\frac{\eta}{2} + \cot\chi_0\cos\frac{\eta}{2}\right| \tag{9.16}$$

Thus we have the matching condition of the outside Schwarzschild coordinates with the inside Friedmann coordinates. With this matching condition we now proceed to determine the double null coordinates for the entire spacetime region.

9.2.2 *Double Null Coordinates*

In this and subsequent sections we will first concentrate on the (1+1) spacetime dimensions by using $dL_{\perp}^2 = 0$ (which corresponds to $d\theta = 0$ and $d\phi = 0$) and quote the results for $(3 + 1)$ dimensions, whenever appropriate. For the purpose of calculating the vacuum expectation values for the regularised stress-energy tensor, it is useful to define and use a coordinate system made up of double null coordinates, constructed from the ingoing and outgoing null rays. Each event x (see Fig. 9.3) can be characterized by two null rays—an incoming ray from $\mathcal{J}^-$, which is labeled as V^+ and an outgoing ray, tracked backward to vertical line $r = 0$ and then reflected to $\mathcal{J}^-$ giving the second null coordinate V^-. To obtain these null coordinates for the global spacetime we proceed as follows: In the interior Friedmann region we define two new coordinates which are both null

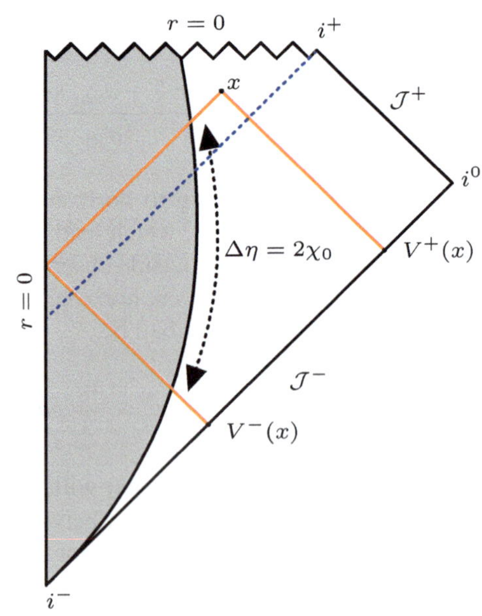

Fig. 9.3 Penrose diagram showing the construction of double null coordinates. Each event x can be characterized by two null rays—an incoming ray from $\mathcal{J}^-$, which is labeled as V^+ and an outgoing ray, tracked backward to the vertical line $r = 0$ and then reflected towards $\mathcal{J}^-$, giving the second null coordinate V^-

$$U = \eta - \chi; \qquad V = \eta + \chi \tag{9.17}$$

Then the line element with the coordinates (V, U) in the interior Friedmann universe reduces to the following form:

$$ds^2_{\text{int}} = -a^2(\eta) dU dV \tag{9.18}$$

In order to describe the exterior structure we define ingoing and outgoing Eddington-Finkelstein coordinates v and u both being null. The exterior line element reduces to:

$$ds^2_{\text{ext}} = -\left(1 - \frac{1}{r}\right) du dv \tag{9.19}$$

in the null coordinates (v, u). Then we can define the double null coordinates V^+ and V^- through the following relations related to the interior Friedmann universe as:

$$V^+ = A(V - \chi_0) \tag{9.20}$$
$$V^- = A(U - \chi_0) \tag{9.21}$$

where we have introduced the function $A(x)$ for future convenience:

$$A(x) \equiv \left\{\cot \chi_0 \left(1 + \frac{a_{\max} \sin \chi_0}{2}\right)\right\} x + 2 \ln \left|\sin \frac{x}{2} + \cot \chi_0 \cos \frac{x}{2}\right|$$
$$+ a_{\max} \sin \chi_0 \left[\cos \frac{x}{2} + \cot \chi_0 \sin \frac{x}{2}\right] \cos \frac{x}{2} \tag{9.22}$$

which connect the interior coordinates (U, V) to the global double null coordinates (V^+, V^-).

From the construction it is easy to verify the following results. The surface $r = 0$, which in the FRW universe, has equation $\chi = 0$ or equivalently (from Eq. (9.17)) $U = V$. Thus from Eqs. (9.20) and (9.21) it is evident that $V^+ = V^-$. Thus the two null coordinates coincide at $r = 0$. Also these null coordinates should be continuous on the surface. Then using the fact that $V = \eta + \chi_0$ on the surface of the star from Eq. (9.20) we get, $V^+ = v$, which it should. For $\chi \neq 0$, the value of V^+ and V^- are never equal for all η. This implies that even though the singularity forms at $\eta = \pi$, in the double null coordinates it is 'spread out'. The value of V^- for the outer surface of the star to reach the singularity is obtained by using $U = \pi - \chi_0$, leading to, $V^- = A(\pi - 2\chi_0)$. By construction this must be equal to V^+ of the point, where the ray enters the dust sphere, which is $A(\eta)$. This helps us to identify, the time at which that particular null ray has entered the sphere as $\pi - 2\chi_0$. This, in turn, fixes the surface of the star completely. Note that this is just a corollary of a more general result: The null rays entering the sphere, getting reflected at $r = 0$ and exiting the sphere satisfy the relation, $d\eta = d\chi$. Thus the ray starts at $\chi = \chi_0$, then goes to $\chi = 0$, again comes out at $\chi = \chi_0$. The difference in the coordinate values for η

between the point of entering and exit of the ray corresponds to $\Delta\eta = 2\chi_0$ (see Fig. 9.3). Hence in the above situation, the final value of η is π, and the entry value of η has to be $\pi - 2\chi_0$, leading back to the previous result.

Another set of relations connecting exterior coordinates (u, v) to the global null coordinates (V^+, V^-) are given by:

$$v = V^+ \tag{9.23}$$

$$\begin{aligned} u = B(U + \chi_0) &= A(U + \chi_0) - a_{\max} \sin\chi_0 \left(1 + \cos(U + \chi_0)\right) \\ &- 2\ln\left|a_{\max}\sin\chi_0\cos^2\left(\frac{U + \chi_0}{2}\right) - 1\right| \end{aligned} \tag{9.24}$$

Thus the interior and exterior line elements can be written in the double null coordinate system (V^+, V^-) in the following manner:

$$ds^2_{\text{int}} = -a^2\left(\frac{U + V}{2}\right)\frac{1}{\frac{dA}{dU}\frac{dA}{dV}}dV^+dV^- \tag{9.25}$$

$$ds^2_{\text{ext}} = -\left(1 - \frac{1}{r}\right)\frac{dB/dU}{dA/dU}dV^+dV^- \tag{9.26}$$

Hence this double null coordinate covers the full spacetime and brings the $(1 + 1)$ sector of the spacetime to the conformally flat form. This is especially suited to evaluate the vacuum expectation value of the regularised stress-energy tensor, which is our next task.

9.3 Regularised Stress-Energy Tensor

We consider a minimally coupled, massless scalar field on the background geometry described by the line elements in Eqs. (9.25) and (9.26). In two dimension the dynamics of the geometry is encoded in the conformal factor which allows us to obtain the vacuum expectation value of the energy momentum tensor for the scalar field. For this we follow the standard procedure [22, 39] and use the following expressions given in terms of the conformal factor:

$$\langle T_{++}\rangle = \frac{1}{12\pi}\left[\frac{1}{2}\frac{\partial_+^2 C}{C} - \frac{3}{4}\left(\frac{\partial_+ C}{C}\right)^2\right] \tag{9.27}$$

$$\langle T_{--}\rangle = \frac{1}{12\pi}\left[\frac{1}{2}\frac{\partial_-^2 C}{C} - \frac{3}{4}\left(\frac{\partial_- C}{C}\right)^2\right] \tag{9.28}$$

$$\langle T_{+-}\rangle = \frac{1}{24\pi}\left[\frac{\partial_+\partial_- C}{C} - \frac{\partial_+ C}{C}\frac{\partial_- C}{C}\right] \tag{9.29}$$

In the above expressions we have introduced a short hand notation: the symbol $\pm$ stands for $V^{\pm}$ respectively. The detailed computation and the explicit expression for the regularised stress-energy tensor, using these relations is given in the Appendix A.1.

9.4 Energy Density and Flux Observed by Different Observers

We shall now compute the energy density and flux at different events in the spacetime using the regularised stress tensor. As we described earlier our primary interest is in the regions B and A inside the event horizon where no static observers can exist. In these regions, we will study the energy density and the flux in the freely-falling frame. But before we do that, it is useful to consider the region C and see how the standard results are reproduced. In this region C (unlike in A and B) we can introduce both static and freely falling observers and study the flux and energy density as measured by both kinds of observers.

9.4.1 *Static Observers in Region C*

We will start with the energy density and flux as measured by the static observers in region C. This will verify our procedure by leading to the standard result of the Hawking radiation with the Tolman redshifted temperature.

A static observer stays outside the event horizon at some fixed radius $r > 1$. Since the observer is not following geodesic he/she has to accelerate (by firing rockets) so as to remain stationary at that radius. The velocity components for this static observer is given by:

$$\dot{V}^{+} = \sqrt{\frac{r}{r-1}} \tag{9.30}$$

$$\dot{V}^{-} = \sqrt{\frac{r}{r-1}} \left(\frac{\cot \chi_0 - \tan\left(\frac{U+\chi_0}{2}\right)}{\cot \chi_0 + \tan\left(\frac{U-\chi_0}{2}\right)} \right) \frac{\cos^2\left(\frac{U-\chi_0}{2}\right)}{\cos^2\left(\frac{U+\chi_0}{2}\right)} \tag{9.31}$$

for a fixed r. The outward normal, determined from the condition $n_a \dot{V}^a = 0$, has the following components:

$$n^{+} = \dot{V}^{+}; \qquad n^{-} = -\dot{V}^{-} \tag{9.32}$$

Thus the energy density and flux can be computed using Eq. (9.2) for the static trajectories as a function of V^{+}. For static observers V^{+} acts as the proper time along their trajectories and we plot both the energy density and flux as a function of V^{+}

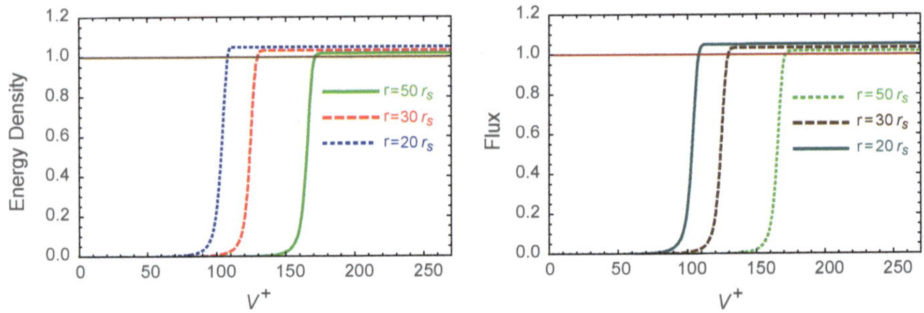

Fig. 9.4 Variation of the energy density and the flux as observed by a static observer as a function of the proper time. The first figure shows the variation of the energy density with proper time V^+. All the static observers at different radii will ultimately observe the standard result for the Hawking radiation with the temperature modified by the Tolman redshift factor. (Note that while plotting the graphs we have normalized them to their values at infinity, i.e., $\pi T_H^2/12$ for convenience.) The same situation is depicted in the second figure for the flux. Both start with a small value and then rise rapidly, ultimately saturating at the appropriate values for the Hawking radiation with the redshifted temperature

in Fig. 9.4. From this figure we see that both energy density and flux shows similar behaviour at large radii. Both of them start from small positive values and finally saturate at the standard thermal spectrum values with the Tolman redshifted value for the temperature.

The key results we want to verify for the static observers are, of course, the late time energy density and flux. The late time limit corresponds to $U \to \pi - 3\chi_0$, under which both the energy density and the flux lead to the following expressions:

$$\mathcal{U} = \frac{\pi T_H^2}{12}\left(1 - \frac{2}{r^4}\right)\left(\frac{r}{r-1}\right) \tag{9.33a}$$

$$\mathcal{F} = \frac{\pi T_H^2}{12}\frac{r}{r-1} \tag{9.33b}$$

We see that, in the asymptotic limit, $r \to \infty$ (corresponding to a static observer at a large distance) above expressions reduce to the standard Hawking energy density and flux. The redshift factor, as is well-known, diverges as the horizon $r = 1$ is approached. From Eq. (9.33a) it is evident that in the near horizon limit the energy density reaches a maximum and then decreases with decreasing r, eventually becoming negative. Thus, even in the Oppenheimer-Snyder dust model there is a region of negative energy density just outside the black hole horizon, just as in the case of a null shell collapse noticed earlier e.g., in [19]. However the flux is always positive and diverges at the horizon.

9.4.2 *Radially In-Falling Observers*

Having reproduced the standard results in region C we now turn our attention to the phenomena inside the event horizon. Since there can be no static observers in this region, it is best to study these effects in the freely falling frame of the radially ingoing observer. To provide the complete picture and to maintain continuity, we will study the results for the freely falling observers in the *entire* spacetime. This can be done from two different perspectives:

(1) We can compute the energy density and flux on the events along the trajectory of a given freely falling observer. This has a clear physical meaning of what such a freely falling observer will see, as she plunges into the singularity.
(2) We can examine the energy density and flux on the events along specific null rays we described in Sect. 9.1 (see Fig. 9.1). This will tell us how the energy density and flux behaves on a selected set of events in the spacetime which are physically well-motivated.

In this section, we shall present the results based on the approach (1), viz. along the trajectories of the observers. We shall describe the behaviour of the energy density and flux on the events along the specific null rays in Sect. 9.5.

Since part of the spacetime is covered by the collapsing dust ball, there exist two types of radially in-falling observers. The first set of observers are (i) those who are comoving with the dust sphere and remain *inside* the collapsing matter, while the second set of observers are (ii) those who are always *outside* the dust sphere. The first set of observers who stay inside the dust sphere ultimately reach the singularity in a finite proper time. These observers are in a Friedmann universe. The observers in a region outside the collapsing matter also hit the singularity in finite time but they live in a Schwarzschild geometry. Thus these two observers are geometrically distinct even though both of them hit the singularity in a finite proper time. We will now consider these two sets of freely falling observers and use them to probe the regions D, A and B.

9.4.2.1 Radially In-Falling Observers: Inside Regions D and A

Let us start with the observers who are inside the dust sphere and are comoving with it. All these observers start at $\eta = 0$ when the dust sphere starts to collapse and reach $\eta = \pi$ in a finite proper time. The trajectory of these observers are characterized by the conformal time η. For such an observer comoving with the dust sphere, the four velocity is given by:

$$u^a = \frac{1}{\sqrt{2}a(\eta)}\left(\frac{dA}{dV}, \frac{dA}{dU}\right) \tag{9.34}$$

$$u_a = -\frac{a}{\sqrt{2}}\left(\frac{1}{dA/dV}, \frac{1}{dA/dU}\right) \tag{9.35}$$

and the unit normal has the following expression:

$$n_a = \frac{a}{\sqrt{2}}\left(\frac{1}{\frac{dA}{dV}}, -\frac{1}{\frac{dA}{dU}}\right) \tag{9.36}$$

$$n^a = \frac{1}{\sqrt{2}a}\left(\frac{dA}{dV}, -\frac{dA}{dU}\right) \tag{9.37}$$

The comoving observer is characterized by the value of χ which remains fixed throughout the trajectory (taken as $\tilde{\chi}$) and hits the singularity as η varies from 0 to π. The variation of energy density for the comoving observer with the proper time η along the trajectory is shown in Fig. 9.5. We note that the energy density for the outermost observer at $\chi = \chi_0$ remains positive and diverges as $\eta \to \pi$. While the energy density as measured by other observers remain finite while reaching a maximum near $\eta = \pi$ and then becomes negative. Finally the energy densities for all the observers diverge in the $\eta \to \pi$ limit. This result is shown in Fig. 9.5.

The energy density diverges as the observer hits the singularity. The ratio of the energy density measured by the observer to that of the energy density of the collapsing matter also diverges. This divergence can be obtained from the leading order behaviour of $\mathcal{U}$ and ρ near $r = 0$, which can be represented, for (1+1) spacetime dimensions, as: (see Eq. (9.85) in the Appendix A.2.2)

$$\mathcal{U}^{(1)} = -\frac{\kappa^2 a_{\max}}{6\pi a^3}; \qquad \rho^{(1)} = \frac{a_{\max}\sin^2\chi_0}{2a}; \qquad \frac{\mathcal{U}^{(1)}}{\rho^{(1)}} \propto \frac{1}{a^2} \tag{9.38}$$

where superscript (1) denote the values of respective quantities in (1+1) spacetime dimensions. Thus, close to the singularity, the energy density of the scalar field dominates over that of the dust sphere.

We next consider the total energy within a small volume with linear dimension ϵ in this (1+1) spacetime both for the scalar field ($\mathcal{E}^{(1)}$) and classical matter ($E^{(1)}$). This energy is given by:

$$\mathcal{E}^{(1)} = \int \mathcal{U}\sqrt{h^{(1)}}dV^{(1)} \sim \frac{1}{\epsilon}; \qquad E^{(1)} = \int \rho\sqrt{h^{(1)}}dV^{(1)} = \epsilon; \qquad \frac{\mathcal{E}^{(1)}}{E^{(1)}} = \frac{1}{\epsilon^2} \tag{9.39}$$

Thus even the total energy of the scalar field within a small volume diverges and is much larger compared to the total energy of the classical matter within that volume. Hence we conclude that energy in the quantum field dominates over the energy of the classical background.

Generalization to (3 + 1) spacetime dimensions Let us now consider the generalization to (3 + 1) spacetime dimensions. Since we are considering s-wave approximation, the energy density for the scalar field in (1+1) spacetime dimensions can be related to that in (3 + 1) spacetime dimensions through the result [7, 22]: $\mathcal{U}^{(1)} \times (1/4\pi r^2) = \mathcal{U}^{(3)}$, where $\mathcal{U}^{(1)}$ and $\mathcal{U}^{(3)}$ are the energy densities in (1+1)

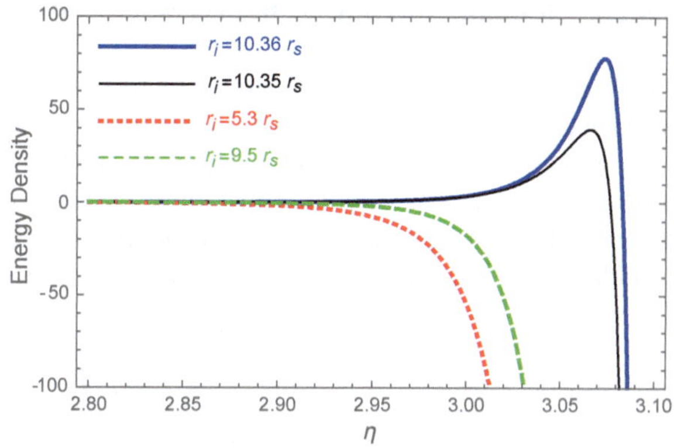

Fig. 9.5 The variation of the energy density as the collapse progresses for an observer inside the sphere and co-moving with the sphere. The energy density diverges as the conformal time η tends to π, the instant when the dust sphere collapses to a singularity

and $(3+1)$ spacetime dimensions respectively. So, inside the dust sphere, we have the following expressions for the energy densities of the scalar field and the dust:

$$\mathcal{U}^{(3)} = -\frac{\kappa^2 a_{\max}}{24\pi^2 a^5 \sin^2\chi_0}; \qquad \rho^{(3)} = \frac{3}{8\pi}\frac{a_{\max}}{a^3}; \qquad \frac{\mathcal{U}^{(3)}}{\rho^{(3)}} \propto \frac{1}{a^2} \tag{9.40}$$

where the superscript (3) denotes the values of the respective quantities in $(3+1)$ spacetime dimensions. Hence the divergence in $\mathcal{U}/\rho$ still persists in $(3+1)$ spacetime dimensions.

As we have argued before, the really important measure is probably not the energy density but the total energy contained in a small volume of size ϵ around the singularity in the $\epsilon \to 0$ limit. From Eq. (9.3) we arrive at the volume element to be: $\sqrt{h}d^3x = a^3 \sin^2\chi \sin\theta \, d\chi d\theta d\phi$. The total energy inside a small volume can be found by integrating the energy density inside a sphere and is given by (with $\epsilon \to 0$):

$$\mathcal{E}^{(3)} = \int_0^{\epsilon} \mathcal{U}^{(3)}\sqrt{h^{(3)}}d^3x \sim \epsilon; \qquad E^{(3)} = \int_0^{\epsilon} \rho^{(3)}\sqrt{h^{(3)}}d^3x \sim \epsilon^3; \qquad \frac{\mathcal{E}^{(3)}}{E^3} = \frac{1}{\epsilon^2} \tag{9.41}$$

It vanishes for both components (as ϵ^3 for the dust and as ϵ for the field), where ϵ represents radius of the small sphere around the singularity. However the total energy in the scalar field dominates over that in the classical background. This suggests that inside the dust sphere, the effect of back-reaction can *not* be neglected in $(3+1)$ spacetime dimensions. (In normal units, the ratio of the energy densities in Eq. (9.41) will go as L_P^2/ϵ^2; so the effect is significant numerically at Planck scales, which is not unexpected.)

9.4.2.2 Radially In-Falling Observers: Outside Regions C and B

The last set of observers we will study are the radially in-falling ones, but outside the dust sphere. They start from some initial radius and then follow a geodesic, ultimately hitting the singularity. They, however, live in the Schwarzschild region of spacetime throughout their life. Any such in-falling observer is parametrised by the energy associated with the geodesic, or—equivalently—by the initial radius from which her suicide mission starts. These two quantities are related via:

$$E = \left(1 - \frac{1}{r_i}\right)^{1/2} \tag{9.42}$$

where r_i stands for the initial radius from which the free-fall begins. From the geodesic equation we can determine the four velocity of these observers, which turns out to be:

$$\dot{V}^+ = \frac{r}{r-1}\left(E - \sqrt{E^2 - \frac{r-1}{r}}\right) \tag{9.43}$$

$$\dot{V}^- = \frac{r}{r-1}\left(E + \sqrt{E^2 - \frac{r-1}{r}}\right)\left(\frac{\cot\chi_0 - \tan\left(\frac{U+\chi_0}{2}\right)}{\cot\chi_0 + \tan\left(\frac{U-\chi_0}{2}\right)}\right)\frac{\cos^2\left(\frac{U-\chi_0}{2}\right)}{\cos^2\left(\frac{U+\chi_0}{2}\right)} \tag{9.44}$$

The components of the normal are determined by the condition $n_a \dot{V}^a = 0$ which leads to the following choice for the outward normal:

$$n^+ = \dot{V}^+; \qquad n^- = -\dot{V}^- \tag{9.45}$$

However we also require the evolution of V^+ as the observer proceeds towards the singularity, i.e., we need V^+ as a function of the observer's conformal time η. This can also be determined from the geodesic equation. For the radially in-falling trajectory the solution to the geodesic equation can be written as:

$$r(\eta) = \frac{r_i}{2}\left(1 + \cos\eta\right) \tag{9.46a}$$

$$\tau = \tau_i + \sqrt{r_i}\frac{r_i}{2}\left(\eta + \sin\eta\right) \tag{9.46b}$$

In this case as well, the conformal time η varies in the same range $0 \leq \eta \leq \pi$. We also have the following initial conditions: $r(\tau_i) = r_i$. From these two relations we can determine $d\tau/d\eta$ and, on using Eq. ((9.43)), we get the differential equation satisfied by $V^+(\eta)$:

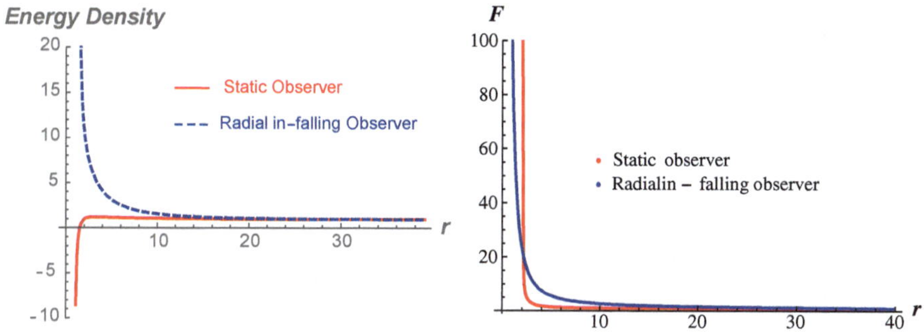

Fig. 9.6 The first figure shows the energy density for the static observers at various radii at late times compared to the radially in-falling observer who crosses those static observers at different times. The second figure depicts the corresponding results for the flux. For static observers near the horizon the flux at late time diverges to infinity as the horizon is approached, while the energy density becomes negative. For radially in-falling observers nothing peculiar happens at $r = 2M$, while both $\mathcal{U}$ and F diverge as $r \to 0$. This feature was also observed earlier for the collapse of a null shell in [19]

$$\frac{dV^+}{d\eta} = r_i \cos\left(\frac{\eta}{2}\right) \frac{\sqrt{r_i}\,[1-(1/r_i)]^{1/2}\cos(\eta/2) - \sin(\eta/2)}{1-\left\{\sec^2(\eta/2)/r_i\right\}} \tag{9.47}$$

The above equation can be integrated to obtain V^+ as a function of the time η. Then both the energy density and flux can be obtained from this prescription. The behaviour of both the energy density and the flux are shown in Fig. 9.6.

We note that both the energy density and the flux show identical behaviour. For $\eta < 0$, when the sphere has not started collapsing, the energy density and flux vanishes. (With our choice $2M = 1$ the asymptotic value of Hawking flux would turn out to be: $(1/192\pi)$. To normalize the figures so that Hawking flux turns out be unity, the figures are drawn with a rescaling of the y-axis.) As the observer moves forward in proper time both the energy density and the flux arise and—as the event horizon is approached—both the energy density and the flux rise rapidly. However the values are finite at the event horizon located at $\eta_H \sim 2.824$. This suggests another interesting aspect to study, which is the comparison between the energy density and flux as measured by a radially in-falling observer and various static observers they encounter along the path. For the static observers the energy density and flux diverge as the event horizon is approached, while for the radially in-falling observer the energy density and flux remain finite at the horizon crossing. We give below the expressions for the energy density and flux as measured by these observers at the horizon:

$$\mathcal{U}_H = \pi T_H^2 \left(\frac{2}{3} - \frac{1}{48E^2} + 2E^2\right) \tag{9.48a}$$

$$\mathcal{F}_H = \pi T_H^2 \left(\frac{1}{48E^2} + 2E^2 \right) \tag{9.48b}$$

For a radially in-falling observer from infinity we have $E = 1$, for which the horizon crossing energy density and flux becomes $\mathcal{U}_H \sim 32\mathcal{U}_\infty$ and $\mathcal{F}_H \sim 24\mathcal{F}_\infty$. (Note that these results were obtained earlier in the context of a null collapse in [19].)

However the energy density and flux finally diverge as the singularity is approached. The behaviour of the energy density and its integral over a small volume has the following expressions near the singularity: (see Eq. (9.87) in Appendix A.2.3)

$$\mathcal{U}^{(1)} = \frac{7\kappa^2}{24\pi}\frac{1}{\epsilon^3}; \qquad \mathcal{E}^{(1)} = \int \mathcal{U}\sqrt{h^{(1)}}dV^{(1)} \sim \frac{1}{\epsilon^2} \tag{9.49}$$

where superscript (1) denotes the expression of the quantities in the (1+1) space-time dimensions. Thus both the energy density and the integrated energy diverges as the singularity is approached. Hence even in region *outside* the dust sphere the backreaction is important near the singularity.

Generalization to $(3+1)$ spacetime dimensions The energy density of the scalar field, as measured by the radially in-falling observers, in (3+1) spacetime dimensions can be obtained by dividing the (1+1) energy density by $4\pi r^2$. (This is a standard procedure adopted in the literature; see e.g., [7, 22]). This leads to the following approximate expression for the energy density near the singularity:

$$\mathcal{U}^{(3)} = \frac{7\kappa^2}{96\pi^2}\frac{1}{r^5} \tag{9.50}$$

This result, at the face of it, shows that the back-reaction effects will be quite significant close to the singularity even *outside* the collapsing dust sphere. However, we need to ensure that the geometrical factor arising from the shrinking of the spatial volume does not over-compensate the divergence. To obtain the proper volume element appropriate for the radially in-falling observer, let us start with the metric in the synchronous coordinates for an observer in free-fall, from a large distance:

$$ds^2 = -d\tau^2 + \frac{1}{r}dR^2 + r^2 d\Omega^2; \qquad r = \left[\frac{3}{2}(R-\tau)\right]^{2/3} \tag{9.51}$$

Thus for a $\tau =$ constant surface the volume element turns out to be:

$$\sqrt{h}d^3x = r^{3/2}\sin\theta \, dRd\theta d\phi = \frac{3}{2}(R-\tau)\sin\theta dRd\theta d\phi \tag{9.52}$$

Integrating this energy density over a sphere of small radius ϵ we get the total energy to be

$$\mathcal{E}^{(3)} = \int_{\tau}^{\epsilon+\tau} \mathcal{U}\sqrt{h}d^3x = \frac{7\kappa^2}{16}\int_{\tau}^{\epsilon+\tau} \frac{1}{\left[\frac{3}{2}(R-\tau)\right]^{10/3}}(R-\tau)\,dR \sim \frac{1}{\epsilon^{4/3}} \quad (9.53)$$

which still diverges in the $\epsilon \to 0$ limit but only as $\epsilon^{-4/3}$. Thus, in the Schwarzschild spacetime region (outside the dust sphere), close to the singularity, the energy density due to scalar field tends to arbitrarily high value. The volume factor does help in (3+1) but not completely.

Thus, on the whole, the results suggest that the back-reaction is important *both* in the outside Schwarzschild regime as well as inside the dust sphere. This has the potential of changing the geometrical structure near the singularity both inside and outside the matter, due to the backreaction.

We will end this section with a few comments on how our results compare with those obtained in some previous attempts [41–44]. In most of these studies, approximate expressions for $\langle T_{ab} \rangle$ have been obtained using a fourth order WKB expansion for the field modes to get unrenormalized $\langle T_{ab} \rangle$ and then eliminating DeWitt-Schwinger counter-terms [45] to get the renormalized value. All these approximate results (which includes both massless and massive fields) for vacuum expectation value of the stress tensor has been obtained outside the event horizon. In [46] the approximate renormalized stress tensor in four-dimensional spacetime was obtained in the interior of the event horizon. Our results follow from the s-wave approximation [7] in which the dominant contribution to the Hawking effect comes from the monopole term ($\ell = 0$) in the multipole expansion. This is also well justified since we are using the radial observers to foliate our spacetime.

As we are interested in the region near the singularity, we will consider dominant terms in the observables when the limit $r \to 0$ is taken. If we calculate the energy density, i.e., $\langle T_{ab} \rangle u^a u^b$ for radially in-falling observer using the approximate stress energy tensor given in [46], it diverges as the singularity is approached (which has been pointed out in [46] itself), but more importantly as $\sim 1/r^5$. This is exactly the divergence we have obtained through our analysis as well. Hence, the energy density expressions obtained under s-wave approximation and energy density calculated using approximate renormalized stress energy tensor given in [46] have similar divergent behavior near the singularity. Moreover, the renormalized energy momentum tensor obtained in [46] includes curvature coupled scalar field as well. Hence the divergent nature of the energy density is a generic feature independent of coupling with curvature.

Using perturbations around the Schwarzschild solution and treating the renormalized energy momentum tensor as a source for this perturbation, it is seen in [46] that Kretschmann scalar diverges more rapidly for the perturbation compared to the classical Kretschmann scalar. This key result also follows and gets verified in our analysis. Along with these, a related fact that curvature for the perturbation grows more rapidly than the background geometry itself is also consistent with our results.

9.5 Energy Density and Flux on Events Along Specific Null Rays

Combining all the results for the static and the radially in-falling observers we can describe the variation of the energy density and flux on events along the null rays introduced previously. For the events along the null rays inside the horizon the best probes are the radially in-falling observers, as they pierce through every spacetime event in the inside region, i.e., both in the regions A and B [see Fig. 9.1]. Thus each radially in-falling observer will intersect the null rays at one unique spacetime event. As the initial radius of the radially in-falling observer is varied, the trajectory will intersect the null ray at a different but still unique spacetime point. All these null rays in the outgoing mode have constant V^- and hence the only parameter that varies along the null rays is V^+. For the events along the null rays outside the horizon we can determine the energy density and flux by using either the freely falling observers or the static observers. We summarize below all the results obtained for these three rays from our analysis.

Events along the null ray just outside the horizon: Let us start the discussion by considering the behaviour of the energy density and flux on the events along the ray which straddles the horizon just outside and ultimately escapes to future null infinity $\mathcal{J}^+$. (This is the ray shown in top-right figure of Fig. 9.1.) As we have discussed earlier, there is nothing peculiar happening in regions C and D, and hence we will not bother to discuss those regions. After its reflection at $r = 0$ the energy density rises, reaches a maxima and then drops back to the Hawking value as the asymptotic infinity is approached [see the left plot in Fig. 9.7]. Similar behaviour is exhibited by the flux, which also reaches a maxima and then decreases ultimately reaching the Hawking value as the asymptotic limit is approached.

Hence we conclude that, on the events along the null ray just outside the horizon, nothing strange happens and the results reproduce the standard black hole radiation, known in the literature.

Events along the null ray just inside the horizon: The second null ray passes through the events which are inside the event horizon (See the bottom-left figure in Fig. 9.1). Initially this ray shares the same geometry as the previous one and hence has almost the same energy density and flux. It also straddles the horizon but—being inside the horizon—its ultimate fate is sealed; it has to hit the singularity in finite time [see Fig. 9.1]. Thus even though the initial events along these two rays share similar physical conditions, the final regions of spacetime encountered by these two rays are very different. One ends at the future null infinity as described earlier, while the other one ends at the singularity.

The energy density and flux in the present case can be calculated using the radially in-falling observers. For different radially in-falling observers (parametrized by the radius from which the free fall starts) the null ray cuts these observers at unique and distinct points. Thus calculating the value of the invariant observables at every point enables us to obtain their variation on the events along the null ray. The energy

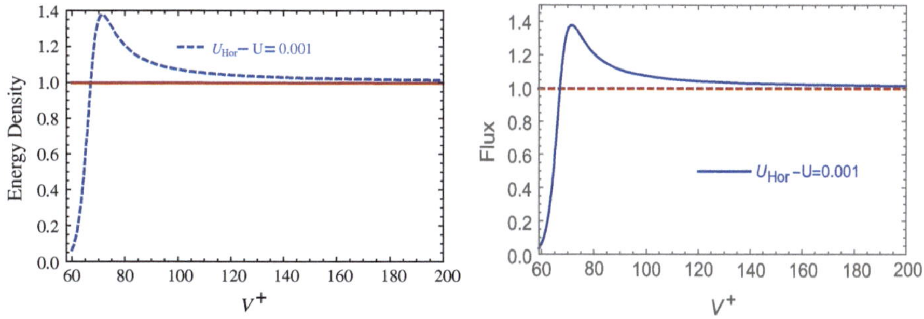

Fig. 9.7 The variation of the energy density and flux on the events along a null ray as a function of time have been shown. Both the curves saturate at late time to the Hawking radiation values implying that all the events on $\mathcal{J}^+$ do receive the standard Hawking flux. The rise of both the flux and energy density at lower values of V^+ corresponds to the near horizon behaviour of these invariant observables. See text for more discussion

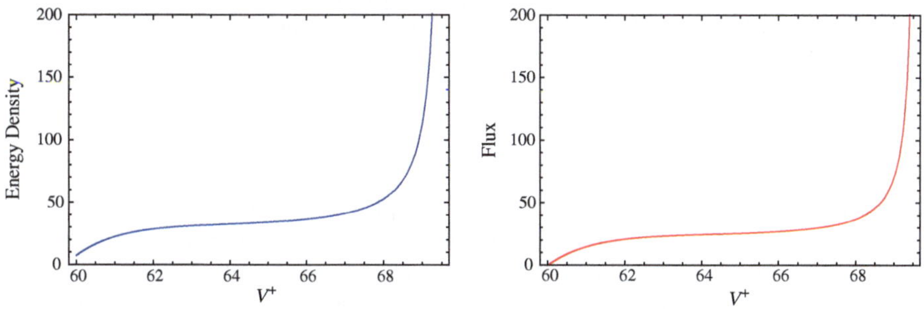

Fig. 9.8 The energy density and flux along null rays very near—but inside—the horizon. These rays come out of the dust sphere after the surface of the dust sphere has crossed the event horizon. Since all these null rays hit the singularity in a finite proper time the energy density and flux diverge. See text for detailed discussion

density along this null ray starts to rise and ultimately diverges as the singularity is approached [see the first figure of Fig. 9.8]. The same feature is also seen in the flux as measured along this null ray.

Thus we can conclude that the final phase of the journey for this null ray shows significant differences compared to the one discussed earlier. For events along this ray close to the singularity, both the invariant observables diverge, in striking contrast to the events along the previous null ray, which ultimately leads to the Hawking flux [see Fig. 9.9]. Naively speaking this makes the back reaction effects quite significant close to the singularity. However, as we said before, such a divergence in the energy *density* can be compensated by the shrinking 3-volume in a region close to the singularity. So we needed to compute the total energy inside a small volume before we can conclude about the divergence. This question has already been addressed in Sect. 9.4.2.2. The final conclusion to be drawn is that these quantities do exhibit a divergence even in $(3+1)$ spacetime dimensions.

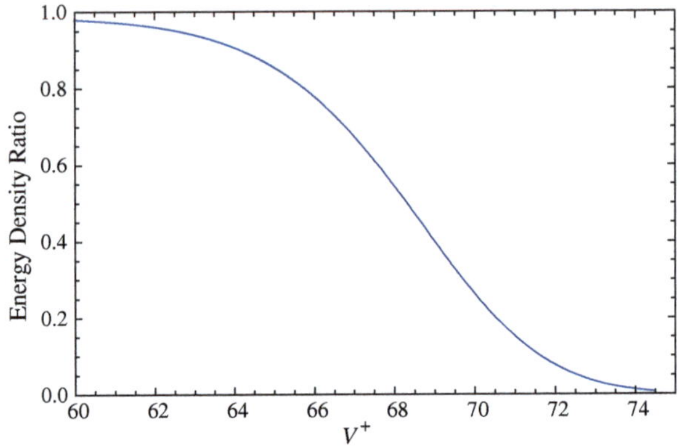

Fig. 9.9 The ratio of the two energy densities along the two null rays straddling the event horizon—one just outside, while another just inside. As the inside ray approaches the singularity this ratio reduces to zero since energy density along the inside ray diverges. However for the rays near the event horizon—but outside—the energy density is finite, making the ratio vanish at late times. For two null rays with almost identical V^+ at the time of reflection at $r = 0$ the energy density turns out to have similar values; it can be seen from the figure that the ratio at the beginning is almost equal to unity

bf Events along the null ray inside the dust sphere: This is the last situation we need to consider. This null ray stays mostly inside the dust sphere and also hits the singularity while it is still inside. On the events along this null ray we can make a direct comparison of the energy density of the quantum field with that of the dust sphere itself. As it moves towards the event horizon the energy density again rises, reaches a maxima and then diverges as the ray hits the singularity [see Fig. 9.1]. The same behaviour was also noted earlier for radially in-falling observers inside the dust sphere. As these observers are used to define the invariant observables along the null rays they exhibit similar behaviour (Fig. 9.10).

In this case as well the ratio $\mathcal{U}/\rho$ diverges as the singularity is approached, i.e., the ratio tends to arbitrarily high value as the null ray reaches arbitrarily close to the singularity. The same result was also obtained earlier in the case of radially in-falling observers inside the dust sphere. Even though the above conclusions were drawn from $(1 + 1)$ spacetime dimensions they carry forward to $(3 + 1)$ spacetime dimensions as well, as shown in Sect. 9.4.2.1. Thus we conclude that the respective energy density for the scalar field and its ratio with dust energy density diverges in $(1 + 1)$ as the singularity is approached. In $(3 + 1)$, even though the energy for the scalar field does not diverge due to the volume factor in $(3+1)$ spacetime dimensions, the ratio $\mathcal{E}^{(3)}/E^{(3)}$ diverges. Thus the backreaction *is* important within the dust sphere as well. As we saw earlier, this is the case at events in region B close to the singularity, where the back reaction effects are significant.

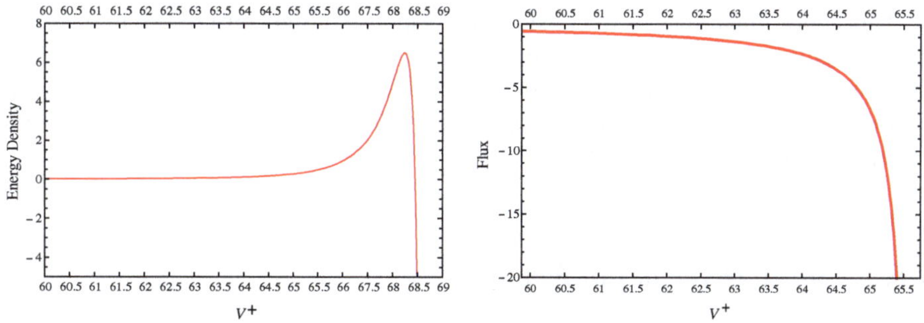

Fig. 9.10 Behaviour of $\mathcal{U}$ and $\mathcal{F}$ at events along the null rays hitting the singularity while they are within the dust sphere. For these rays the value of the U coordinate is greater than the value of the U coordinate at which the dust ball hits the singularity. The null ray initially shows a growth as seen for the radially in-falling observers inside the dust sphere. After reaching a maximum the energy density ultimately decreases, finally it diverges as it hits the singularity. Identical divergence can also be seen for the flux. See text for detailed discussion

9.6 Effective Temperature Measured by Detectors

9.6.1 Effective Temperature as a Measure of Detector Response

We believe, for reasons described in Sect. 9.3, that the regularised stress-tensor expectation value is the appropriate quantity to study in our case. Nevertheless, for the sake of completeness and to compare our results with those obtained earlier in the literature, we will also consider the detector response in this Section.

Let $\gamma(\tau)$ be the trajectory of an asymptotic stationary detector expressed in terms of its proper time τ. Then, under certain conditions, one can associate [21, 47, 48] a temperature T_- with this detector by:

$$T_- = \frac{1}{2\pi}\left|\frac{\ddot{V}^-}{\dot{V}^-}\right| \tag{9.54}$$

(One can also associate a second temperature T_+ in an analogous fashion; but we will not need it for our discussion.) The usefulness of the above definition arises from the fact that it can be applied to non-stationary, non-asymptotic observers as well. It can be easily verified [21] that $T_-(\tau)$ indeed leads to the Tolman redshifted temperature [40] as measured by static Unruh-DeWitt detectors. The approximate constancy of T_- can be expressed through the *adiabaticity condition* [47, 48] which can be expressed as:

$$\eta_- \equiv \left|\frac{\dot{T}_-}{T_-}\right| \ll 1 \tag{9.55}$$

This condition allows us to consider the response of the Unruh-DeWitt detectors in a straightforward way, by evaluating $T_-(\tau)$. Hence with every trajectory we can introduce a temperature $T_-(\tau)$ along with its adiabaticity parameter η_-. This setup will provide another handle on the quantum field theory of the scalar field in the collapsing background geometry.

9.6.2 *Static Detectors in Region C*

For a static detector the effective temperature corresponding to V^- exists which, in the late time limit, $(U \to \pi - 3\chi_0)$ reduces to:

$$T_-^{\rm late} = \frac{1}{4\pi}\sqrt{\frac{r}{r-1}} = T_H\sqrt{\frac{r}{r-1}} \tag{9.56}$$

(the general expression is given in Eq. (9.89) in the Appendix F.) This is precisely the Tolman redshifted temperature, showing the validity and use of this effective temperature formalism. The variation of the effective temperature with V^+, proportional to the proper time of the static detector has been shown in Fig. 9.11. We have also illustrated the variation of the effective temperature at late time with radii, showing the divergence as the horizon is approached.

Incidentally, the detector temperature allows us to study another question which is of interest by itself. Consider a collapsing structure which—unlike the pressure-free dust studied so far—exhibits the following behaviour. The dust sphere starts to collapse from certain radius, continues to collapse until it reaches $r = 2M + \epsilon$ (with $\epsilon \ll 2M$) and then stops collapsing due to, say, internal dynamics and becomes static. What kind of radiation will be detected by a static detector at large distances? Previous studies, based on quantum field theory [49], has shown that the collapsing body (i)

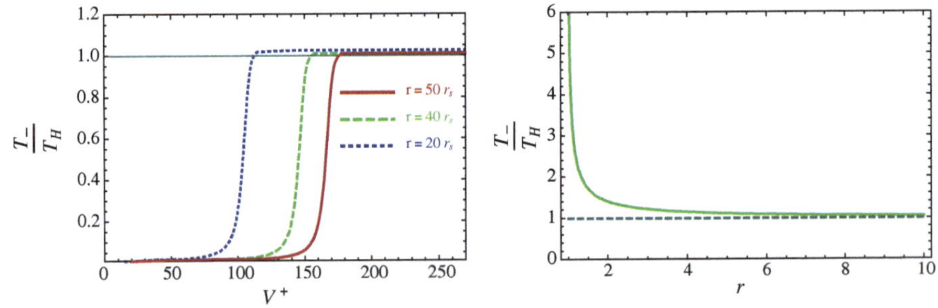

Fig. 9.11 The effective temperature T_- reaches the Tolman redshifted Hawking temperature T_H at late V^+ (*left figure*). In the near horizon regime the effective temperature T_- is positive and diverges as the horizon is approached (*right figure*)

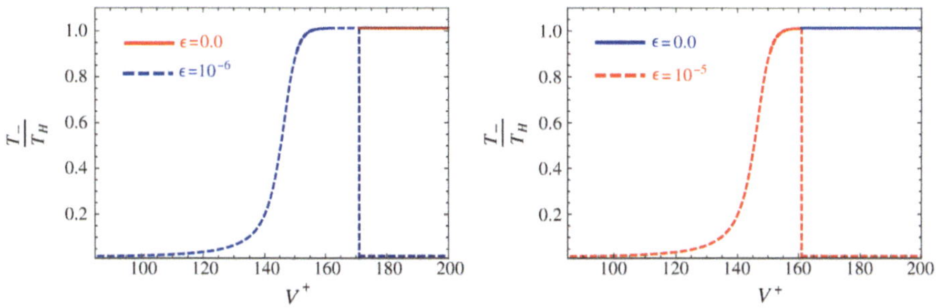

Fig. 9.12 The variation of the effective temperature with V^+ as the surface of the dust sphere approaches the horizon. If, for some reason, the collapse is halted at $r = 2M + \epsilon$ (before forming the horizon) the effective temperature—which initially raises towards the standard Hawking value—drops to zero when the dust sphere becomes static. This situation is shown for two values of the final radii in the two plots

will emit radiation closely resembling the Hawking effect during the collapsing phase and (ii) the effective temperature will drop down to zero at late times. It is worthwhile to study this situation using the detector response.

We will consider the detector response, i.e., effective temperature measured by static detectors in this collapse scenario. The relevant collapsing geometry can be easily achieved by applying the same equations as given by Eq. (9.11) with η being restricted to the range: $0 \leq \eta \leq \pi - 2\chi_0 - 2\epsilon$. Here ϵ is connected to the quantity $r_{\text{stop}} - r_s$, i.e., the radial distance from the horizon at which the collapse stops. This relation can be easily obtained as: $r_{\text{stop}} - r_s = 2\,(\cot \chi_0)\,\epsilon$. Hence, given the value of ϵ, we can obtain the effective temperature at all times. We find it to be nonzero and approaching the saturated Tolman redshifted Hawking value during the collapse. While after the collapse stops, there is no formation of trapped region and thus the effective temperature drops to zero. This situation is being depicted in Fig. 9.12 for two different choices of ϵ. It turns out that as ϵ decreases the effective temperature more closely resembles the Hawking temperature. As $\epsilon \to 0$, we recover the usual Hawking evaporation result.

Let us now return to our main theme. Having discussed the energy density, flux and effective temperature for static observers, we will now proceed to determine the same for radially in-falling observers, both inside the dust sphere and outside.

9.6.3 *Radially In-Falling Detectors: Inside Regions D and A*

Let us first consider the effective temperature measured by the observers inside the dust sphere. As the spacetime can be globally mapped by the coordinates (V^+, V^-), we can follow the same procedure as adopted before to calculate the effective

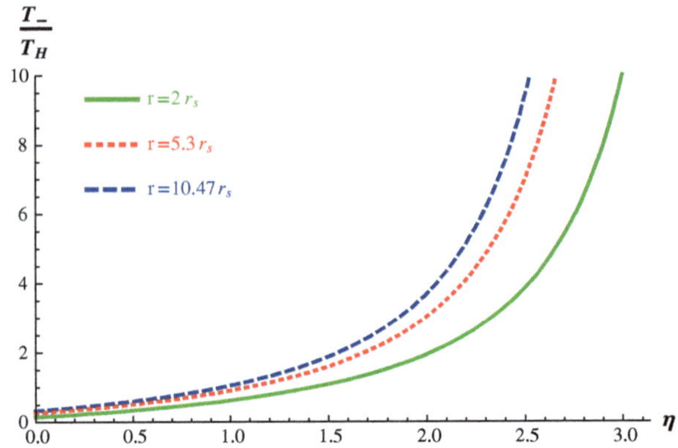

Fig. 9.13 The effective temperature T_- for radially in-falling detectors comoving with the dust sphere and remaining inside it. It is evident that the effective temperature diverges as the detector approaches the singularity

temperature T_-. For completeness we provide the expression for the effective temperature T_- along the trajectory of the radially in-falling detector in Eq. (643) in the Appendix F. The effective temperature depends on the proper time η along the trajectory of the radially in-falling observer and is plotted in Fig. 9.13. It becomes arbitrarily large as $\eta \to \pi$, i.e., as the radially in-falling detector approaches the singularity arbitrarily close.

9.6.4 *Radially In-Falling Detectors: Outside Regions C and B*

Let us now consider the case of Unruh-DeWitt detectors outside the dust sphere moving on a radially in-falling trajectory. The effective temperature they measure along their radially in-falling trajectory is specified by the energy E. This is a non-stationary phenomenon, since—as the detector approaches the singularity the local curvature grows rapidly. Thus this trajectory allows us to probe the time dependent Hawking temperature in detail. The expression is given by Eq. (9.91) in Appendix F.

Here we are interested in two limits: the asymptotic one and the near horizon limit. In the asymptotic limit we have $\ddot{u} = 0$ and $(r-1)/r = 1$, such that the effective temperature reduces to:

$$T_-^{\text{Asym}} = T_H \left(E + \sqrt{E^2 - 1} \right) \tag{9.57}$$

which is consistent with what we expect. For a radially in-falling observer who starts her journey from spatial infinity has energy $E = 1$. Then this observer will detect a temperature $T_-^{\text{Asym}} = T_H$. While for observers with $E \neq 1$, the asymptotic temperature would be different from the Hawking value, as obtained earlier in the context of a null collapse in [21]. In this regime the adiabatic parameter is negligibly small.

On the other hand, at horizon crossing, we have the following expressions: $\dot{r} = -E$, $\dot{V}^+ = (1/2E)$ and $\ddot{r} = -(1/2)$. Then we obtain the following expression:

$$\frac{\ddot{V}^-}{\dot{V}^-} = \frac{2\ddot{u} - \dot{u}^2}{2\dot{u}} = \lim_{r \to 1} \frac{4\dot{r}^2(r+1) - 4r(-\ddot{r} + \dot{r}\dot{v})}{4r\dot{r}} = -2E \tag{9.58}$$

Thus the near horizon effective temperature is given by :

$$T_-^{\rm Hor} = 4ET_H \tag{9.59}$$

Thus the effective temperature is not only non zero, but for states with $E \geq 1$, it exceeds the Hawking value. For an Unruh-DeWitt detector dropped from infinity we have $E = 1$ and thus the detector will perceive four times the Hawking temperature at the horizon crossing. (This result was obtained earlier in Ref. [21, 47]). However at this stage the adiabaticity parameter η_- has large value, and hence the interpretation of T_- as the temperature has some ambiguity, and one needs to consider progressively larger frequency modes.

The behaviour of T_- has been plotted against the conformal time η for the detector in Fig. 9.14. The effective temperature remains finite at the horizon crossing, $\eta_H \sim 2.82$, while diverges later as the detector hits the singularity. We have also plotted the behavior of $T_-^{\rm Asym}$ and $T_-^{\rm Hor}$ as a function of the energy of the in-falling detector. We note that for unbound trajectories ($E \geq 1$), the horizon crossing temperature is always greater than the asymptotic value.

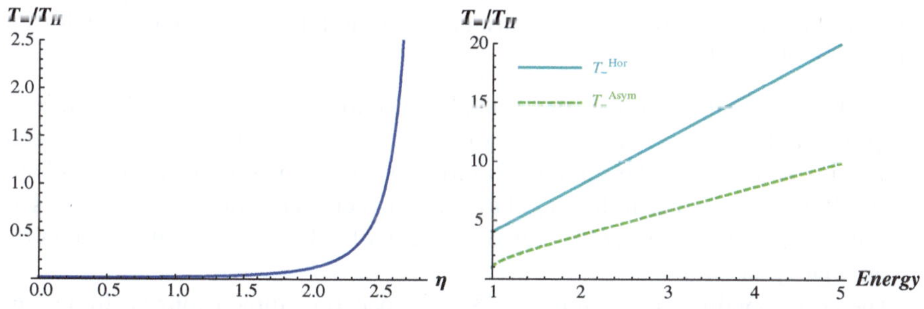

Fig. 9.14 The first figure shows the behaviour of T_-, normalized to the Hawking value T_H, as the radially in-falling observer approaches the singularity. The next figure describes the variation of the effective temperature with the energy gap of the detector in the asymptotic and near horizon regime. The near horizon value is greater than the asymptotic value for unbound trajectories i.e. trajectories with $E \geq 1$

9.7 Conclusion

While the quantum field theory outside the black hole event horizon is well studied in the literature, the corresponding issues in the region inside the event horizon have not attracted sufficient attention. This was the key motivation for this chapter. We have considered a massless scalar field in the background geometry of a collapsing dust ball (Oppenheimer-Snyder model) in the in-vacuum state defined on $\mathcal{J}^-$. We focused on the (regularised) stress-energy tensor in this vacuum state as a physically relevant diagnostic of the quantum effects. The stress-energy tensor, in turn, leads to two scalar observables—energy density and flux in the normal direction—related to an observer at any point in the spacetime. The key results obtained through this analysis are summarized below (Fig. 9.15):

- Let us first start with the radially in-falling observers *inside* the dust sphere and the invariant observables they measure along their trajectory.

1. In (1+1) spacetime dimensions the radially in-falling observers *inside* the dust sphere find that the corresponding energy density and flux becomes arbitrarily large as they approach the singularity (arbitrarily close) in a finite proper time. Even though the energy density of the dust ball (ρ) itself diverges as the singularity is approached, scalar field energy density ($\mathcal{U}$) diverges faster. The same conclusion holds for the total energy within an infinitesimal volume as well. Hence in (1+1) spacetime dimensions the scalar field dominates over the classical source.
2. In $(3+1)$ spacetime dimensions also the energy densities of both the scalar field and classical matter becomes arbitrarily large close to the singularity. Just as in (1+1), in $(3+1)$ as well the scalar field energy density dominates over the classical source. Thus backreaction is *important* inside the dust sphere.

- The second class of observers we consider are the radially in-falling observers *outside* the dust sphere. For them, we arrive at the following conclusions for the invariant observables:

1. In (1+1) spacetime dimensions, radially in-falling observers *outside* the dust sphere find that the energy density of the scalar field becomes arbitrarily large near the singularity. The same is also true for fluxes measured by these radially in-falling observers near the singularity. Since there is no classical matter present in this region this divergent energy of the scalar field has the potential to alter the singularity structure.
2. The corresponding situation in the $(3+1)$ spacetime dimensions is similar and both the energy density and flux diverge as the singularity is approached. The divergence in the energy density of the scalar field can *prohibit* the formation of singularity itself due to existence of back-reaction near the singularity.

   Thus, the energy density in the scalar field dominates over the classical background energy density making the backreaction important both inside and outside the dust sphere. All these conclusions hold for energy density and flux as measured along the events in the null rays hitting the singularity either inside the dust sphere or outside.

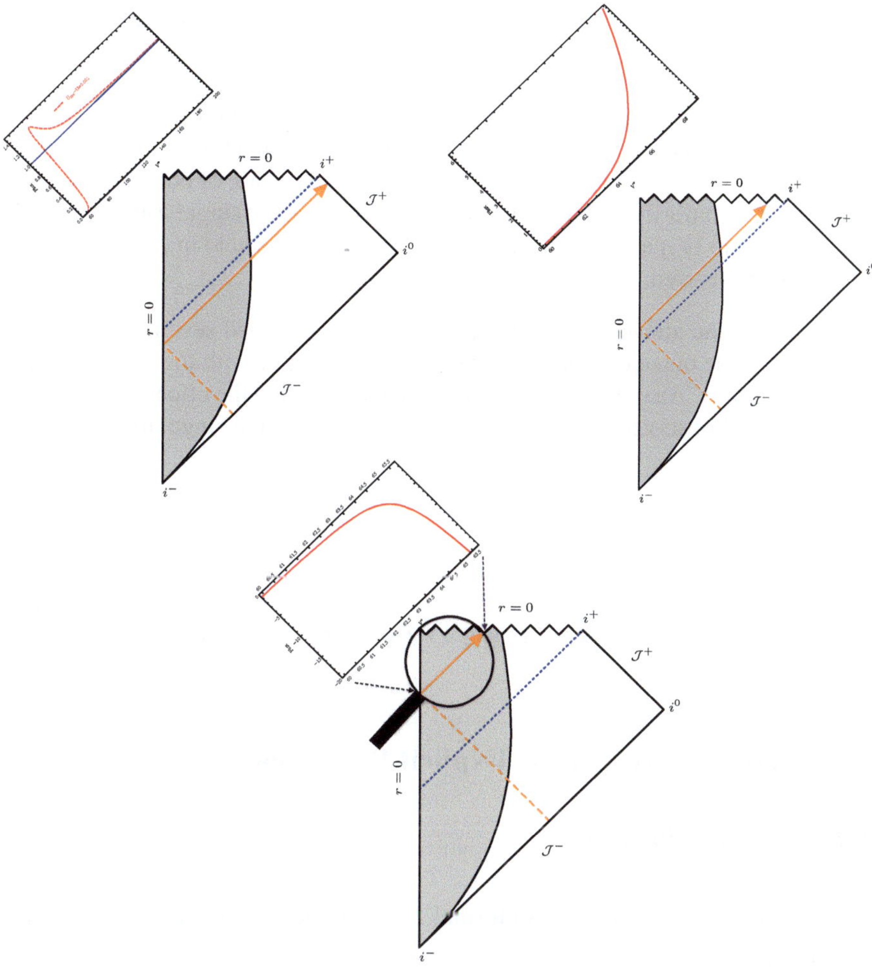

Fig. 9.15 Penrose diagrams (along with the associated graphs) illustrate how the flux along the null rays vary. In the *top-left* figure we consider a ray straddling the event horizon but remains outside. The flux in the near horizon regime first shows an increment, but soon tends to the Hawking value as the ray approaches the future null infinity. The behaviour of flux along the null ray straddling the event horizon just inside, but remaining *outside* matter at late times is shown in the *top-right* figure. This null ray hits the singularity in a finite proper time and the flux diverges. The bottom figure shows the divergent nature of the flux for null rays *inside* the dust sphere. The flux diverges near the singularity as expected

- The radially in-falling observers outside the dust sphere observe that there is nothing peculiar at the event horizon; neither the energy density nor the flux diverges there.
- The study of the events along the null ray straddling on the outside of the event horizon leads to the standard Hawking energy density and flux at late times, while the events along the ray inside the horizon sees an increase of both energy density and flux to arbitrarily high values as it approaches near the singularity. Even though derived in (1+1) spacetime dimensions, the same results hold in $(3+1)$ spacetime dimensions as well, with milder divergences.

In addition to the above main results, we have also obtained several results for the observers on the outside of the event horizon which agree with the previous studies and expectations. Also, for completeness, we have studied the effective temperature formalism and observed similar effects of divergence at the singularity in this case as well.

Appendix A

In this Appendix, we present the steps for deriving various results mentioned in Chap. 9.

A.1 Stress-Energy Tensor: Explicit Derivation

A.1.1 Exterior Region

The various derivatives of the conformal factor in the exterior region have the following expressions:

$$\partial_+ C = \frac{r-1}{2r^3}\frac{dB/dU}{dA/dU}; \qquad \partial_+^2 C = \frac{(r-1)(3-2r)}{4r^5}\frac{dB/dU}{dA/dU} \tag{9.60}$$

$$\partial_- C = \frac{r-1}{r}\partial_-\left(\frac{dB/dU}{dA/dU}\right) - \frac{r-1}{2r^3}\left(\frac{dB/dU}{dA/dU}\right)^2 \tag{9.61}$$

$$\begin{aligned}\partial_-^2 C = &-\frac{3}{2}\frac{r-1}{r^3}\frac{dB/dU}{dA/dU}\partial_-\left(\frac{dB/dU}{dA/dU}\right) + \frac{r-1}{r}\partial_-^2\left(\frac{dB/dU}{dA/dU}\right)\\ &+\frac{(3-2r)(r-1)}{4r^5}\left(\frac{dB/dU}{dA/dU}\right)^3\end{aligned} \tag{9.62}$$

$$\partial_-\partial_+ C = \frac{r-1}{2r^3}\partial_-\left(\frac{dB/dU}{dA/dU}\right) - \left(\frac{dB/dU}{dA/dU}\right)^2 \frac{(3-2r)(r-1)}{4r^5} \tag{9.63}$$

With the following expressions for energy momentum tensor:

$$\langle T_{++}\rangle = \frac{\kappa^2}{48\pi}\left(\frac{3}{r^4} - \frac{4}{r^3}\right); \qquad \langle T_{+-}\rangle = \frac{\kappa^2}{12\pi}\frac{r-1}{r^4}\frac{dB/dU}{dA/dU} \tag{9.64}$$

$$\begin{aligned}
\langle T_{--}\rangle &= \frac{\kappa^2}{48\pi}\Bigg[\left(\frac{dB/dU}{dA/dU}\right)^2\left(\frac{3}{r^4} - \frac{4}{r^3}\right) \\
&+ 16\left\{\frac{1}{2}\frac{\partial_-^2\left(\frac{dB/dU}{dA/dU}\right)}{\frac{dB/dU}{dA/dU}} - \frac{3}{4}\left(\frac{\partial_-\left(\frac{dB/dU}{dA/dU}\right)}{\frac{dB/dU}{dA/dU}}\right)^2\right\}\Bigg] \\
&= \frac{\kappa^2}{48\pi}\left(\frac{dB/dU}{dA/dU}\right)^2\Bigg[\left(\frac{3}{r^4} - \frac{4}{r^3}\right) + \frac{16}{\left(\frac{dB}{dU}\right)^2} \\
&\times\Bigg[\left\{\frac{1}{2}\frac{\partial_U^2\,(dB/dU)}{dB/dU} - \frac{3}{4}\left(\frac{\partial_U\,(dB/dU)}{dB/dU}\right)^2\right\} \\
&- \left\{\frac{1}{2}\frac{\partial_U^2\,(dA/dU)}{dA/dU} - \frac{3}{4}\left(\frac{\partial_U\,(dA/dU)}{dA/dU}\right)^2\right\}\Bigg]\Bigg]
\end{aligned} \tag{9.65}$$

these relations can be simplified to arrive at,

$$\langle T_{+-}\rangle = \frac{\kappa^2}{12\pi}\frac{r-1}{r^4}\left(\frac{\cot\chi_0 + \tan\left(\frac{U-\chi_0}{2}\right)}{\cot\chi_0 - \tan\left(\frac{U+\chi_0}{2}\right)}\right)\frac{\cos^2\left(\frac{U+\chi_0}{2}\right)}{\cos^2\left(\frac{U-\chi_0}{2}\right)} \tag{9.66}$$

$$\begin{aligned}
\langle T_{--}\rangle &= \frac{\kappa^2}{48\pi}\left(\left(\frac{\cot\chi_0 + \tan\left(\frac{U+\chi_0}{2}\right)}{\cot\chi_0 - \tan\left(\frac{U+\chi_0}{2}\right)}\right)\frac{\cos^2\left(\frac{U+\chi_0}{2}\right)}{\cos^2\left(\frac{U-\chi_0}{2}\right)}\right)^2\Bigg[\left(\frac{3}{r^4} - \frac{4}{r^3}\right) \\
&+ \left(\frac{a_{max}\cos^2\left(\frac{U+\chi_0}{2}\right)}{\sin\chi_0\left(\cot\chi_0 - \tan\left(\frac{U+\chi_0}{2}\right)\right)}\right)^{-2} \\
&\times\Bigg(-15\left[\tan^2\left(\frac{U+\chi_0}{2}\right) - \tan^2\left(\frac{U-\chi_0}{2}\right)\right] \\
&- 6\cot\chi_0\left[\tan\left(\frac{U+\chi_0}{2}\right) + \tan\left(\frac{U-\chi_0}{2}\right)\right]
\end{aligned}$$

$$+\frac{4\cot\chi_0}{\sin^2\chi_0}\left[\frac{1}{\cot\chi_0-\tan\left(\frac{U+\chi_0}{2}\right)}-\frac{1}{\cot\chi_0+\tan\left(\frac{U-\chi_0}{2}\right)}\right]+\frac{a_{max}}{\sin\chi_0}\left[\frac{1}{\left(\cot\chi_0-\tan\left(\frac{U+\chi_0}{2}\right)\right)^2}-\frac{1}{\left(\cot\chi_0+\tan\left(\frac{U-\chi_0}{2}\right)\right)^2}\right]\Bigg)\Bigg] \tag{9.67}$$

In arriving at the above relations we have used the following expressions for the various derivatives dA/dU and dB/dU are respectively:

$$\frac{dA}{dU}=\frac{a_{max}\cos^2\left(\frac{U-\chi_0}{2}\right)}{\sin\chi_0\left(\cot\chi_0+\tan\left(\frac{U-\chi_0}{2}\right)\right)};\quad \frac{dB}{dU}=\frac{a_{max}\cos^2\left(\frac{U+\chi_0}{2}\right)}{\sin\chi_0\left(\cot\chi_0-\tan\left(\frac{U+\chi_0}{2}\right)\right)} \tag{9.68}$$

$$\frac{d^2A}{dU^2}=-\frac{1}{2}\frac{\left[1+2\sin^2\left(\frac{U-\chi_0}{2}\right)+\cot\chi_0\sin\left(U-\chi_0\right)\right]}{\sin^4\chi_0\left(\cot\chi_0+\tan\left(\frac{U-\chi_0}{2}\right)\right)^2} \tag{9.69}$$

$$\frac{d^2B}{dU^2}=\frac{1}{2}\frac{\left[1+2\sin^2\left(\frac{U+\chi_0}{2}\right)-\cot\chi_0\sin\left(U+\chi_0\right)\right]}{\sin^4\chi_0\left(\cot\chi_0-\tan\left(\frac{U+\chi_0}{2}\right)\right)^2} \tag{9.70}$$

as well as the following derivatives:

$$\begin{aligned}\frac{1}{2}\frac{\partial_U^2\left(dA/dU\right)}{dA/dU}-\frac{3}{4}\left(\frac{\partial_U\left(dA/dU\right)}{dA/dU}\right)^2&=-\frac{1}{16}\Bigg[15\tan^4\left(\frac{U-\chi_0}{2}\right)\\&+24\cot\chi_0\tan^3\left(\frac{U-\chi_0}{2}\right)+10\tan^2\left(\frac{U-\chi_0}{2}\right)\\&+4\cot^2\chi_0\left\{1+2\tan^2\left(\frac{U-\chi_0}{2}\right)\right\}\\&+16\cot\chi_0\tan\left(\frac{U-\chi_0}{2}\right)-1\Bigg]\left[\cot\chi_0+\tan\left(\frac{U-\chi_0}{2}\right)\right]^{-2}\\&=-\frac{1}{16}\Bigg[15\tan^2\left(\frac{U-\chi_0}{2}\right)-6\cot\chi_0\tan\left(\frac{U-\chi_0}{2}\right)+5\left(1+\sin^{-2}\chi_0\right)\\&-\frac{4\sin^{-2}\chi_0\cot\chi_0}{\cot\chi_0+\tan\left(\frac{U-\chi_0}{2}\right)}-\frac{a_{max}}{\sin\chi_0\left(\cot\chi_0+\tan\left(\frac{U-\chi_0}{2}\right)\right)^2}\Bigg]\end{aligned} \tag{9.71}$$

and

$$\frac{1}{2}\frac{\partial_U^2\,(dB/dU)}{dB/dU}-\frac{3}{4}\left(\frac{\partial_U\,(dB/dU)}{dB/dU}\right)^2=-\frac{1}{16}\Bigg[15\tan^4\left(\frac{U+\chi_0}{2}\right)$$
$$-24\cot\chi_0\tan^3\left(\frac{U+\chi_0}{2}\right)-10\tan^2\left(\frac{U+\chi_0}{2}\right)$$
$$+4\cot^2\chi_0\left\{1+2\tan^2\left(\frac{U+\chi_0}{2}\right)\right\}$$
$$-16\cot\chi_0\tan\left(\frac{U+\chi_0}{2}\right)-1\Bigg]\left[\cot\chi_0-\tan\left(\frac{U+\chi_0}{2}\right)\right]^{-2}$$
$$=-\frac{1}{16}\Bigg[15\tan^2\left(\frac{U+\chi_0}{2}\right)+6\cot\chi_0\tan\left(\frac{U+\chi_0}{2}\right)+5\left(1+\sin^{-2}\chi_0\right)$$
$$-\frac{4\sin^{-2}\chi_0\cot\chi_0}{\cot\chi_0-\tan\left(\frac{U+\chi_0}{2}\right)}-\frac{a_{max}}{\sin\chi_0\left(\cot\chi_0-\tan\left(\frac{U+\chi_0}{2}\right)\right)^2}\Bigg] \tag{9.72}$$

A.1.2 Interior Region

For the interior region the various derivatives of the conformal factors are:

$$\begin{aligned}\frac{1}{C}\partial_+C&=\frac{1}{(dA/dV)}\left[\frac{1}{a^2}\frac{da^2}{dV}-\frac{d^2A/dV^2}{dA/dV}\right]\\ \frac{1}{C}\partial_-C&=\frac{1}{(dA/dU)}\left[\frac{1}{a^2}\frac{da^2}{dU}-\frac{d^2A/dU^2}{dA/dU}\right]\end{aligned} \tag{9.73}$$

$$\begin{aligned}\frac{1}{C}\partial_+^2C=\frac{1}{(dA/dV)^2}&\left[\frac{1}{a^2}\frac{d^2a^2}{dV^2}-\frac{3}{a^2}\frac{da^2}{dV}\frac{1}{dA/dV}\frac{d^2A}{dV^2}\right.\\&\left.-\frac{1}{dA/dV}\frac{d^3A}{dV^3}+3\left(\frac{1}{dA/dV}\frac{d^2A}{dV^2}\right)^2\right]\end{aligned} \tag{9.74}$$

$$\begin{aligned}\frac{1}{C}\partial_-^2C=\frac{1}{(dA/dU)^2}&\left[\frac{1}{a^2}\frac{d^2a^2}{dU^2}-\frac{3}{a^2}\frac{da^2}{dU}\frac{1}{dA/dU}\frac{d^2A}{dU^2}-\frac{1}{dA/dU}\frac{d^3A}{dU^3}\right.\\&\left.+3\left(\frac{1}{dA/dU}\frac{d^2A}{dU^2}\right)^2\right]\end{aligned} \tag{9.75}$$

$$\frac{1}{C}\partial_-\partial_+ C = \frac{1}{\frac{dA}{dV}\frac{dA}{dU}}\left[\frac{1}{a^2}\frac{d^2a^2}{dUdV} - \frac{1}{a^2}\frac{da^2}{dU}\frac{1}{dA/dV}\frac{d^2A}{dV^2} - \frac{1}{a^2}\frac{da^2}{dV}\frac{1}{dA/dU}\frac{d^2A}{dU^2}\right.$$
$$\left. + \frac{1}{dA/dV}\frac{d^2A}{dV^2}\frac{1}{dA/dU}\frac{d^2A}{dU^2}\right] \tag{9.76}$$

Then the components of the stress energy tensor in the inside region are:

$$\langle T_{++}\rangle = \frac{1}{12\pi}\frac{1}{(dA/dV)^2}\left[\left[\left\{\frac{1}{2a^2}\frac{d^2a^2}{dV^2} - \frac{3}{4}\left(\frac{1}{a^2}\frac{da^2}{dV}\right)^2\right\} - \left\{\frac{1}{2}\frac{1}{dA/dV}\frac{d^3A}{dV^3}\right.\right.\right.$$
$$\left.\left.\left. - \frac{3}{4}\frac{1}{(dA/dV)^2}\left(\frac{d^2A}{dV^2}\right)^2\right\}\right]\right]$$
$$= \frac{1}{12\pi}\frac{1}{(dA/dV)^2}\left[-\frac{1}{8}\left(\frac{3a_{max}\sin\chi_0}{r\left(\frac{U+V}{2}\right)} - 2\right)\right.$$
$$\left. - \left\{\frac{1}{2}\frac{\partial_V^2(dA/dV)}{(dA/dV)} - \frac{3}{4}\left(\frac{\partial_V(dA/dV)}{(dA/dV)}\right)^2\right\}\right] \tag{9.77}$$

$$\langle T_{--}\rangle = \frac{1}{12\pi}\frac{1}{(dA/dU)^2}\left[\left[\left\{\frac{1}{2a^2}\frac{d^2a^2}{dU^2} - \frac{3}{4}\left(\frac{1}{a^2}\frac{da^2}{dU}\right)^2\right\} - \left\{\frac{1}{2}\frac{1}{dA/dU}\frac{d^3A}{dU^3}\right.\right.\right.$$
$$\left.\left.\left. - \frac{3}{4}\frac{1}{(dA/dU)^2}\left(\frac{d^2A}{dU^2}\right)^2\right\}\right]\right]$$
$$= \frac{1}{12\pi}\frac{1}{(dA/dU)^2}\left[-\frac{1}{8}\left(\frac{3a_{max}\sin\chi_0}{r\left(\frac{U+V}{2}\right)} - 2\right) - \left\{\frac{1}{2}\frac{\partial_U^2(dA/dU)}{(dA/dU)}\right.\right.$$
$$\left.\left. - \frac{3}{4}\left(\frac{\partial_U(dA/dU)}{(dA/dU)}\right)^2\right\}\right] \tag{9.78}$$

$$\langle T_{+-}\rangle = \frac{1}{24\pi}\frac{1}{dA/dV}\frac{1}{dA/dU}\left[\frac{1}{a^2}\frac{d^2a^2}{dUdV} - \frac{1}{a^2}\frac{da^2}{dU}\frac{1}{a^2}\frac{da^2}{dV}\right]$$
$$= -\frac{1}{48\pi}\frac{1}{dA/dV}\frac{1}{dA/dU}\frac{1}{1+\cos\left(\frac{U+V}{2}\right)} \tag{9.79}$$

where various derivatives of the quantity A are given in Appendix F.1.1.

A.2 Energy Density and Flux For Various Observers

A.2.1 Static Observer

Below we provide the full expressions for energy density and flux calculated for static observer:

$$\begin{aligned}
\mathcal{U} &= \langle T_{++}\rangle \left(\dot{V}^{+}\right)^2 + \langle T_{--}\rangle \left(\dot{V}^{-}\right)^2 + 2\langle T_{+-}\rangle \dot{V}^{+}\dot{V}^{-} \\
&= \frac{\kappa^2}{48\pi}\left(\frac{r}{r-1}\right)\Bigg[\left(-\frac{2}{r^4}\right) + \frac{\sin^2\chi_0}{a_{max}^2\cos^4\left(\frac{U+\chi_0}{2}\right)}\left(\cot\chi_0 - \tan\left(\frac{U+\chi_0}{2}\right)\right)^2 \\
&\times \Bigg\{ -15\left[\tan^2\left(\frac{U+\chi_0}{2}\right) - \tan^2\left(\frac{U-\chi_0}{2}\right)\right] \\
&- 6\cot\chi_0\left[\tan\left(\frac{U+\chi_0}{2}\right) + \tan\left(\frac{U-\chi_0}{2}\right)\right] \\
&+ \frac{4\cot\chi_0}{\sin^2\chi_0}\left[\frac{1}{\cot\chi_0 - \tan\left(\frac{U+\chi_0}{2}\right)} - \frac{1}{\cot\chi_0 + \tan\left(\frac{U-\chi_0}{2}\right)}\right]\Bigg\} \\
&+ \frac{\sin\chi_0}{a_{max}\cos^4\left(\frac{U+\chi_0}{2}\right)}\left[1 - \frac{\left(\cot\chi_0 - \tan\left(\frac{U+\chi_0}{2}\right)\right)^2}{\left(\cot\chi_0 + \tan\left(\frac{U-\chi_0}{2}\right)\right)^2}\right]\Bigg]
\end{aligned} \tag{9.80}$$

and the expression for flux turns out to be:

$$\begin{aligned}
\mathcal{F} &= -\langle T_{ab}\rangle u^a n^b = -\langle T_{++}\rangle \left(\dot{V}^{+}\right)^2 + \langle T_{--}\rangle \left(\dot{V}^{-}\right)^2 \\
&= \frac{\kappa^2}{48\pi}\left(\frac{r}{r-1}\right)\Bigg[\frac{\sin^2\chi_0}{a_{max}^2\cos^4\left(\frac{U+\chi_0}{2}\right)}\left(\cot\chi_0 - \tan\left(\frac{U+\chi_0}{2}\right)\right)^2 \\
&\times \Bigg\{ -15\left[\tan^2\left(\frac{U+\chi_0}{2}\right) - \tan^2\left(\frac{U-\chi_0}{2}\right)\right] \\
&- 6\cot\chi_0\left[\tan\left(\frac{U+\chi_0}{2}\right) + \tan\left(\frac{U-\chi_0}{2}\right)\right] \\
&+ \frac{4\cot\chi_0}{\sin^2\chi_0}\left[\frac{1}{\cot\chi_0 - \tan\left(\frac{U+\chi_0}{2}\right)} - \frac{1}{\cot\chi_0 + \tan\left(\frac{U-\chi_0}{2}\right)}\right]\Bigg\} \\
&+ \frac{\sin\chi_0}{a_{max}\cos^4\left(\frac{U+\chi_0}{2}\right)}\left[1 - \frac{\left(\cot\chi_0 - \tan\left(\frac{U+\chi_0}{2}\right)\right)^2}{\left(\cot\chi_0 + \tan\left(\frac{U-\chi_0}{2}\right)\right)^2}\right]\Bigg]
\end{aligned} \tag{9.81}$$

A.2.2 Radially In-Falling Observers: Inside

The energy density for radially in-falling observer has the following expression:

$$
\begin{aligned}
\mathcal{U} = \frac{\kappa^2}{48\pi}\frac{1}{a^2(\eta)}\Bigg[& -8\sec^2\frac{\eta}{2} + 4 + \frac{1}{2}\Bigg\{15\tan^2\left(\frac{\eta-\chi_0-\tilde{\chi}}{2}\right) \\
& - 6\cot\chi_0\tan\left(\frac{\eta-\chi_0-\tilde{\chi}}{2}\right) + 5\left(1+\sin^{-2}\chi_0\right) \\
& - \frac{4\sin^{-2}\chi_0\cot\chi_0}{\cot\chi_0+\tan\left(\frac{\eta-\chi_0-\tilde{\chi}}{2}\right)} - \frac{a_{max}}{\sin\chi_0\left(\cot\chi_0+\tan\left(\frac{\eta-\chi_0-\tilde{\chi}}{2}\right)\right)^2} \\
& + 15\tan^2\left(\frac{\eta-\chi_0+\tilde{\chi}}{2}\right) - 6\cot\chi_0\tan\left(\frac{\eta-\chi_0+\tilde{\chi}}{2}\right) + 5\left(1+\sin^{-2}\chi_0\right) \\
& - \frac{4\sin^{-2}\chi_0\cot\chi_0}{\cot\chi_0+\tan\left(\frac{\eta-\chi_0+\tilde{\chi}}{2}\right)} - \frac{a_{max}}{\sin\chi_0\left(\cot\chi_0+\tan\left(\frac{\eta-\chi_0+\tilde{\chi}}{2}\right)\right)^2}\Bigg\}\Bigg]
\end{aligned}
\tag{9.82}
$$

while the flux has the following expression:

$$
\begin{aligned}
\mathcal{F} = \frac{\kappa^2}{48\pi}\frac{1}{a^2(\eta)}\frac{1}{2}\Bigg\{ & 15\tan^2\left(\frac{\eta-\chi_0-\tilde{\chi}}{2}\right) - 6\cot\chi_0\tan\left(\frac{\eta-\chi_0-\tilde{\chi}}{2}\right) \\
& + 5\left(1+\sin^{-2}\chi_0\right) - \frac{4\sin^{-2}\chi_0\cot\chi_0}{\cot\chi_0+\tan\left(\frac{\eta-\chi_0-\tilde{\chi}}{2}\right)} \\
& - \frac{a_{max}}{\sin\chi_0\left(\cot\chi_0+\tan\left(\frac{\eta-\chi_0-\tilde{\chi}}{2}\right)\right)^2} - 15\tan^2\left(\frac{\eta-\chi_0+\tilde{\chi}}{2}\right) \\
& + 6\cot\chi_0\tan\left(\frac{\eta-\chi_0+\tilde{\chi}}{2}\right) - 5\left(1+\sin^{-2}\chi_0\right) \\
& + \frac{4\sin^{-2}\chi_0\cot\chi_0}{\cot\chi_0-\tan\left(\frac{\eta-\chi_0+\tilde{\chi}}{2}\right)} + \frac{a_{max}}{\sin\chi_0\left(\cot\chi_0+\tan\left(\frac{\eta-\chi_0+\tilde{\chi}}{2}\right)\right)^2}\Bigg\}
\end{aligned}
\tag{9.83}
$$

A.2.3 Radially In-falling Observers: Outside

For radially in-falling observer outside the dust ball has the following expression for energy density:

$$
\begin{aligned}
\mathcal{U} &= \langle T_{++}\rangle \left(\dot{V}^{+}\right)^2 + \langle T_{--}\rangle \left(\dot{V}^{-}\right)^2 + 2\langle T_{+-}\rangle \dot{V}^{+}\dot{V}^{-} \\
&= \frac{\kappa^2}{48\pi}4E^2\left(\frac{r}{r-1}\right)^2\left(\frac{3}{r^4}-\frac{4}{r^3}\right) + \frac{\kappa^2}{24\pi}\left(\frac{r}{r-1}\right)\left(-\frac{7}{r^4}+\frac{8}{r^3}\right) \\
&+ \frac{\kappa^2}{48\pi}\left(\frac{r}{r-1}\right)^2\left(E+\sqrt{E^2-\frac{r-1}{r}}\right)^2 \\
&\times \Bigg[\frac{\sin^2\chi_0}{a_{max}^2\cos^4\left(\frac{\eta+\chi_0-\tilde{\chi}}{2}\right)}\left(\cot\chi_0 - \tan\left(\frac{\eta+\chi_0-\tilde{\chi}}{2}\right)\right)^2 \\
&\times \Bigg\{-15\left[\tan^2\left(\frac{\eta+\chi_0-\tilde{\chi}}{2}\right) - \tan^2\left(\frac{\eta-\chi_0-\tilde{\chi}}{2}\right)\right] \\
&- 6\cot\chi_0\left[\tan\left(\frac{\eta+\chi_0-\tilde{\chi}}{2}\right) + \tan\left(\frac{\eta-\chi_0-\tilde{\chi}}{2}\right)\right] \\
&+ \frac{4\cot\chi_0}{\sin^2\chi_0}\left[\frac{1}{\cot\chi_0-\tan\left(\frac{\eta+\chi_0-\tilde{\chi}}{2}\right)} - \frac{1}{\cot\chi_0+\tan\left(\frac{\eta-\chi_0-\tilde{\chi}}{2}\right)}\right]\Bigg\} \\
&+ \frac{\sin\chi_0}{a_{max}\cos^4\left(\frac{\eta+\chi_0-\tilde{\chi}}{2}\right)}\left[1-\frac{\left(\cot\chi_0-\tan\left(\frac{\eta+\chi_0-\tilde{\chi}}{2}\right)\right)^2}{\left(\cot\chi_0+\tan\left(\frac{\eta-\chi_0-\tilde{\chi}}{2}\right)\right)^2}\right]\Bigg]
\end{aligned}
\tag{9.84}
$$

and the flux has the following expression:

$$
\begin{aligned}
\mathcal{F} &= -\langle T_{ab}\rangle u^a n^b = -\langle T_{++}\rangle \left(\dot{V}^{+}\right)^2 + \langle T_{--}\rangle \left(\dot{V}^{-}\right)^2 \\
&= \frac{\kappa^2}{48\pi}4E\sqrt{E^2-\frac{r-1}{r}}\left(\frac{r}{r-1}\right)^2\left(\frac{3}{r^4}-\frac{4}{r^3}\right) + \frac{\kappa^2}{48\pi}\left(\frac{r}{r-1}\right)^2 \\
&\times \left(E+\sqrt{E^2-\frac{r-1}{r}}\right)^2\Bigg[\frac{\sin^2\chi_0}{a_{max}^2\cos^4\left(\frac{\eta+\chi_0-\tilde{\chi}}{2}\right)}\left(\cot\chi_0 - \tan\left(\frac{\eta+\chi_0-\tilde{\chi}}{2}\right)\right)^2 \\
&\times \Bigg\{-15\left[\tan^2\left(\frac{\eta+\chi_0-\tilde{\chi}}{2}\right) - \tan^2\left(\frac{\eta-\chi_0-\tilde{\chi}}{2}\right)\right] \\
&- 6\cot\chi_0\left[\tan\left(\frac{\eta+\chi_0-\tilde{\chi}}{2}\right) + \tan\left(\frac{\eta-\chi_0-\tilde{\chi}}{2}\right)\right] \\
&+ \frac{4\cot\chi_0}{\sin^2\chi_0}\left[\frac{1}{\cot\chi_0-\tan\left(\frac{\eta+\chi_0-\tilde{\chi}}{2}\right)} - \frac{1}{\cot\chi_0+\tan\left(\frac{\eta-\chi_0-\tilde{\chi}}{2}\right)}\right]\Bigg\} \\
&+ \frac{\sin\chi_0}{a_{max}\cos^4\left(\frac{\eta+\chi_0-\tilde{\chi}}{2}\right)}\left[1-\frac{\left(\cot\chi_0-\tan\left(\frac{\eta+\chi_0-\tilde{\chi}}{2}\right)\right)^2}{\left(\cot\chi_0+\tan\left(\frac{\eta-\chi_0-\tilde{\chi}}{2}\right)\right)^2}\right]\Bigg]
\end{aligned}
\tag{9.85}
$$

A.3 Effective temperature for various observers

In this section we present the full expression for effective temperature measured by detectors in various trajectories. For static observer the effective temperature turns out to be:

$$\begin{aligned}
T_- &= \frac{1}{2\pi}\left|\frac{\ddot{V}^-}{\dot{V}^-}\right| \\
&= \frac{1}{2\pi}\left|\dot{u}\left(\frac{\frac{d^2A}{dU^2}}{\frac{dA}{dU}\frac{dB}{dU}} - \frac{\frac{d^2B}{dU^2}}{\left(\frac{dB}{dU}\right)^2}\right)\right| \\
&= \frac{1}{4\pi}\sqrt{\frac{r}{r-1}}\sin^4\chi_0 \\
&\times\left\{\frac{\left[1+2\sin^2\left(\frac{U-\chi_0}{2}\right)+\cot\chi_0\sin\left(U-\chi_0\right)\right]\left(\cot\chi_0-\tan\left(\frac{U+\chi_0}{2}\right)\right)}{\left(\cot\chi_0+\tan\left(\frac{U-\chi_0}{2}\right)\right)\cos^2\left(\frac{U+\chi_0}{2}\right)\cos^2\left(\frac{U-\chi_0}{2}\right)}\right. \\
&+\left.\frac{\left[1+2\sin^2\left(\frac{U+\chi_0}{2}\right)-\cot\chi_0\sin\left(U+\chi_0\right)\right]}{\cos^4\left(\frac{U+\chi_0}{2}\right)}\right\}
\end{aligned} \tag{9.86}$$

For radially in-falling observes inside the dust sphere the effective temperature turns out to be:

$$T_- = \frac{1}{4\pi}\left|\frac{1+2\sin^2\left(\frac{\eta-\chi_0-\tilde{\chi}}{2}\right)+\cot\chi_0\sin\left(\frac{\eta-\chi_0-\tilde{\chi}}{2}\right)}{\cos^2\left(\frac{\eta-\chi_0-\tilde{\chi}}{2}\right)\left\{\cot\chi_0+\tan\left(\frac{\eta-\chi_0-\tilde{\chi}}{2}\right)\right\}}\right| \tag{9.87}$$

Finally, for radially in-falling observers in the Schwarzschild spacetime effective temperature takes the following expression:

$$\begin{aligned}
T_- &= \frac{1}{2\pi}\left|\frac{\ddot{u}}{\dot{u}} - \dot{u}\frac{\sin^4\chi_0}{2}\right. \\
&\left\{\frac{\left[1+2\sin^2\left(\frac{\eta-\chi_0-\tilde{\chi}}{2}\right)+\cot\chi_0\sin\left(\eta-\chi_0-\tilde{\chi}\right)\right]\left(\cot\chi_0-\tan\left(\frac{\eta+\chi_0-\tilde{\chi}}{2}\right)\right)}{\left(\cot\chi_0+\tan\left(\frac{\eta-\chi_0-\tilde{\chi}}{2}\right)\right)\cos^2\left(\frac{\eta+\chi_0-\tilde{\chi}}{2}\right)\cos^2\left(\frac{\eta-\chi_0-\tilde{\chi}}{2}\right)}\right. \\
&+\left.\left.\frac{\left[1+2\sin^2\left(\frac{\eta+\chi_0-\tilde{\chi}}{2}\right)-\cot\chi_0\sin\left(\eta+\chi_0-\tilde{\chi}\right)\right]}{\cos^4\left(\frac{\eta+\chi_0-\tilde{\chi}}{2}\right)}\right\}\right|
\end{aligned} \tag{9.88}$$

References

1. S. Hawking, Black hole explosions. Nature **248**, 30–31 (1974)
2. S. Hawking, Particle creation by Black Holes. Commun. Math. Phys. **43**, 199–220 (1975)
3. N. Birrell, P. Davies, *Quantum Fields in Curved Space* (Cambridge Monogr. Math, Phys, 1982)
4. T. Padmanabhan, Gravity and the thermodynamics of horizons. Phys. Rep. **406**, 49–125 (2005). arXiv:gr-qc/0311036 [gr-qc]
5. T. Padmanabhan, Thermodynamical aspects of gravity: new insights. Rept. Prog. Phys. **73**, 046901 (2010). arXiv:0911.5004 [gr-qc]
6. A.D. Helfer, Do black holes radiate? Rept. Prog. Phys. **66**, 43–1008 (2003). arXiv:gr-qc/0304042 [gr-qc]
7. A. Fabbri, J. Navarro-Salas, *Modeling Black Hole Evaporation* (2005)
8. V. Mukhanov, S. Winitzki, *Introduction to Quantum Effects in Gravity*, 1st edn (Cambridge University Press, Cambridge, 2007)
9. L.E. Parker, D. Toms, *Quantum Field Theory in Curved Spacetime*. Cambridge Monographs on Mathematical Physics (Cambridge University Press, Cambridge, 2009), http://www.cambridge.org/de/knowledge/isbn/item2327457
10. T. Padmanabhan, *Gravitation: Foundations and Frontiers* (Cambridge University Press, Cambridge, UK, 2010)
11. M. Visser, Essential and inessential features of Hawking radiation. Int. J. Mod. Phys. D **12**, 649–661 (2003). arXiv:hep-th/0106111 [hep-th]
12. S. Takagi, Vacuum noise and stress induced by uniform accelerator: Hawking-Unruh effect in Rindler manifold of arbitrary dimensions. Prog. Theor. Phys. Suppl. **88**, 1–142 (1986)
13. G. Acquaviva, G.F.R. Ellis, R. Goswami, A.I.M. Hamid, Constructing black hole entropy from gravitational collapse. Phys. Rev. D **91**(6), 064017 (2015). arXiv:1411.5708 [gr-qc]
14. M. Wielgus, M.A. Abramowicz, G.F.R. Ellis, F.H. Vincent, Cosmic background radiation in the vicinity of a Schwarzschild black hole: no classic firewall. Phys. Rev. D **90**(12), 124024 (2014). arXiv:1406.6551 [gr-qc]
15. G.F.R. Ellis, Astrophysical black holes may radiate, but they do not evaporate. arXiv:1310.4771 [gr-qc]
16. G. Polhemus, A.J. Hamilton, Large entropy production inside black holes: A Simple model. JHEP **1008**, 093 (2010). arXiv:1006.4195 [gr-qc]
17. A.J. Hamilton, G. Polhemus, Stereoscopic visualization in curved spacetime: seeing deep inside a black hole. New J. Phys. **12**, 123027 (2010). arXiv:1012.4043 [gr-qc]
18. K. Banerjee, A. Paranjape, Semiclassical environment of collapsing shells. Phys. Rev. D **80**, 124006 (2009). arXiv:0909.4668 [gr-qc]
19. S. Singh, S. Chakraborty, Black hole kinematics: the âinâ -vacuum energy density and flux for different observers. Phys. Rev. D **90**(2), 024011 (2014). arXiv:1404.0684 [gr-qc]
20. S. Singh, Excavations at the gravitationally collapsed site: recent findings. J. Phys. Conf. Ser. **600**(1), 012035 (2015). arXiv:1412.1028 [gr-qc]
21. M. Smerlak, S. Singh, New perspectives on Hawking radiation. Phys. Rev. D **88**(10), 104023, (2013). arXiv:1304.2858 [gr-qc]
22. R. Brout, S. Massar, R. Parentani, P. Spindel, A Primer for black hole quantum physics. Phys. Rep. **260**, 329–454 (1995). arXiv:0710.4345 [gr-qc]
23. W. Unruh, Notes on black hole evaporation. Phys. Rev. D **14**, 870 (1976)
24. P. Langlois, Causal particle detectors and topology. Ann. Phys. **321**, 2027–2070 (2006). arXiv:gr-qc/0510049 [gr-qc]
25. J. Louko, A. Satz, Transition rate of the Unruh-DeWitt detector in curved spacetime. Class. Quant. Grav. **25**, 055012 (2008). arXiv:0710.5671 [gr-qc]
26. A. Satz, Then again, how often does the Unruh-DeWitt detector click if we switch it carefully? Class. Quant. Grav. **24**, 1719–1732 (2007). arXiv:gr-qc/0611067 [gr-qc]
27. L. Ford, T.A. Roman, Motion of inertial observers through negative energy. Phys. Rev. D **48**, 776–782. arXiv:gr-qc/9303038 [gr-qc]

28. L. Ford, T.A. Roman, Averaged energy conditions and quantum inequalities. Phys. Rev. D **51**, 4277–4286 (1995). arXiv:gr-qc/9410043 [gr-qc]
29. L. Ford, T.A. Roman, Averaged energy conditions and evaporating black holes. Phys. Rev. D **53**, 1988–2000 (1996). arXiv:gr-qc/9506052 [gr-qc]
30. M. Visser, Gravitational vacuum polarization. 1: Energy conditions in the Hartle-Hawking vacuum. Phys. Rev. D **54**, 5103–5115 (1996). arXiv:gr-qc/9604007 [gr-qc]
31. M. Visser, Gravitational vacuum polarization. 2: Energy conditions in the Boulware vacuum. Phys. Rev. D **54**, 5116–5122 (1996). arXiv:gr-qc/9604008 [gr-qc]
32. M. Visser, Gravitational vacuum polarization. 3: Energy conditions in the (1+1) Schwarzschild space-time. Phys. Rev. D **54**, 5123–5128 (1996). arXiv:gr-qc/9604009 [gr-qc]
33. T. Padmanabhan, T. Singh, Response of an accelerated detector coupled to the stress—Energy tensor. Class. Quant. Grav. **4**, 1397–1407 (1987)
34. J. Oppenheimer, H. Snyder, On Continued gravitational contraction. Phys. Rev. **56**, 455–459 (1939)
35. C. Vaz, S. Gutti, C. Kiefer, T. Singh, Quantum gravitational collapse and Hawking radiation in 2+1 dimensions. Phys. Rev. D **76**, 124021 (2007). arXiv:0710.2164 [gr-qc]
36. S. Ross, R.B. Mann, Gravitationally collapsing dust in (2+1)-dimensions. Phys. Rev. D **47**, 3319–3322 (1993). arXiv:hep-th/9208036 [hep-th]
37. R. Casadio, Semiclassical collapse of a sphere of dust and Hawking radiation. Nucl. Phys. Proc. Suppl. **57**, 177–180 (1997). arXiv:gr-qc/9611062 [gr-qc]
38. E. Poisson, *A Relativist's Toolkit: The Mathematics of Black-Hole Mechanics*, 1st edn (Cambridge University Press, Cambridge, 2007)
39. P. Davies, S. Fulling, W. Unruh, Energy momentum tensor near an evaporating Black Hole. Phys. Rev. D **13**, 2720–2723 (1976)
40. R.C. Tolman, P. Ehrenfest, Temperature equilibrium in a static gravitational field. Phys. Rev. **36**(12), 1791–1798 (1930)
41. K.W. Howard, Vacuum $<t(\mu)(\nu)>$ in Schwarzschild space-time. Phys. Rev. D **30**, 2532–2547 (1984)
42. B. Jensen, A. Ottewill, Renormalized electromagnetic stress tensor in Schwarzschild space-time. Phys. Rev. D **39**, 1130 (1989)
43. P.R. Anderson, W.A. Hiscock, D.A. Samuel, Stress energy tensor of quantized scalar fields in static black hole space-times. Phys. Rev. Lett. **70**, 1739–1742 (1993)
44. P.R. Anderson, W.A. Hiscock, D.J. Loranz, Semiclassical stability of the extreme Reissner-Nordstrom black hole. Phys. Rev. Lett. **74**, 4365–4368 (1995). arXiv:gr-qc/9504019 [gr-qc]
45. S.M. Christensen, Vacuum expectation value of the stress tensor in an arbitrary curved background: the covariant point separation method. Phys. Rev. D **14**, 2490–2501 (1976)
46. W.A. Hiscock, S.L. Larson, P.R. Anderson, Semiclassical effects in black hole interiors. Phys. Rev. D **56**, 3571–3581 (1997). arXiv:gr-qc/9701004 [gr-qc]
47. L.C. Barbado, C. Barcelo, L.J. Garay, Hawking radiation as perceived by different observers. Class. Quant. Grav. **28**, 125021 (2011). arXiv:1101.4382 [gr-qc]
48. C. Barcelo, S. Liberati, S. Sonego, M. Visser, Minimal conditions for the existence of a Hawking-like flux. Phys. Rev. D **83**, 041501 (2011). arXiv:1011.5593 [gr-qc]
49. A. Paranjape, T. Padmanabhan, Radiation from collapsing shells, semiclassical backreaction and black hole formation. Phys. Rev. D **80**, 044011 (2009). arXiv:0906.1768 [gr-qc]

Chapter 10
Information Retrieval from Black Holes

10.1 Introduction: The Black Hole Information Loss Paradox

Evaporation of black holes threatens the most basic and fundamental feature of the standard quantum theory, namely unitary evolution as one tries extrapolating the results obtained at the semi-classical level [1]. Existence of black hole entropy [2, 3] as well as evaporation of black holes by emission of a (nearly) thermal radiation of positive energy due to quantum effects [1] leads to a thermodynamical description of black holes. Further the flux of negative energy flowing into the black hole decreases its mass and the corresponding mass lost by the black hole appears in the form of energy of the thermal radiation. Even though this is essential for the celebrated thermodynamic description of black holes, such a process, together with other properties of black holes, appears to violate the standard unitary quantum mechanics giving rise to the black hole information loss paradox [4]. The crux of the paradox can be summarized as follows—as the black hole evaporates completely without leaving any remnant behind, one is justified in assuming that the entire information content of the collapsing body is either destroyed or must be encoded in the resulting radiation. However, residual radiation in this process being (dominantly) thermal, is unable to contain much of the information of what has fallen into the black hole and hence incapable of making the theory unitary. Therefore, most of the information content of the matter, forming the black hole becomes unavailable to the future asymptotic observers (see also [5–12]).

In this chapter, we will argue that the version of the paradox concerning the information content of the initial data originates due to a hybrid analysis of a process which should be fully quantum mechanical in nature. The matter which forms a black hole in the first place must be treated as fundamentally quantum mechanical and hence must follow a quantum evolution. We expect that the classical description at lowest order is adequate, leading to formation of an event horizon but the information about the quantum nature of the collapsing material must not be completely ignored while studying this process. The quantum matter forming the black hole will populate its

S. Chakraborty, *Classical and Quantum Aspects of Gravity in Relation to the Emergent Paradigm*, Springer Theses, DOI 10.1007/978-3-319-63733-4_10

modes at future asymptotia in a non thermal manner and hence the presence of this effect can be felt even in semi-classical level.

We will see that the symmetry characteristics of the initial state forming the black hole will determine whether asymptotic observers can reconstruct the initial state completely or partially. Therefore, analysis of the allowed set of symmetries present in the characterization of the initial state will play a central role in the recovery of the initial information content, just from the spectrum function when the black hole is evaporated completely. Recently, it has been shown [13] how to reconstruct a qubit state which is thrown into a black hole by measuring changes in the black hole characteristics due to such a process. Our scheme, even though somewhat similar in spirit, but calculates the projection of states in an infinite dimensional Hilbert space.

In the semi-classical approximation, the formation of the classical event horizon is unavoidable, as the part forming the black hole follows classical equations of motion. Due to the presence of the horizon, pure to mixed state transition is also imminent. Thus a part of the field modes always remain hidden from the asymptotic observer, resulting in a mixed state description. We show that, at the semi-classical level itself, there exists additional quantum hairs in the resulting radiation profile. At this stage we must emphasize that *we are not focusing on the non-thermal distortions originating from the vacuum itself, but on the part which is originating from non-vacuum component of the quantum state.* As discussed earlier [14, 15], there can be various sources of non-thermal corrections to the Hawking radiation, even when the initial state is a vacuum. We henceforth mention the total Hawking radiation endowed with all these corrections, as *the vacuum response*. We show that if there exist corrections over and above the vacuum response, information regarding the initial sector of Hilbert space, which formed the black hole, can be extracted.

Information in an initial state of a collapsing field carrying non-zero stress energy can be stored using a superposed state in the corresponding Hilbert space. The initial non-vacuum state we will be using are of the following form:

$$|\Psi\rangle_{\rm in} = \int_0^\infty \frac{d\omega}{\sqrt{4\pi\omega}} f(\omega)\hat{a}^\dagger(\omega)|0\rangle_{\rm in}. \tag{10.1}$$

This is an excited state which is also a superposition of one-particle states. The function $f(\omega)$ completely encodes the information about the initial state and the idea is to reconstruct this function from the spectrum of the black hole radiation as perceived by the asymptotic observers.

We will first demonstrate how this reconstruction of initial state works, for a collapse model forming Schwarzschild black hole in $(3+1)$ dimension as a quasi-static process. In this case we will consider a spherically symmetric scalar field undergoing a collapse process to form a black hole. The complete quantum analysis of this process requires the study of the quantum evolution of the field as well as that of the "quantum geometry" and the back-reaction. However, lack of good control over either the quantum sector of the geometry or the back-reaction in $(3+1)$ dimension, compels us to adopt a semi-classical approach, where we take the matter field to be described by:

$$\hat{\phi}(x) = \phi_0 \mathbb{I} + \delta\hat{\phi}, \tag{10.2}$$

where ϕ_0 corresponds to the part which dominantly describes the evolution of geometry. That is,

$$\langle\Psi|T_{\mu\nu}[\phi]|\Psi\rangle \sim \langle\Psi|T_{\mu\nu}[\phi_0]|\Psi\rangle,$$

will act as the source for the spacetime geometry and will lead to the gravitational collapse. We can also think of a process in which some of the highly excited modes ϕ_0 of the field, acting as classical matter, collapse to form a black hole, while some other modes $\delta\hat{\phi}$ evolve quantum mechanically as a *test field* in this background. These *quantum* modes are populated, i.e., such modes will be in a non-vacuum state of the field and hence will carry some small amount of energy into the black hole. Classically, the test field modes, as they cross the horizon, will make the black hole larger and then will become inaccessible to future asymptotic observers. However, quantum mechanically, we show that such a process will lead to a *non-vacuum distortion* in the late time Hawking radiation, resulting in the recovery of information about the initial quantum state possible. The late time radiation will have a frequency space correlation which becomes non-zero if the initial state was not in a vacuum configuration.

We will also discuss a $(1+1)$ dimensional dilatonic black hole solution also known as the CGHS model. This model includes the effect of the back-reaction, in which the problem of forming a black hole with quantum matter as the source can be solved exactly. Therefore, in this model, we will be able to not only account for the back-reaction of the test field modes $\delta\hat{\phi}$ but also recover the information about the field ϕ_0 which forms the black hole itself in the first place. Further, being a conformally flat spacetime, the Bogoliubov coefficients can be calculated exactly and hence the information content in the out-going modes can be calculated at any stage of evolution, not only at late times. Thus, one can track the loss of information, if any, during the course of evolution and evaporation.

In Sects. 10.2 and 10.3, we present the characterization of the initial state we will be dealing with in this chapter. We will also briefly discuss quantum field modes in a spherically symmetric collapse scenario forming a black hole. These modes will be used to specify the non vacuum state of the test quantum field in the spherically symmetric case. In Sect. 10.4, we discuss the resulting correlation spectrum arising from such non-vacuum, single particle states. Since accommodating the back-reaction in the $(3+1)$ black hole formation is technically challenging, we go to lower dimensions to get a handle on the back-reaction analytically. For this purpose we will consider the evaporation of a $(1+1)$ dimensional dilatonic black hole solution. We will briefly describe this model in Sect. 10.5, and shall connect up with the corresponding $(3+1)$ results. In Sect. 10.6, we will show how the concept of the non-vacuum distortion can be utilized to harness information about the matter falling into the black hole, which normally would have been invisible to the asymptotic observer. We discuss the class of symmetry characterizing the initial state, for which such an asymptotic observer can reconstruct the initial data. In Sect. 10.7, we discuss the implication of our scheme of retrieval of information and the scope for further generalization. The detailed calculations have been presented in Appendix A.

10.2 Initial State of Spherical Collapse

The case we discuss first is that of a real scalar quantum field living in a collapsing spacetime background, which eventually would harbor a black hole. The initial state of the field has been specified at the past null infinity ($\mathcal{J}^-$). The geometry at $\mathcal{J}^-$ is Minkowski-like and hence the corresponding modes describing the quantum field will be the flat spacetime modes. For the flat spacetime free-field theory, the in-falling field decomposition is given by

$$\hat{\phi}(\mathbf{x}) = \int \frac{d^3\mathbf{k}}{\sqrt{2\omega_\mathbf{k}}}(\hat{a}_\mathbf{k} e^{ik\cdot x} + \hat{a}^\dagger_\mathbf{k} e^{-ik\cdot x}), \tag{10.3}$$

$$= \int d^3\mathbf{k}\hat{\phi}_\mathbf{k}(t)e^{i\mathbf{k}\cdot\mathbf{x}}, \tag{10.4}$$

where

$$\hat{\phi}_\mathbf{k}(t) = \frac{(\hat{a}_\mathbf{k} e^{-i\omega_\mathbf{k} t} + \hat{a}^\dagger_{-\mathbf{k}} e^{i\omega_\mathbf{k} t})}{\sqrt{2\omega_\mathbf{k}}}, \tag{10.5}$$

satisfying $\hat{\phi}_\mathbf{k} = \hat{\phi}^*_{-\mathbf{k}}$ for a real field. We further define $\hat{\hat{\phi}}_\mathbf{k} = \hat{a}_\mathbf{k} e^{-i\omega_\mathbf{k} t}$, such that,

$$\sqrt{2\omega_\mathbf{k}}\hat{\phi}_\mathbf{k} = \hat{\hat{\phi}}_\mathbf{k} + \hat{\hat{\phi}}^\dagger_{-\mathbf{k}}. \tag{10.6}$$

Therefore, specifying $\hat{\hat{\phi}}_\mathbf{k}$ is equivalent to specifying $\hat{\phi}_\mathbf{k}$. This operator describes the field configuration in terms of the corresponding Fourier momentum modes at $\mathcal{J}^-$. We define an observable of the momentum correlation by introducing the following Hermitian operator:

$$\begin{aligned}\hat{N}_{\mathbf{k}_1\mathbf{k}_2} &\equiv \hat{\hat{\phi}}_{\mathbf{k}_1}\hat{\hat{\phi}}^\dagger_{\mathbf{k}_2} + \hat{\hat{\phi}}_{\mathbf{k}_2}\hat{\hat{\phi}}^\dagger_{\mathbf{k}_1} \\ &= \hat{a}_{\mathbf{k}_1}\hat{a}^\dagger_{\mathbf{k}_2} e^{-i(\omega_{\mathbf{k}_1}-\omega_{\mathbf{k}_2})t} + \hat{a}_{\mathbf{k}_2}\hat{a}^\dagger_{\mathbf{k}_1} e^{i(\omega_{\mathbf{k}_1}-\omega_{\mathbf{k}_2})t}.\end{aligned} \tag{10.7}$$

For a massless field in a spherically symmetric configuration, this operator can also measure the frequency correlation if one suppress the angular dependence. We will later concentrate specifically on these cross-correlators in order to retrieve the information about what went into the black hole. Let us consider a field which undergoes spherically symmetric (s-wave) collapse, slowly [14, 15] to form a large mass black hole (see Fig. 10.1). The relevant positive frequency modes describing the initial state will be

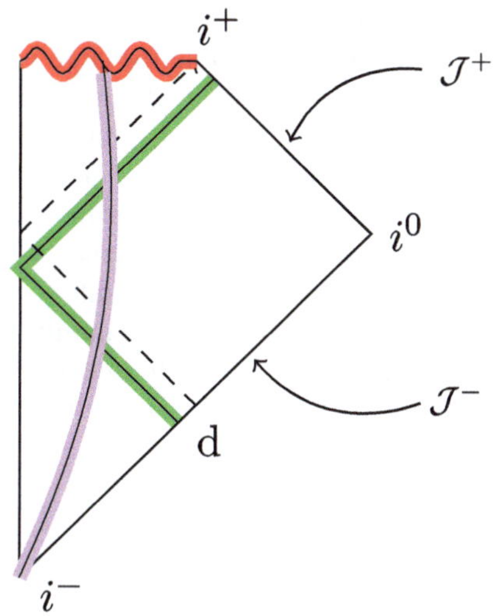

Fig. 10.1 Penrose diagram For schwarzschild collapse

$$u_\omega(t, r, \theta, \phi) \sim \frac{1}{r\sqrt{\omega}} e^{-i\omega(t+r)} S(\theta, \phi) \tag{10.8}$$

where $S(\theta, \phi)$ gives a combination of spherical harmonics $Y_{lm}(\theta, \phi)$. For this collapsing scenario we take the initial state to be in-moving at $\mathcal{J}^-$ which is totally spherically symmetric, i.e., $l = 0$. Once an event horizon is formed through the collapse process, the full state can again be described using a combined description at the event horizon $\mathcal{H}$ as well as on the future null infinity ($\mathcal{J}^+$) [16, 17], i.e., the field configuration of spacetime can also be described using positive and negative frequency modes compatible to $\mathcal{J}^+$ and on the horizon $\mathcal{H}$. For an asymptotic observer, the end configuration of the field will be the out-state, described using modes at $\mathcal{J}^+$, which are again flat spacetime modes owing to the asymptotic flatness of the model. The field content of the out-state can be obtained using the Bogoliubov coefficients between the modes at $\mathcal{J}^-$ and $\mathcal{J}^+$ respectively [16, 17]. The asymptotic form of these Bogoliubov coefficients are given as [1]

$$\begin{aligned}
\alpha_{\Omega\omega} &= \frac{1}{2\pi\kappa}\sqrt{\frac{\Omega}{\omega}} \exp\left[\frac{\pi\Omega}{2\kappa}\right] \exp\left[i(\Omega-\omega)d\right] \exp\left[\frac{i\Omega}{\kappa}\log\frac{\omega}{C}\right]\Gamma\left[-\frac{i\Omega}{\kappa}\right], \\
\beta_{\Omega\omega} &= -\frac{1}{2\pi\kappa}\sqrt{\frac{\Omega}{\omega}} \exp\left[-\frac{\pi\Omega}{2\kappa}\right] \exp\left[i(\Omega+\omega)d\right] \exp\left[\frac{i\Omega}{\kappa}\log\frac{\omega}{C}\right]\Gamma\left[-\frac{i\Omega}{\kappa}\right].
\end{aligned} \tag{10.9}$$

where Ω is the frequency of the out-modes at $\mathcal{J}^+$, the parameter κ being the surface gravity of the black hole, while C is a product of affine parametrization of incoming and outgoing null rays [16, 17] and d is an arbitrary constant marking the last null ray reaching $\mathcal{J}^+$.

One can set $d = 0$ through co-ordinate transformations on $\mathcal{J}^-$. These Bogoliubov coefficients are accurate for large values of ω. At small ω values, the expressions in Eq. (10.9) will receive additional corrections. However, when we are interested in the late-time radiation at future null infinity ($\mathcal{J}^+$), it turns out that the dominating spectra will come from those modes which have just narrowly escaped the black hole, i.e., which were scattered just before the formation of the event horizon. Such modes are the ones with high frequencies at the past null infinity. So the calculations done with Eq. (10.9) will be accurate to the leading order. We will concentrate on states which are eigenstates of the number operator defined with the help of the in-modes, but are not the energy or momentum eigenstates. Let the state of the field undergoing collapse to a black hole spacetime, be a (superposition of) single particle excitation state

$$|\Psi\rangle_{in} = \int_0^\infty \frac{d\omega}{\sqrt{4\pi\omega}} f(\omega)\hat{a}^\dagger(\omega)|0\rangle. \tag{10.10}$$

(We will generalize this analysis for higher excited states in the appendices) We introduce a new function $g(z)$, using a dimensionless variable z, which is related to the frequency ω of mode functions at $\mathcal{J}^-$, as

$$\log \frac{\tilde{\omega}}{C} = z \Rightarrow f(Ce^z) = g(z), \tag{10.11}$$

to re-write the state as

$$|\Psi\rangle_{\text{in}} = \int_{-\infty}^\infty \frac{dz}{\sqrt{4\pi}} g(z)\hat{a}^\dagger(z)|0\rangle. \tag{10.12}$$

In order to specify the state we need to specify either $f(\omega)$ in Eq. (10.10) or equivalently $g(z)$ in Eq. (10.12). We will see how much of information about this function can be retrieved from the outgoing modes. Before proceeding to the black hole emission spectra, we introduce the function

$$F(y) = \int_{-\infty}^\infty dz g(z) e^{iyz}, \tag{10.13}$$

which will be used to characterize the initial state in Eq. (10.10) or Eq. (10.12). This is an equally good measure for encoding the information about the state, since it is just a Fourier transform of an $\mathbb{L}^2$ function. It is useful to construct yet another function

$$\tilde{F}\left(\frac{\Omega}{\kappa}\right) = \exp\left[\frac{\pi\Omega}{2\kappa}\right] F\left(\frac{\Omega}{\kappa}\right), \tag{10.14}$$

from Eq. (10.13). Then one can obtain the distribution $g(t)$ from Eq. (10.14) as

$$g(z) = \frac{1}{2\pi} \int_{-\infty}^{\infty} d\left(\frac{\Omega}{\kappa}\right) \tilde{F}\left(\frac{\Omega}{\kappa}\right) e^{-i\frac{\Omega}{\kappa}z} e^{-\frac{\pi\Omega}{2\kappa}}. \tag{10.15}$$

We emphasize once again that the one particle state is completely specified once we have complete knowledge of the function $\tilde{F}(\Omega/\kappa)$. We will discuss the information about $\tilde{F}(\Omega/\kappa)$ available in the out-going modes, by analyzing the frequency space correlations. Such correlations will turn out to be definitive tools for the recovery of the information. After discussing these correlations briefly in the next section, we will go on to study a special case of this correlation, the self-correlation, which gives the emission profile of the black hole, and we will show that the emission profile develops a non-vacuum, non-thermal, part capable of storing information.

10.3 Information of Black Hole Formation: Correlation Function

We wish to obtain the information regarding the quantum states falling into the black hole, or more elaborately, regarding the quantum states which formed the black hole itself. For this purpose, we start with an observable which measures the frequency space correlation in the out-going modes. The annihilation operator $\hat{b}_\Omega$ associated with the outgoing modes is related to the creation and annihilation operator $\hat{a}_\omega$ and $\hat{a}_\omega^\dagger$ of the ingoing mode as,

$$\hat{b}_\Omega = \int d\omega \left(\alpha_{\Omega\omega}^* \hat{a}_\omega - \beta_{\Omega\omega}^* \hat{a}_\omega^\dagger\right) \tag{10.16}$$

where, $\alpha_{\Omega\omega}$ and $\beta_{\Omega\omega}$ are the associated Bogoliubov coefficients. Now following Eq. (10.7), the frequency space correlation for the outgoing modes turns out to yield

$$\hat{N}_{\Omega_1\Omega_2} = \hat{b}_{\Omega_1}^\dagger \hat{b}_{\Omega_2} e^{-i(\Omega_1-\Omega_2)t} + \hat{b}_{\Omega_2}^\dagger \hat{b}_{\Omega_1} e^{i(\Omega_1-\Omega_2)t} \tag{10.17}$$

Recently, an analogous operator was used in [18] for studying the growth of loop corrections in an interacting field theory. For the initial state (in-state) of the field, being vacuum $|0\rangle$ or one with a definite momentum $|\mathbf{k}\rangle$ (and hence for all the Fock basis states), the expectation value of this correlation operator identically vanishes. We now consider this quantum correlation of the field in the out-going modes. The correlation operator $\hat{N}_{\Omega_1\Omega_2}$ defined in Eq. (10.17) for the out-going modes is related to that of the in-moving modes through Bogoliubov transformations, as presented in Eq. (10.16). If the test field $\delta\hat{\phi}$ starts in the in-vacuum state $|0\rangle_{\text{in}}$, then the expectation value of the frequency correlation reads,

$$
\begin{aligned}
{}_{\rm in}\langle 0|\hat{N}_{\Omega_1\Omega_2}|0\rangle_{\rm in} &= \delta(\Omega_1-\Omega_2)\times e^{\frac{-\pi(\Omega_1+\Omega_2)}{2\kappa}}\frac{\sqrt{\Omega_1\Omega_2}}{4\pi^2\kappa^2} \\
&\times\left\{\Gamma\left[-i\frac{\Omega_1}{\kappa}\right]\Gamma\left[i\frac{\Omega_2}{\kappa}\right]e^{-i(\Omega_1-\Omega_2)t}+\text{c.c.}\right\}
\end{aligned}
\tag{10.18}
$$

which vanishes identically for the off-diagonal elements and hence *the asymptotic future observer will also measure no frequency correlation* in the out-going modes. On the other hand, the diagonal elements of this observable gives the number spectrum. Such an observer measures the out-going spectrum to be a thermal one which can be verified by taking $\Omega_1 = \Omega_2$.

However, when the test field starts in a non-vacuum state, the out-going modes will develop a frequency correlation and as a result the expectation of Eq. (10.17) will become non-zero. The correction to the expectation of the frequency correlator $\hat{N}_{\Omega_1\Omega_2}$ in a non-vacuum in-state leads to,

$$
\begin{aligned}
{}_{\rm in}\langle\psi|\hat{N}_{\Omega_1\Omega_2}|\psi\rangle_{\rm in} &= \Bigg[\left(\int\frac{d\omega}{\sqrt{4\pi\omega}}f(\omega)\alpha^*_{\Omega_2\omega}\right) \\
&\times\left(\int\frac{d\bar{\omega}}{\sqrt{4\pi\bar{\omega}}}f^*(\bar{\omega})\alpha_{\Omega_1\bar{\omega}}\right)+\left(\int\frac{d\omega}{\sqrt{4\pi\omega}}f^*(\omega)\beta^*_{\Omega_2\omega}\right) \\
&\times\left(\int\frac{d\bar{\omega}}{\sqrt{4\pi\bar{\omega}}}f(\bar{\omega})\beta_{\Omega_1\bar{\omega}}\right)\Bigg]e^{-i(\Omega_1-\Omega_2)t}+\text{c.c}
\end{aligned}
\tag{10.19}
$$

Using the expressions for $\alpha_{\Omega\omega}$ and $\beta_{\Omega\omega}$ from Eq. (10.9), we finally obtain,

$$
\begin{aligned}
{}_{\rm in}\langle\psi|\hat{N}_{\Omega_1\Omega_2}|\psi\rangle_{\rm in} &= \frac{1}{4\pi}\left[A(\Omega_1)A(\Omega_2)^*+\text{c.c.}\right] \\
&+\frac{1}{4\pi}\left[B(\Omega_1)B(\Omega_2)^*+\text{c.c.}\right].
\end{aligned}
\tag{10.20}
$$

with

$$
A(\Omega) = e^{-\frac{\pi\Omega}{2\kappa}}\frac{\sqrt{\Omega}}{2\pi\kappa}\Gamma\left[-i\frac{\Omega}{\kappa}\right]F\left(\frac{\Omega}{\kappa}\right)e^{-i\Omega t},
\tag{10.21}
$$

and

$$
B(\Omega) = e^{\frac{\pi\Omega}{2\kappa}}\frac{\sqrt{\Omega}}{2\pi\kappa}\Gamma\left[-i\frac{\Omega}{\kappa}\right]F^*\left(-\frac{\Omega}{\kappa}\right)e^{-i\Omega t}
\tag{10.22}
$$

expressed in terms of the time co-ordinate of out-going observers. It is clear that the frequency correlation for two distinctly separated frequencies (i.e., $\Omega_1 \neq \Omega_2$), as discussed above, remains zero for all the field configurations which were in the vacuum (incidentally, also for configurations in individual Fock basis elements) in the in-state. However, as for the out-state, the frequency correlation remains zero *only if* the in-state was a vacuum. The out-going modes develop frequency correlation, even if the in-state was a non-vacuum Fock basis state but with zero correlation. Alternatively, those fields which carried some amount of stress energy into the black hole definitely develop some non-zero frequency correlation at late times, while only those fields which were in a vacuum state develop no late time frequency correlation. Therefore, by just measuring this operator, a late time observer will be able to tell if some non-zero stress-energy has entered the black hole or not.

Further the observer can also decipher the state that entered into the black hole, by reconstructing $F(\Omega/\kappa)$ from this non-zero expectation value of the correlation. We will demonstrate the technique below. We can also consider a special case of this correlation, e.g., the self-correlation for simplicity. We measure the change in self-correlation, which is just the spectrum operator [19], once a non-vacuum state perturbs the black hole configuration and we will see that this change encodes the information of interest (see Appendix A). Expectedly, other correlations carry information about the in-going states in a much more efficient way.

10.4 Radiation from Black Hole: Information About the Initial State

In this section we will explicitly show that Eq. (10.20) can be used by out-moving observers for retrieving information regarding $F(\Omega)$ and thus for reconstructing the state presented in Eq. (10.1). From the off-diagonal elements of Eq. (10.20) one can construct a complex quantity

$$\mathcal{D}_{\Omega_1\Omega_2} \equiv N_{\Omega_1\Omega_2} + \frac{i}{\Delta\Omega}\frac{\partial}{\partial t}N_{\Omega_1\Omega_2}, \tag{10.23}$$

where $\Delta\Omega = \Omega_1 - \Omega_2$. In terms of the function $F(\Omega/\kappa)$, the above expression can be re-written as

$$\mathcal{D}_{\Omega_1\Omega_2} = \frac{1}{2\pi}\frac{\sqrt{\Omega_1\Omega_2}}{4\pi^2\kappa^2}\Gamma\left[-i\frac{\Omega_1}{\kappa}\right]\Gamma\left[i\frac{\Omega_2}{\kappa}\right]\left\{e^{\frac{\pi(\Omega_1+\Omega_2)}{2\kappa}}F\left(-\frac{\Omega_2}{\kappa}\right)F^*\left(-\frac{\Omega_1}{\kappa}\right)\right.$$
$$\left.+\,e^{-\frac{\pi(\Omega_1+\Omega_2)}{2\kappa}}F\left(\frac{\Omega_1}{\kappa}\right)F^*\left(\frac{\Omega_2}{\kappa}\right)\right\}e^{-i(\Omega_1-\Omega_2)t}. \tag{10.24}$$

For a real initial state, we have $F(\Omega/\kappa) = F^*(-\Omega/\kappa)$ and therefore, Eq. (10.24) can be used to extract the function $F(\Omega/\kappa)$ as,

$$\begin{aligned} S_{\Omega_1\Omega_2} &\equiv \frac{4\pi^3\kappa^2}{\sqrt{\Omega_1\Omega_2}} \frac{\mathcal{D}_{\Omega_1\Omega_2} e^{i(\Omega_1-\Omega_2)t}}{\Gamma\left[-i\frac{\Omega_1}{\kappa}\right]\Gamma\left[i\frac{\Omega_2}{\kappa}\right]\cosh\left(\frac{\pi(\Omega_1+\Omega_2)}{2\kappa}\right)} \\ &= F\left(\frac{\Omega_1}{\kappa}\right) F^*\left(\frac{\Omega_2}{\kappa}\right). \end{aligned} \tag{10.25}$$

Clearly, the left hand side of Eq. (10.25) can be determined by observing the emission spectrum and hence is under our control completely. From the above relation, we observe that this quantity has to be separable as a product in terms of the frequencies Ω_1 and Ω_2. Using this property, we can obtain the function $F(\Omega/\kappa)$, upto an irrelevant constant phase, from the symmetric sum

$$\log S_{\Omega_1\Omega_2} = \log F\left(\frac{\Omega_1}{\kappa}\right) + \log F^*\left(\frac{\Omega_2}{\kappa}\right), \tag{10.26}$$

or, alternatively, by fixing one of the frequencies and varying the other one. Therefore, for the real initial state, the state can be identically and completely reconstructed from frequency correlations in the out-going modes. This is similar in spirit to [20], where we devised a formalism to deal with the analysis of field content of a non-vacuum pure state corresponding to a particular observer with respect to another set of observers, using the correlation functions. The information about the state through the function $f(\omega)$, along with the Bogoliubov coefficients, completely characterize the deviations form the standard vacuum response. The analysis of the spectrum operator [9, 21] will also capture this distortion. The extraction of information about initial data using the spectral distortion $\hat{N}_\Omega = \langle\Psi|\hat{b}^\dagger_\Omega\hat{b}_\Omega|\Psi\rangle - \langle 0|\hat{b}^\dagger_\Omega\hat{b}_\Omega|0\rangle$, as reported in [19] has been presented in detail in Appendix A.1. Further, in Appendix A.2, we show that for a particular class of symmetric initial states, interesting quantities can be obtained from the out-states. In particular, one hopes that higher order correlation functions may yield the information regarding the many particle sectors subsequently. However, in this chapter, we only focus on the single particle case, for simplicity.

We have thus demonstrated the existence of semi-classical "quantum" hairs in the case of spherically symmetric collapse which would have formed a Schwarzschild black hole classically. We learn that if we are aware of the symmetries of the system which is going to form a black hole, from some general principles, we can figure out how the non-vacuum response would look like. The test field which contributes infinitesimal energy to the formation of the hole will reflect its non-vacuum character in the late time radiation. That is to say, its spectra will show deviations from the expected vacuum response, in accordance with the symmetries of initial data. Measurements of such non-vacuum distortion will reveal partial or complete character of the state of the field depending on the knowledge of the symmetry of initial profile.

The cases discussed above were all based on test field approximations. We can extrapolate this idea further to conjecture that if we have correct account of the back-reaction, or quantum gravity corrected Bogoliubov coefficients, they will still provide a handle for initial data as in Eq. (10.24) (see also Eq. (9.60) in Appendix A). It will be worthwhile to demonstrate these ideas for a set-up which includes the back-reaction. In a general $(3 + 1)$ collapse scenario the handling of back-reaction remains an open problem. Even for the spherically symmetric collapse case, which we discussed above, accounting for the back-reaction is a tedious task. We will instead be focussing on a $(1 + 1)$ dimensional dilatonic CGHS black hole model. In this case the issue of back-reaction can be exactly handled and even the full quantum gravity calculations can be implemented in a perturbative manner. However, in this chapter, we content ourselves with the semi-classical scheme wherein the back-reaction of the test field has been accounted for. We will see how these non-vacuum distortions lead to additional quantum hairs, which would have been missed in a classical treatment.

10.5 CGHS Model: Introduction

The CGHS black hole solution [22, 23] is a $(1 + 1)$ dimensional gravity model of a dilatonic field ϕ (along with possibly various other matter fields). The theory is being described by the action,

$$\mathcal{A} = \frac{1}{2\pi}\int d^2x\sqrt{-g}\left[e^{-2\phi}\left(R + 4(\nabla\phi)^2 + 4\lambda^2\right) - \frac{1}{2}\sum_{i=1}^{N}(\nabla f_i)^2\right] \tag{10.27}$$

where λ^2 stands for the cosmological constant and f_i denotes for $i-$ th matter field; N such total fields may be present. Since all two dimensional space-times are conformally flat the metric ansatz essentially involves a single unknown function, the conformal factor, which is written in double null coordinates as,

$$ds^2 = -e^{2\rho}dx^+dx^-. \tag{10.28}$$

For the matter fields, the classical solutions are those in which, $f_i(x^+, x^-) = f_{i+}(x^+) + f_{i-}(x^-)$. Thus given some particular matter fields, one can obtain corresponding solutions for ϕ and ρ respectively. The simplest among all of them corresponds to the vacuum solution in which $e^{-2\rho} = e^{-2\phi} = (M/\lambda) - \lambda^2x^+x^-$. This represents a black hole of mass M, with the line element,

$$ds^2 = -\frac{dx^+dx^-}{\frac{M}{\lambda} - \lambda^2x^+x^-}, \tag{10.29}$$

while in absence of any mass, i.e., $M = 0$ one obtains the linear dilatonic vacuum solution as

$$ds^2 = -\frac{dx^+dx^-}{-\lambda^2x^+x^-}. \tag{10.30}$$

A more realistic and dynamical situation corresponds to the case when an in-coming matter collapses and subsequently forms a singularity. If the matter starts at x_i^+ and extends up to x_f^+, then the line element turns out to be,

$$ds^2 = -\frac{dx^+dx^-}{\frac{M(x^+)}{\lambda} - \lambda^2 x^+x^- - P^+(x^+)x^+}, \tag{10.31}$$

where $M(x^+)$ and $P^+(x^+)$ correspond to the following integrals,

$$M(x^+) = \int_{x_i^+}^{x^+} dy^+ y^+ T_{++}(y^+), \tag{10.32}$$

$$P^+(x^+) = \int_{x_i^+}^{x^+} dy^+ T_{++}(y^+). \tag{10.33}$$

The region outside x_f^+ corresponds to a black hole of mass $M \equiv M(x_f^+)$. One can easily check that there is a curvature singularity where the conformal factor diverges.

The singularity is hidden behind an event horizon for future null observers receiving the out-moving radiation. The location of the event horizon can be obtained starting from the location of the apparent horizon. This can, in turn, be obtained using $\partial_+ A \leq 0$, where A is the transverse area. Using the four dimensional analog we end up getting, $\partial_+ e^{-2\phi} \leq 0$. Here the equality leads to the location of the event horizon beyond x_f^+, which happens to be at $x_h^- = -P^+/\lambda^2$.

Thermodynamics, as well as Hawking evaporation of such a black hole solution (also with different matter couplings), have been extensively studied in the literature [23–27]. Even higher order loop corrections to the thermal character of vacuum response have been studied [28]. We will demonstrate our ideas for the simplest setting as depicted in Eq. (10.27).

For completeness, we first study the genesis of Hawking radiation in such a black hole formation. The field equations for this action, at the classical level can be written as

$$-\partial_+\partial_- e^{-2\phi} - \lambda^2 e^{2\rho-2\phi} = 0, \tag{10.34}$$

$$2e^{-2\phi}\partial_+\partial_-(\rho-\phi) + \partial_+\partial_- e^{-2\phi} + \lambda^2 e^{2\rho-2\phi} = 0, \tag{10.35}$$

since the stress energy tensor decouples into left-moving and right-moving parts. On the other hand, the two constraint equations involve the energy momentum tensor components on both the initial slices and lead to,

$$\partial_+^2 e^{-2\phi} + 4\partial_+\phi\partial_+(\rho-\phi)\, e^{-2\phi} + T_{++} = 0, \tag{10.36a}$$

$$\partial_-^2 e^{-2\phi} + 4\partial_-\phi\partial_-(\rho-\phi)\, e^{-2\phi} + T_{--} = 0. \tag{10.36b}$$

Solving the field equations, Eqs. (10.34) and (10.35), we obtain the conformal gauge $\rho = \phi$, where the constraint equations are expressed as

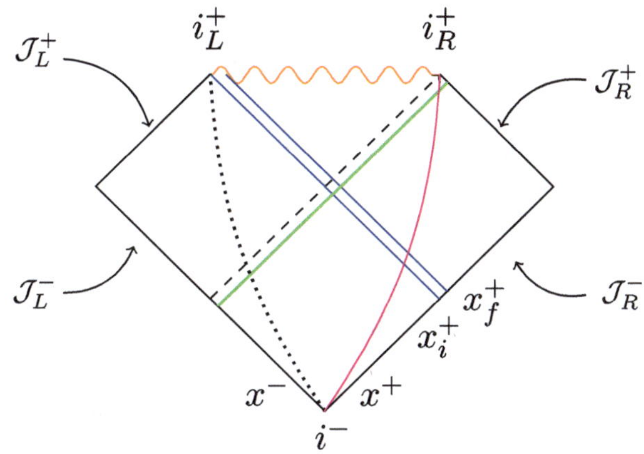

Fig. 10.2 A CGHS black hole

$$\partial_+^2 e^{-2\phi} + T_{++} = 0, \tag{10.37a}$$

$$\partial_-^2 e^{-2\phi} + T_{--} = 0. \tag{10.37b}$$

The solution of these equations provide the classical geometry which is illustrated in Fig. 10.2. The spacetime prior to x_i^+ is flat while the spacetime beyond x_f^+ is being described by the black hole geometry.

For carrying out the semi-classical analysis on this spacetime, we introduce co-ordinate systems which are suited for $\mathcal{J}_L^-$ and $\mathcal{J}_R^+$ respectively. We introduce the co-ordinate set $z^\pm$

$$\pm \lambda x^\pm = e^{\pm\lambda z^\pm}, \tag{10.38}$$

which maps the entire $\mathcal{J}_L^-$ into $z^- \in (-\infty, \infty)$. We also obtain another co-ordinate system suited for $\mathcal{J}_R^+$ as $\sigma_{\text{out}}^\pm$. Let us first discuss the transformation between (z^+, z^-) and $(\sigma_{\text{out}}^+, \sigma_{\text{out}}^-)$ defined as

$$z^+ = \sigma_{\text{out}}^+; \qquad z^- = -\frac{1}{\lambda}\ln\left[e^{-\lambda\sigma_{\text{out}}^-} + \frac{P^+}{\lambda}\right]. \tag{10.39}$$

The horizon, located at $x^- = -P^+/\lambda^2$, will get mapped to $z^- = z_i^- = -\frac{1}{\lambda}\log(P^+/\lambda)$ in these new co-ordinates. 'In' state modes are defined on the asymptotically flat region $\mathcal{J}_L^-$ moving towards $\mathcal{J}_R^+$ and the convenient basis modes on $\mathcal{J}_L^-$ can be taken to be,

$$u_\omega = \frac{1}{\sqrt{2\omega}} e^{-i\omega z^-}, \tag{10.40}$$

where $\omega > 0$. The 'out' region corresponds to $\mathcal{J}_R^+$ which receives the state from $\mathcal{J}_L^-$ after the black hole has been formed. The basis modes in the out region at $\mathcal{J}_R^+$ are given by

$$v_\omega = \frac{1}{\sqrt{2\omega}} e^{-i\omega\sigma^-_{\rm out}} \Theta(z_i^- - z^-), \tag{10.41}$$

where Θ is the usual step function arising from the fact that the out modes are supported by states on $\mathcal{J}_L^-$ only in the interval $(-\infty, 0)$. Again the state of the field can be specified fully on $\mathcal{J}_L^-$ or jointly on $\mathcal{J}_R^+$ and on the event horizon $\mathcal{H}_R$. Since the mode functions at $\mathcal{H}_R$ correspond to part of the field falling into the singularity, which cannot be detected by observers at $\mathcal{J}_R^+$, they must be traced over. Thus, the precise form of mode decomposition on $\mathcal{H}_R$ does not affect physical results for $\mathcal{J}_L^-$. Therefore, we can expand the dilaton field in a different mode basis as,

$$f = \int_0^\infty d\omega \left[a_\omega u_\omega + a_\omega^\dagger u_\omega^*\right], \quad \text{(in)} \tag{10.42}$$

$$= \int_0^\infty d\omega \left[b_\omega v_\omega + b_\omega^\dagger v_\omega^* + \hat{b}_\omega \hat{v}_\omega + \hat{b}_\omega^\dagger \hat{v}_\omega^*\right], \quad \text{(out)} \tag{10.43}$$

where $a_\omega^\dagger$ corresponds to creation operator associated with the 'in' region. In an identical manner $b_\omega^\dagger$ and $\hat{b}_\omega^\dagger$ stand for the creation operators for the 'out' region and the black hole interior region respectively. The inner product between v_Ω and u_ω^* corresponds to,

$$\begin{aligned}
\alpha_{\Omega\omega} &= -\frac{i}{\pi}\int_{-\infty}^{z_i^-} dz^- v_\Omega \partial_- u_\omega^* \\
&= \frac{1}{2\pi}\sqrt{\frac{\omega}{\Omega}}\int_{-\infty}^{z_i^-} dz^- \exp\left[\frac{i\Omega}{\lambda}\ln\left\{\left(e^{-\lambda z^-} - \frac{P^+}{\lambda}\right)\right\} + i\omega z^-\right] \\
&= \frac{1}{2\pi\lambda}\sqrt{\frac{\omega}{\Omega}}\left(\frac{P^+}{\lambda}\right)^{i(\Omega-\omega)/\lambda} B\left(-\frac{i\Omega}{\lambda} + \frac{i\omega}{\lambda}, 1 + \frac{i\Omega}{\lambda}\right),
\end{aligned} \tag{10.44}$$

while the inner product between v_Ω and u_ω yields

$$\begin{aligned}
\beta_{\Omega\omega} &= \frac{i}{\pi}\int_{-\infty}^{z_i^-} dz^- v_\Omega \partial_- u_\omega \\
&= \frac{1}{2\pi}\sqrt{\frac{\omega}{\Omega}}\int_{-\infty}^{z_i^-} dz^- \exp\left[\frac{i\Omega}{\lambda}\ln\left\{\left(e^{-\lambda z^-} - \frac{P^+}{\lambda}\right)\right\} - i\omega z^-\right] \\
&= \frac{1}{2\pi\lambda}\sqrt{\frac{\omega}{\Omega}}\left(\frac{P^+}{\lambda}\right)^{i(\Omega+\omega)/\lambda} B\left(-\frac{i\Omega}{\lambda} - \frac{i\omega}{\lambda}, 1 + \frac{i\Omega}{\lambda}\right),
\end{aligned} \tag{10.45}$$

where $B(x, y)$ is the Beta function.

The vacuum response can be obtained from Eq. (10.45), which has a thermal profile in the limit of large frequencies, which will correspond to the late time limit for observers at $\mathcal{J}_R^+$. In general the vacuum response for the CGHS black hole also comes with a non-thermal part. This late-time thermal response originates from the vacuum state at $\mathcal{J}_L^-$, but, due to asymmetry in the left- and right-moving modes, the modes which form the black hole, actually originate only from $\mathcal{J}_R^-$, rather than from $\mathcal{J}_L^-$. If one has to retrieve the information associated with the quantum states that formed the black hole in the first place, the observers moving left are the relevant ones. Therefore, we will concentrate on the future, left moving observers who are moving in the flat spacetime throughout the journey. We want to study the black hole evaporation at the semi-classical level for one such observer, which we will do in the subsequent section.

10.6 Information Regarding the Collapsing Matter

The left-moving matter is introduced for $x^+ \geq x_i^+$ and hence the spacetime is in a vacuum configuration prior the introduction of the matter shell. Hence the spacetime, in the region $x^+ < x_i^+$ is flat and the metric is given as in the Eq. (10.30), which, on using the co-ordinates

$$x^+ = -\frac{1}{\lambda y^+}; \qquad x^- = -\frac{1}{\lambda y^-}, \tag{10.46}$$

remains the same

$$ds^2 = -\frac{dy^+ dy^-}{-\lambda^2 y^+ y^-}. \tag{10.47}$$

The location x_i^+ is marked by $y_i^+ = -1/\lambda x_i^+$. Using another set of co-ordinate transformation the metric on $\mathcal{J}_R^-$ can be brought into the desired flat form. A left-moving observer who completely stays in the region to the past of $\mathcal{J}_L^+$ always remains in flat spacetime.

However, such an observer is able to access only a portion of the initial data on $\mathcal{J}_R^-$. Therefore, the Bogoliubov coefficients between a complete set of mode functions u_ω defined on $\mathcal{J}_R^-$ and the complete set of mode functions v_ω defined on $\mathcal{J}_L^+$, are given by

$$\begin{aligned} \alpha_{\Omega\omega} &= -\frac{i}{\pi}\int_{-\infty}^{\chi_i^+} d\chi^+ v_\Omega \partial_- u_\omega^* \\ &= \frac{1}{2\pi}\sqrt{\frac{\omega}{\Omega}}\int_{-\infty}^{\chi_i^+} d\chi^+ \exp\left[\frac{i\Omega}{\lambda}\ln\left\{\left(e^{-\lambda\chi^+} - |y_i^+|\right)\right\} + i\omega\chi^+\right] \\ &= \frac{1}{2\pi\lambda}\sqrt{\frac{\omega}{\Omega}}|y_i^+|^{\frac{i(\Omega-\omega)}{\lambda}} B\left(-\frac{i\Omega}{\lambda} + \frac{i\omega}{\lambda}, 1 + \frac{i\Omega}{\lambda}\right), \end{aligned} \tag{10.48}$$

and

$$\begin{aligned}\beta_{\Omega\omega} &= \frac{i}{\pi}\int_{-\infty}^{\chi_i^+} d\chi^+ v_\Omega \partial_+ u_\omega \\ &= \frac{1}{2\pi}\sqrt{\frac{\omega}{\Omega}}\int_{-\infty}^{\chi_i^+} d\chi^+ \exp\left[\frac{i\Omega}{\lambda}\ln\left\{\left(e^{-\lambda\chi^+} - |y_i^+|\right)\right\} - i\omega\chi^+\right] \\ &= \frac{1}{2\pi\lambda}\sqrt{\frac{\omega}{\Omega}}|y_i^+|^{\frac{i(\Omega+\omega)}{\lambda}} B\left(-\frac{i\Omega}{\lambda} - \frac{i\omega}{\lambda}, 1 + \frac{i\Omega}{\lambda}\right).\end{aligned} \tag{10.49}$$

Thus, as before, the observer at $\mathscr{J}_L^+$ will observe Bogoliubov coefficients similar to the ones observed by their right moving counterparts but with the parameter exchange $|y_i^+| \leftrightarrow P^+/\lambda$ [29]. However, it should be emphasized that these nontrivial Bogoliubov coefficients are totally due to the tracing over of the modes which lie in the future of $\mathscr{J}_L^+$ and not due to any geometry change. One can check that if the fraction of the tracing over vanishes, which would corresponds to the limit $|y_i^+| \to 0$, the Bogoliubov coefficients will assume a trivial form. Along identical lines, the Bogoliubov coefficients in Eqs. (10.44) and (10.45) will also assume a trivial form in the limit $P^+ \to 0$. For these observers the effect of tracing over is indistinguishable from that of change in geometry and both of these effects vanish simultaneously in the above limit. However, surprisingly the vacuum response for both these observers is indistinguishable. In particular, the late time radiation for such observers on $\mathscr{J}_L^+$ for the vacuum state (of a test field) on $\mathscr{J}_R^-$ is also thermal *with the same temperature as measured by their right-moving counterparts!*

Further, any non vacuum state on $\mathscr{J}_R^-$ will lead to non-vacuum distortions in the radiation as we have elaborated earlier. Thus, we will now use the Bogoliubov coefficients for extracting information regarding the matter that formed the black hole.

10.6.1 Test Field Approximation

To start with, we can first do a quick demonstration, under the test field approximation, in order to connect up with the earlier case of the Schwarzschild black hole, by assuming that the matter field ϕ forming the black hole is classical. Then we add a little more matter $\delta\hat{\phi}$ to the black hole in a perturbative manner. That is to say, we add another small matter pulse to the collapse process with support in the region $x^+ > x_f^+$. This small chunk must be treated perturbatively as quantum matter. Given the form of the Bogoliubov coefficients as in Eqs. (10.48) and (10.49), we note that they remain independent of the matter content in the region $x^+ > x_i^+$ and solely depend on the co-ordinate at which the first matter shell was introduced, serving as a horizon for the left-moving observers. Therefore, an asymptotically left-moving observer will use Eqs. (10.48) and (10.49) to compute the spectral distortion and hence reconstruct the state of the test matter which is thrown in, through a procedure similar to what was done for the Schwarzschild black hole. Thus, for asymptotic observers at late times, the high frequency approximation of Eqs. (10.48) and (10.49) leads to a form exactly as in Eq. (10.9). Hence, for such observers, the results developed in Sect. 10.4 and Appendix A.2 are exactly applicable.

10.6.2 Adding Back-Reaction: Extracting Information About the Matter Forming the Black Hole

In the CGHS case, we can do away with the high frequency approximation, since it is possible to exactly calculate the results for observers at any finite time (not necessarily at late times). Moreover, the results derived for the left-moving asymptotic observers remains oblivious to the matter introduced beyond their perceived horizon and hence does not care about the geometry beyond the horizon at all. Therefore, all the back-reaction of the collapsing matter can now be accounted for, since they *do not* change the geometry profile of left-moving observers. Being a conformally flat spacetime, we know the exact mode functions irrespective of the back-reaction in the entire spacetime which includes the relevant left portion. Therefore, the form of 10.48 and 10.49 are exact, even in the presence of back-reaction of the matter field, or even when we take the collapsing matter ϕ itself to be quantum matter, which we will do next.

For a complete semi-classical treatment, we take the field $\hat{\phi}$ to be quantum mechanical. At the semi-classical level, the stress energy tensor components are replaced by their expectation values $\langle T_{\pm\pm}\rangle$ and $\langle T_{+-}\rangle$. Being a two dimensional spacetime the expectation values get an additional contribution from the conformal anomaly term. Therefore, the classical equations (for $N = 1$) get modified to

$$- \partial_+\partial_- e^{-2\phi} - \lambda^2 e^{2\rho-2\phi} = \frac{1}{12\pi}\partial_+\partial_-\rho, \tag{10.50}$$

$$2e^{-2\phi}\partial_+\partial_-(\rho-\phi) + \partial_+\partial_- e^{-2\phi} + \lambda^2 e^{2\rho-2\phi} = 0, \tag{10.51}$$

whereas the constraint equations also pick up conformal anomaly corrections as

$$\partial_+^2 e^{-2\phi} + 4\partial_+\phi\partial_+(\rho-\phi)\, e^{-2\phi} + \langle T_{++}\rangle = 0, \tag{10.52a}$$

$$\partial_-^2 e^{-2\phi} + 4\partial_-\phi\partial_-(\rho-\phi)\, e^{-2\phi} + \langle T_{--}\rangle = 0. \tag{10.52b}$$

In order to remain true to the classical geometry, the state of the matter field should be the one in which the classical flat geometry is realized prior to x_i^+. Thus, we require $\exp(\rho) = \exp(\rho_{\text{flat}}) = 1/\lambda^2 x^+ x^-$ in that region suggesting that the matter support is only in the region given by $x^+ > x_i^+$. Therefore, the classical values of the $T_{\perp\perp}$ are realized by the expectation values $\langle T_{\perp\perp}\rangle$. We can also judiciously choose the boundary conditions for the set of initial states such that the contribution arising from the conformal anomaly can be canceled in the region of interest, giving rise to a flat spacetime semi-classically, see for instance [30–32]. We further define $\mathcal{I}_L^+$ as the line $x^+ = x_i^+$. For our consideration, we will need the part of the spacetime in the causal past of $\mathcal{I}_L^+$, which remains unaffected by the conformal anomaly term with such a judicious choice of family of quantum states.

Supported by such quantum states, the geometry of the spacetime remains as elaborated above and we can use the expressions for the Bogoliubov coefficients as earlier. The asymptotic expressions for the Bogoliubov coefficients as in 10.48 and 10.49 resemble those of the spherically symmetric collapse model 10.9. Therefore, the spectral distortion as measured by the late time observers will be exactly as discussed in Sect. 10.4 and Appendix A.2.

We can now obtain the exact expression and the symmetry profile of the initial data necessary for the retrieval of complete information by a generic observer on $\mathcal{J}_L^+$. Using the Bogoliubov coefficients in Eqs. (10.48) and (10.49) we can obtain the non-vacuum correction to the correlator and hence the vacuum spectrum through Eq. (10.7) (see also Eq. (9.60)). Again, we will consider the case of a single particle state as in Eq. (10.1), which has the stress energy support as discussed above. The symmetry profile required in the initial state of the quantum field, for the retrieval of information by the late time observers, remains exactly the same as that for the spherically symmetric Schwarzschild model. Therefore, such symmetry profiles appear uniquely for all late time observers and *any real initial data can be reconstructed by the late time observers.*

In this case, since we have exact expressions for the Bogoliubov coefficients, we can also obtain symmetry condition for all observers on $\mathcal{J}_L^+$ and not only for the late time observers. This will demonstrate the ability to reconstruct the information at any stage of evolution, which can then be used to track the information content throughout the evolution and quantify whether there is any information loss for a more general initial data. We further define another function $\tilde{g}(\omega)$, as

$$f(\bar{\omega}') = \bar{\omega}'\tilde{g}(\bar{\omega}')e^{-\frac{\pi}{2}\bar{\omega}'}\Gamma[-i\bar{\omega}']\left|y_i^+\right|^{-i\bar{\omega}'} . \tag{10.53}$$

The transformation presented in Eq. (10.53) relates the frequency correlation to the symmetries of $\tilde{g}(\bar{\omega})$. The scheme of information retrieval from the frequency correlator can then be implemented exactly as before, for $\tilde{g}(\bar{\omega})$. Reconstruction of $\tilde{g}(\bar{\omega})$ is equivalent to reconstruction of $f(\bar{\omega})$. (Analysis for the self-correlator, i.e., the spectrum operator has been discussed in Appendix A.4.) Different set of observers will require different set of symmetries in order to extract maximum amount of information from the non-vacuum correlator. For a given initial state we can also follow the information content during the course of evaporation, culminating in the late time result.

Thus, using a consistent semi-classical treatment one can recover the matter quantum state using the distortion of the thermal radiation as detected by the asymptotic observers. However any asymptotic observer would have measured a thermal radiation which is independent of the black hole's mass—and decided entirely by the cosmological constant—in the CGHS model. This remains true throughout, even during the evaporation of the black hole. For the left moving asymptotic observer, surprisingly the black hole region does not shrink as an outcome of this radiation emission, since the location of horizon is not being decided by the mass content inside the horizon. Still such an observer will be able to reconstruct all the information about the mass/energy content beyond her horizon. Therefore, it is reasonable to expect that such observers associate a notion of entropy to the black hole which can be significantly different from what a right moving observer will do. Hence it is expected that the notion of entropy for left-moving observers should also be different from the standard black hole entropy expression. In the Schwarzschild black hole formation on the other hand, due to spherical symmetry, left-moving or right moving observers not only measure the same temperature, but witness an identical geometry change during the formation as well as evaporation of the black hole. Therefore, unlike the current situation, those observers should be associating the same entropy expression for the hole.

10.7 Conclusions

In this chapter, we have focused on the reconstruction of initial data which formed the black hole by observing a particular kind of distortion to the Hawking radiation. For some matter to end up inside the horizon (carrying some energy, charge etc. with it), the state of the matter field must be non-vacuum. Moreover, the non-vacuum nature of the state manifests itself in the correlation between the modes escaping the capture of the black hole and the modes entering the horizon and eventually hitting the singularity. Observing the portion which narrowly escaped the horizon, we can re-construct the correlation and hence the state of the field at the initial slice. We have constructed an observable capable of capturing this correlation. Using this observable judiciously, one can extract information pertaining to the initial condition, which otherwise remains hidden from the asymptotic late time observers. The diagonal elements of this correlation matrix yield the distortion of the emission spectrum, which becomes non-thermal for non-vacuum in-states. It is noteworthy that these non-vacuum distortions will always be present in the most general case including quantum gravity, back-reaction etc. Only the corresponding Bogoliubov coefficients will be generalized for the particular situation under consideration.

In $(3 + 1)$ dimensional spacetime, the inclusion of the semiclassical back-reaction is not under control. For the purpose of demonstration of our ideas we first discussed a semi-classical case of a spherically symmetric collapse which forms a black hole through a slow, s-wave process. On top of this evolving geometry we introduced another test field, which is being described by a massless scalar field, initially set in a particular field configuration at asymptotic, past, null infinity. The complete recovery of the initial data corresponds to the deciphering of the quantum state of the field in an unambiguous manner. For this simplified set up, the vacuum response is thermal, and the non-vacuum response corresponds to non-thermal distortions of the late time Hawking radiation. Such distortions constrain some operators acting on the conjugate space of the frequency representation of the quantum matter field. Typically such operators will provide information about the single particle sectors of the Hilbert space. The spectral distortion being just a single function of the frequency, it is expected to capture the finite correlations of an arbitrary, general, initial state. Higher order correlations may be required in order to obtain additional information about the initial state.

If (i) the state of the field corresponds to a single particle state and (ii) we have access to the symmetries of the distribution in the frequency space, we can recover very many information about the initial state through such non-vacuum distortions. In particular, we show that the existence of a class of symmetries of non-zero measure must encode all the information about the initial state in the non-vacuum distortion. In particular, for real initial data, we can recover the state of a single excitation, *completely*, using frequency correlations between the out-going modes. Although the non-vacuum distortions do not make the out-state a pure one, still there are enough correlations available even in the mixed state, to reconstruct the initial state. Also we did not need to study higher order correlations in order to obtain further information about the initial state.

The simple case we discussed first, was without the incorporation of back-reaction in a non-vacuum configuration. In a general setting, we need to account for the back-reaction as well, howsoever small they might be. The Bogoliubov coefficients incorporating the back-reaction can be obtained through the mode functions of in- and out-configurations in such a modified geometry. In general it is a very difficult and tedious job. However, even in some simplified scenarios addressing the issue of back-reaction should throw some light on the concepts involved. For this purpose we considered the case of back-reaction in a $(1 + 1)$ dimensional dilatonic black hole solution, viz., the CGHS model. The dilatonic vacuum solution is perturbatively unstable towards forming a black hole if the perturbation is around the in-moving modes. Even at the classical level the exact Hawking radiation of this model is thermal at the high frequency regime (i.e., late time) with corrections at low energies. We call such a radiation profile collectively as the vacuum response. We further showed that the non-vacuum distortions for the left-moving observers reveal information about the initial left-moving distribution which collapsed to form the black hole in the first place. In this CGHS black hole spacetime there is always a part of past null infinities, which is causally disconnected to the asymptotic future observers. Classically no information pertaining to the matter field configuration in this disconnected sector will ever be available to such observers due to causal structure of thespacetime. However, we see that the non-vacuum distortions are capable of revealing the information about such configurations through frequency correlations. The modes appearing as distortions at $\mathcal{J}^+$ did have, in the past, some correlations, with the modes which had entered the horizon, at $\mathcal{J}^-$. Measurement of such distortions can, in principle, also tell us about the white hole region if it were accompanying the black hole region as suggested in some earlier works [33–35]. A simple extrapolation from this idea can also be used in order to study the fate of such information regarding the matter field which formed the black hole. The retrieval process for the late time observers is *exactly the same* as for the Schwarzschild late time observers, yielding the same result. We also discuss how the information about the in-state can be tracked throughout the evolution and evaporation process, and not only at late times, for the CGHS case.

Appendix A

A.1 Spectrum Operator

Using the expression for the correction term over the vacuum thermal spectrum, we can obtain the distortion from thermal Hawking radiation for one particle initial state of the field which is undergoing the collapse as,

$$N_\Omega = \left[\left| \int_0^\infty \frac{d\tilde{\omega}'}{\sqrt{4\pi\tilde{\omega}'}} \alpha^*_{\Omega\tilde{\omega}'} f(\tilde{\omega}') \right|^2 + \left| \int_0^\infty \frac{d\tilde{\omega}}{\sqrt{4\pi\tilde{\omega}}} \beta_{\Omega\tilde{\omega}} f(\tilde{\omega}) \right|^2 \right]. \quad (10.54)$$

It must be noted that the expression in Eq. (10.54) is general enough to include cases when the Bogoliubov coefficients as in Eq. (10.9) are modified by back-reaction, angular momentum, quantum gravity etc. In any case, the non-vacuum part of the radiation spectra provides a constraint for $f(\omega)$ in form of Eq. (10.54). Using Eq. (10.14), we can rewrite Eq. (10.54) as

$$N_\Omega = \frac{1}{4\pi} \frac{1}{4\pi\kappa} \frac{1}{\sinh \frac{\pi\Omega}{\kappa}} \left[\left| \tilde{F}\left(\frac{\Omega}{\kappa}\right) \right|^2 + \left| \tilde{F}\left(-\frac{\Omega}{\kappa}\right) \right|^2 \right]. \quad (10.55)$$

We can decompose $|\tilde{F}(\Omega/\kappa)|^2$ into symmetric $\tilde{S}(\Omega/\kappa)$ and an anti-symmetric part $\tilde{A}(\Omega/\kappa)$

$$\left| \tilde{F}\left(\frac{\Omega}{\kappa}\right) \right|^2 = \tilde{S}\left(\frac{\Omega}{\kappa}\right) + \tilde{A}\left(\frac{\Omega}{\kappa}\right). \quad (10.56)$$

With this decomposition, we realize from Eq. (10.55) that the symmetric part of $|\tilde{F}(\Omega/\kappa)|^2$ is entirely characterized by the distribution function N_Ω of the radiation,

$$\tilde{S}\left(\frac{\Omega}{\kappa}\right) = 8\pi^2 \kappa N_\Omega \sinh \frac{\pi\Omega}{\kappa}. \quad (10.57)$$

Further, if the in-state is normalized to unity, we have

$$\int_{-\infty}^{\infty} d\left(\frac{\Omega}{\kappa}\right) e^{-\frac{\pi\Omega}{\kappa}} \left[\tilde{S}\left(\frac{\Omega}{\kappa}\right) + \tilde{A}\left(\frac{\Omega}{\kappa}\right) \right] = 8\pi^2, \quad (10.58)$$

which together with Eq. (10.57) regulates the integral (and hence the asymptotic behavior) of $\tilde{A}(\kappa)$. Apart from this constraint, $\tilde{A}(\kappa)$ is a completely arbitrary anti-symmetric function. Therefore, the radiation spectra fixes the symmetric part of the probability density in the Fourier space corresponding to z. However, the anti-symmetric part of this probability density remains largely unspecified.

In terms of the function $g(z)$ defined in Eq. (10.15), the symmetric part $\tilde{S}\left(\frac{\Omega}{\kappa}\right)$ can be written as

$$\begin{aligned} e^{\pi\frac{\Omega}{\kappa}} \times \int_{-\infty}^{\infty} dy \int_{-\infty}^{\infty} dz g(z-y/2) g^*(z+y/2) e^{-i\frac{\Omega}{\kappa}y} \\ + e^{-\pi\frac{\Omega}{\kappa}} \int_{-\infty}^{\infty} dy \int_{-\infty}^{\infty} dz g(z-y/2) g^*(z+y/2) e^{i\frac{\Omega}{\kappa}y} = 2\tilde{S}\left(\frac{\Omega}{\kappa}\right). \end{aligned} \tag{10.59}$$

As, we see that $F(\kappa)$ is "momentum space representation" conjugate to $g(z)$, the above expression can be written in terms of the Wigner function corresponding to the phase space of $(z, \Omega/\kappa)$,

$$e^{\pi\frac{\Omega}{\kappa}} \int_{-\infty}^{\infty} dz \mathcal{W}_g\left(z, \frac{\Omega}{\kappa}\right) + e^{-\pi\frac{\Omega}{\kappa}} \int_{-\infty}^{\infty} dz \mathcal{W}_g\left(z, -\frac{\Omega}{\kappa}\right) = 2\tilde{S}\left(\frac{\Omega}{\kappa}\right), \tag{10.60}$$

where the Wigner function is defined as,

$$\mathcal{W}_g\left(z, \frac{\Omega}{\kappa}\right) = \int_{-\infty}^{\infty} dy g(z-y/2) g^*(z+y/2) e^{-i\frac{\Omega}{\kappa}y}. \tag{10.61}$$

Also, with the relation

$$\left|F\left(\frac{\Omega}{\kappa}\right)\right|^2 = \int_{-\infty}^{\infty} dz \mathcal{W}_g\left(z, \frac{\Omega}{\kappa}\right), \tag{10.62}$$

we obtain,

$$e^{\pi\frac{\Omega}{\kappa}} \left|F\left(\frac{\Omega}{\kappa}\right)\right|^2 + e^{-\pi\frac{\Omega}{\kappa}} \left|F\left(-\frac{\Omega}{\kappa}\right)\right|^2 = 2\tilde{S}\left(\frac{\Omega}{\kappa}\right), \tag{10.63}$$

which is an obvious illustration of Eq. (10.57). Therefore, integrating the relation Eq. (10.63) over the frequency range at $\mathcal{J}^+$ we obtain the relation

$$\begin{aligned} 2\int_0^{\infty} d\left(\frac{\Omega}{\kappa}\right) \tilde{S}\left(\frac{\Omega}{\kappa}\right) &= \int_0^{\infty} d\left(\frac{\Omega}{\kappa}\right) e^{\frac{\pi\Omega}{\kappa}} \int_{-\infty}^{\infty} dz \mathcal{W}_g\left(z, \frac{\Omega}{\kappa}\right) \\ &\quad + \int_0^{\infty} d\left(\frac{\Omega}{\kappa}\right) e^{-\frac{\pi\Omega}{\kappa}} \int_{-\infty}^{\infty} dz \mathcal{W}_g\left(z, -\frac{\Omega}{\kappa}\right) \\ &= \int_{-\infty}^{\infty} dy e^{\pi y} \left|F(y)\right|^2. \end{aligned} \tag{10.64}$$

Although, the state which would be completely specified, if we know $F(y)$, remains arbitrary apart form this constraint, the symmetric part $\tilde{S}\,(\Omega/\kappa)$ which is completely specified through the non-vacuum distortion, fixes the expectation of the exponentiated momenta conjugate to $z(=\log\omega/C)$.

For a n-particle state

$$|\Psi\rangle = \int_0^\infty \prod_{i=1}^{n} \frac{d\omega_i}{\sqrt{2\pi\omega_i}} f(\omega_1, ...\omega_n)\hat{a}^\dagger(\omega_i)|0\rangle_M, \tag{10.65}$$

the radiation profile over the thermal component fixes the expectation of a single particle exponentiated momentum, i.e.,

$$16\int_0^\infty d\left(\frac{\Omega}{\kappa}\right)\pi^2\kappa\left(\sinh\frac{\pi\Omega}{\kappa}\right)N_\Omega = \frac{1}{n}\langle\Psi|e^{\frac{\pi\hat{\Omega}}{\kappa}}|\Psi\rangle. \tag{10.66}$$

Additional information about the initial state can only be obtained from the spectrum if the initial state has some symmetries. We will discuss few interesting cases below.

- If $F(y)$ is a real and symmetric function, then we see from Eq. (10.63) that it gets completely specified in terms of $\tilde{S}(\Omega/\kappa)$. As, a result the initial state also gets completely specified as $g(z)$ can be obtained by the inverse Fourier transform. However, by virtue of the properties of Fourier transform, $g(z)$ also happens to be real and symmetric. This symmetry corresponds to a duality in the frequency space distribution about the surface gravity parameter κ. These states are very special class of initial states whose information get coded entirely in the radiation from the black hole within the framework of standard unitary quantum mechanics.
- For a slightly more general case, the reality condition on $g(z)$ can be traded for by imposing relation between $F(\Omega/\kappa)$ and $F(-\Omega/\kappa)$, which is to specify the symmetry of $F(y)$ in the positive and negative half planes. Such a specification of symmetry constrains the distribution $F(y)$ to remain arbitrary in one of the half planes and amounts to reducing the degrees of freedom by half. Let us assume $F(\Omega/\kappa)$ is real, that means

$$g(z) = g^*(-z). \tag{10.67}$$

Now additionally if we impose,

$$F(-y) = K(y)\,F(y)\,, \tag{10.68}$$

for a specified function $K(y)$, then

$$\int_{-\infty}^{\infty} dz g(z) e^{-iyz} = \int_{-\infty}^{\infty} dz g(z) e^{iyz} K(y)\,. \tag{10.69}$$

Therefore, using the condition Eq. (10.67), we can obtain from Eq. (10.69)

$$g^*(z) = \int_{-\infty}^{\infty} dz' g(z') \int_{-\infty}^{\infty} dy K(y) e^{iy(z'-z)}$$
$$= \int_{-\infty}^{\infty} dz' g(z') \tilde{K}(z'-z), \tag{10.70}$$

where $\tilde{K}(q)$ is the inverse Fourier transform of $K(y)$. Therefore, for such a symmetry in the probability amplitude, the state can be recovered from

$$F^2\left(\frac{\Omega}{\kappa}\right) = 16\pi^2\kappa \frac{\left(\sinh \frac{\pi\Omega}{\kappa}\right)}{e^{\frac{\pi\Omega}{\kappa}} + \left(K\left(\frac{\Omega}{\kappa}\right)\right)^2 e^{-\frac{\pi\Omega}{\kappa}}} N_\Omega. \tag{10.71}$$

Therefore, we see that the symmetry of the prescribed class for one particle sate encodes the entire information of the in-state in the resulting radiation from the black hole. If the initial condition of the collapse demands symmetry of such kinds, the resulting mixed state has enough information in the spectra to completely specify the state. We will further consider some other classes of symmetries in the initial data for spherically symmetric collapse models and their imprints in the non-vacuum distortions.

A.2 Real Initial Distribution

For real distributions, the Fourier transform will satisfy

$$|F(y)|^2 = |F(-y)|^2 \tag{10.72}$$

Therefore, $|K(y)| = 1$ and we have the relation

$$\left|F\left(\frac{\Omega}{\kappa}\right)\right|^2 + \left|F\left(-\frac{\Omega}{\kappa}\right)\right|^2 = 8\pi^2\kappa \tanh \frac{\pi\Omega}{\kappa} N_\Omega. \tag{10.73}$$

For a symmetric algebraic operator of y

$$\hat{\mathcal{O}}_{\text{even}}(y) = \hat{\mathcal{O}}_{\text{even}}(-y), \tag{10.74}$$

the expression

$$8\pi^2\kappa \times \tanh\frac{\pi\Omega}{\kappa}\mathcal{O}_{\text{even}}\left(\frac{\Omega}{\kappa}\right)N_\Omega$$
$$= \mathcal{O}_{\text{even}}\left(\frac{\Omega}{\kappa}\right)\left|F\left(\frac{\Omega}{\kappa}\right)\right|^2 + \mathcal{O}_{\text{even}}\left(-\frac{\Omega}{\kappa}\right)\left|F\left(-\frac{\Omega}{\kappa}\right)\right|^2 \tag{10.75}$$

which on integration over the whole frequency range gives the expectation value of the operator

$$\int_0^\infty d\left(\frac{\Omega}{\kappa}\right)8\pi^2\kappa\tanh\frac{\pi\Omega}{\kappa}\mathcal{O}_{\text{even}}\left(\frac{\Omega}{\kappa}\right)N_\Omega = \int_0^\infty d\left(\frac{\Omega}{\kappa}\right)\left[\mathcal{O}_{\text{even}}\left(\frac{\Omega}{\kappa}\right)\left|F\left(\frac{\Omega}{\kappa}\right)\right|^2 + \mathcal{O}_{\text{even}}\left(-\frac{\Omega}{\kappa}\right)\left|F\left(-\frac{\Omega}{\kappa}\right)\right|^2\right]$$
$$= \int_{-\infty}^\infty \mathcal{O}_{\text{even}}(y)|F(y)|^2. \tag{10.76}$$

By similar logic, one can argue that expectation of all odd algebraic operators vanish in this case, i.e., with a symmetric $|F(y)|^2$,the expectation value for an odd observable

$$\langle\hat{\mathcal{O}}_{\text{odd}}(y)\rangle = \int_{-\infty}^\infty dy\mathcal{O}_{\text{odd}}(y)|F(y)|^2 = 0. \tag{10.77}$$

Thus in this scenario, expectation of all algebraic operators in y will be given in terms of spectral distortion. Any general operator $\hat{\mathcal{O}}(y)$ can be decomposed in terms of its even and odd parts

$$\hat{\mathcal{O}}(y) = \hat{\mathcal{O}}_{\text{even}}(y) + \hat{\mathcal{O}}_{\text{odd}}(y). \tag{10.78}$$

Therefore, to obtain $\langle\hat{\mathcal{O}}(y)\rangle$ one only requires $\langle\hat{\mathcal{O}}_{\text{even}}(y)\rangle$, which can be easily obtained from Eq. (10.76). Similarly for the generalized symmetry class

$$\langle\hat{\mathcal{O}}(y)\rangle = \int_0^\infty dy\left[\mathcal{O}\left(y\right)\left|F\left(y\right)\right|^2 + \mathcal{O}\left(-y\right)\left|F\left(-y\right)\right|^2\right]$$
$$= \int_0^\infty dy(\mathcal{O}\left(y\right) + |K(y)|^2\mathcal{O}\left(-y\right))\left|F\left(y\right)\right|^2$$
$$= 16\pi^2\kappa\int_0^\infty d\bar{\Omega}\frac{[\mathcal{O}\left(\bar{\Omega}\right) + |K(\bar{\Omega})|^2\mathcal{O}\left(-\bar{\Omega}\right)]\sinh\bar{\Omega}}{e^{\frac{\pi\Omega}{\kappa}} + \left|K\left(\bar{\Omega}\right)\right|^2 e^{-\frac{\pi\Omega}{\kappa}}}N_\Omega, \tag{10.79}$$

where we have used the expression of $|F\left(y\right)|^2$ in the range $y \in (0, \infty)$ from Eq. (10.71), in the third equality. Thus even with the specified symmetry class $K(y)$, all the algebraic operators on the momentum space become fixed.

A.3 State for Step Function Support

Let us excite some right-moving modes beyond x_i^+ (for simplicity we work with single particle states), such that the normal ordered operator $\hat{T}_{++}(x^+)$ has support only in the region inside the horizon, i.e.,

$$\langle \hat{T}_{++}(x^+) \rangle_{\text{Regularized}} = h(x^+)\Theta(x^+ - x_i^+), \tag{10.80}$$

for some well behaved function $h(x^+)$ and the step function $\Theta(x^+)$.

If the single particle state is taken to be in the frame of observers which would have described the linear dilaton vacuum, then

$$|\Psi\rangle = \int_\omega f(\omega)\hat{a}_\omega^\dagger |0\rangle, \tag{10.81}$$

where $\int_\omega$ stands for $\int d\omega/\sqrt{4\pi\omega}$ and the right-moving quantum field is given on $\mathcal{J}_L^+$ as

$$\hat{f}_+(y^+) = \int_\omega (\hat{a}_\omega u_\omega(y^+) + \hat{a}_\omega^\dagger u_\omega^*(y^+)), \tag{10.82}$$

with mode functions $u_\omega(y^+)$. Then the equation Eq. (10.80) can be re-written as

$$\left| \int_\omega f(\omega) u'_\omega(y^+) \right|^2 = h_1(y^+)\Theta(y^+ - y_i^+), \tag{10.83}$$

where$'$ denotes a derivative with respect to y^+ and y_i^+ marking the location corresponding to x_i^+. The function $h_1(y^+)$ absorbs the Jacobian of transformation from $x^\pm$ basis to $y^\pm$ basis,

$$T_{++}(x^+) = \frac{\partial y^\mu}{\partial x^+}\frac{\partial y^\nu}{\partial x^+} T_{\mu\nu}(y^+) = \frac{\partial y^+}{\partial x^+}\frac{\partial y^+}{\partial x^+} T_{++}(y^+) \tag{10.84}$$

The condition Eq. (10.83) can be realized by

$$\int_\omega f(\omega) u'_\omega(y^+) = \tilde{h}(y^+)\Theta(y^+ - y_i^+), \tag{10.85}$$

with some other well behaved function $\tilde{h}(y^+)$. Owing to the conformal flatness of the two dimensional spacetime and the conformal nature of minimally coupled massless scalar field, the mode functions can be written as

$$u'_\omega(y^+) = -i\omega u_\omega(y^+). \tag{10.86}$$

Therefore, we only require to have

$$\int_{\omega} f(\omega)\omega u_{\omega}(x^+) = \tilde{h}(x^+)\Theta(x^+ - x_i^+) = \zeta(x^+), \tag{10.87}$$

where we have absorbed the factor i in the redefinition of $\tilde{h}(x^+)$. Taking the inner product of the Eq. (10.87) with itself and using the orthonormal properties of the mode functions, we write

$$\int_{\omega} \omega^2 |f(\omega)|^2 = (\zeta(x^+), \zeta(x^+)). \tag{10.88}$$

Fo r the states satisfying Eq. (10.88), the expression Eq. (10.87) can be inverted for given $\zeta(x^+)$, using the completeness of mode functions to obtain a consistent state. Therefore, the one particle states respecting Eq. (10.88) will have mode excitations beyond x_i^+.

A.4 Information Retrieval for the CGHS Black Hole

Using a particular representation of the Gamma function

$$\Gamma[z] = i^z \int_{-\infty}^{\infty} dq e^{zq} e^{-ie^q}, \tag{10.89}$$

we can write down a product formula

$$\Gamma[i(\bar{\omega} - \bar{\omega}')]\Gamma[-i(\bar{\omega} - \bar{\omega}'')] - e^{-\pi\bar{\omega}} e^{\frac{\pi}{2}(\bar{\omega}' + \bar{\omega}'')} \times \int_{-\infty}^{\infty} dq_1 dq_2 e^{i\bar{\omega}(q_1 - q_2)} e^{i(\bar{\omega}' q_1 - \bar{\omega}'' q_2)} e^{-ie^{q_1} + ie^{q_2}}, \tag{10.90}$$

which appears in the spectrum operator expression. The correction in the vacuum thermal radiation as received by asymptotic left moving observer can be expressed as

$$2\pi\lambda \sinh \pi\bar{\omega} N_{\bar{\omega}} = \int_0^{\infty}\int_0^{\infty} \frac{d\bar{\omega}'}{\bar{\omega}'}\frac{d\bar{\omega}''}{\bar{\omega}''} \mathrm{Sym}_{\bar{\omega}}\left[\frac{\Gamma[i(\bar{\omega} - \bar{\omega}')]\Gamma[-i(\bar{\omega} - \bar{\omega}'')]}{\Gamma[-i\bar{\omega}']\Gamma[i\bar{\omega}'']}\right] \times \left|y_i^+\right|^{i(\bar{\omega}' - \bar{\omega}'')} f(\bar{\omega}') f^*(\bar{\omega}''), \tag{10.91}$$

where $\mathrm{Sym}_x[f(x)] = (f(x) + f(-x))/2$ with $\bar{\omega'} = \omega'/\lambda$. Using Eq. (10.53), the spectral distortion can be re-written as,

$$2\pi\lambda \sinh \pi\bar{\omega} N_{\bar{\omega}} = \int_0^\infty d\bar{\omega}' d\bar{\omega}'' \int_{-\infty}^\infty dq_1 dq_2 \mathrm{Sym}_{\bar{\omega}}[e^{-\pi\bar{\omega}} e^{i\bar{\omega}(q_1-q_2)}] \times e^{i(\bar{\omega}' q_1 - \bar{\omega}'' q_2)} e^{-ie^{q_1}+ie^{q_2}} g(\omega') g^*(\omega''). \quad (10.92)$$

Using, $g(\bar{\omega})$, introduce yet another function

$$\chi(q) = e^{-ie^q} \int_0^\infty d\bar{\omega} e^{-i\bar{\omega}q} g(\bar{\omega}), \quad (10.93)$$

to express the spectral distortion as

$$2\pi\lambda \sinh \pi\bar{\omega} N_{\bar{\omega}} = \int_{-\infty}^\infty dq_1 dq_2 \mathrm{Sym}_{\bar{\omega}}[e^{-\pi\bar{\omega}} e^{i\bar{\omega}(q_1-q_2)}] \chi(q_1)\chi^*(q_2). \quad (10.94)$$

$$4\pi\lambda \sinh \pi\bar{\omega} N_{\bar{\omega}} = \int_{-\infty}^\infty dq_1 dq_2 [e^{-\pi\bar{\omega}} e^{i\bar{\omega}(q_1-q_2)} + e^{\pi\bar{\omega}} e^{-i\bar{\omega}(q_1-q_2)}] \chi(q_1)\chi^*(q_2), \quad (10.95)$$

which simply gives

$$4\pi\lambda \sinh \pi\bar{\omega} N_{\bar{\omega}} = e^{\pi\bar{\omega}} |\mathcal{F}_\chi(\bar{\omega})|^2 + e^{-\pi\bar{\omega}} |\mathcal{F}_\chi(-\bar{\omega})|^2, \quad (10.96)$$

with $\mathcal{F}_\chi(\bar{\omega})$ being the Fourier transform of $\chi(q)$ w.r.t. $\bar{\omega}$

$$\mathcal{F}_\chi(\bar{\omega}) = \int_{-\infty}^\infty dq e^{-i\bar{\omega}q} \chi(q), \quad (10.97)$$

which gives Eq. (10.96) as the analogue of the Eq. (10.63) for the spherical symmetric collapse. Therefore, we can follow the same steps as outlined in Sect. 10.3 and Sect. 10.4 to recover informations regarding $\mathcal{F}_\chi(\bar{\omega})$. Using the inverse transformations Eqs. (10.97), (10.93) and (10.53) we can recover the information regarding the field state $f(\bar{\omega})$ using the moments of $\mathcal{F}_\chi(\bar{\omega})$.

References

1. S. Hawking, Particle creation by Black Holes. Commun. Math. Phys. **43**, 199–220 (1975)
2. J. Bekenstein, Black holes and the second law. Lett. Nuovo Cimento Soc. Ital. Fis. **4**, 737–740 (1972)
3. W.G. Unruh, R.M. Wald, What happens when an accelerating observer detects a Rindler particle. Phys. Rev. D **29**, 1047–1056 (1984)
4. S.D. Mathur, The Information paradox: a pedagogical introduction. Class. Quant. Grav. **26**, 224001 (2009). arXiv:0909.1038 [hep-th]
5. R.M. Wald, Stimulated emission effects in particle creation near Black Holes. Phys. Rev. D **13**, 3176–3182 (1976)

6. P. Panangaden, R.M. Wald, Probability distribution for radiation from a Black Hole in the presence of incoming radiation. Phys. Rev. D **16**, 929–932 (1977)
7. R. Sorkin, A simple derivation of stimulated emission by Black Holes. Class. Quant. Grav. **4**, L149 (1987)
8. J. Audretsch, R. Muller, Amplification of the Black Hole Hawking radiation by stimulated emission. Phys. Rev. D **45**, 513–519 (1992)
9. D.N. Page, Information in black hole radiation. Phys. Rev. Lett. **71** 3743–3746 (1993). arXiv:hep-th/9306083 [hep-th]
10. M. Schiffer, Is it possible to recover information from the black hole radiation? Phys. Rev. D **48**, 1652–1658 (1993). arXiv:hep-th/9303011 [hep-th]
11. R. Muller, C.O. Lousto, Recovery of information from black hole radiation by considering stimulated emission. Phys. Rev. D **49**, 1922–1928 (1994). arXiv:gr-qc/9307001 [gr-qc]
12. C. Adami, G.L. Ver Steeg, Classical information transmission capacity of quantum black holes. Class. Quant. Grav. **31**, 075015 (2014). arXiv:gr-qc/0407090 [gr-qc]
13. A. Chatwin-Davies, A.S. Jermyn, S.M. Carroll, *How to Recover a Qubit That Has Fallen Into a Black Hole*. arXiv:1507.03592 [hep-th]
14. M. Visser, Thermality of the Hawking flux. JHEP **07**, 009 (2015). arXiv:1409.7754 [gr-qc]
15. F. Gray, S. Schuster, A. Van-Brunt, M. Visser, The Hawking cascade from a black hole is extremely sparse. arXiv:1506.03975 [gr-qc]
16. T. Padmanabhan, Thermodynamical aspects of gravity: new insights. Rept. Prog. Phys. **73**, 046901 (2010). arXiv:0911.5004 [gr-qc]
17. L.E. Parker, D. Toms, *Quantum Field Theory in Curved Spacetime*. Cambridge Monographs on Mathematical Physics (Cambridge University Press, 2009), http://www.cambridge.org/de/knowledge/isbn/item2327457
18. E.T. Akhmedov, H. Godazgar, F.K. Popov, Hawking radiation and secularly growing loop corrections. Phys. Rev. D **93**(2), 024029 (2016). arXiv:1508.07500 [hep-th]
19. K. Lochan, T. Padmanabhan, Extracting information about the initial state from the black hole radiation. Phys. Rev. Lett. **116**(5), 051301 (2016). arXiv:1507.06402 [gr-qc]
20. K. Lochan, T. Padmanabhan, Inertial nonvacuum states viewed from the Rindler frame. Phys. Rev. D **91**(4), 044002 (2015). arXiv:1411.7019 [gr-qc]
21. N. Birrell, P. Davies, *Quantum Fields in Curved Space* (Cambridge Monogr. Math, Phys, 1982)
22. C.G. Callan, Jr., S.B. Giddings, J.A. Harvey, A. Strominger, Evanescent black holes. Phys. Rev. D **45**, 1005–1009 (1992). arXiv:hep-th/9111056 [hep-th]
23. A. Fabbri, J. Navarro-Salas, *Modeling Black Hole Evaporation* (2005)
24. B. Birnir, S.B. Giddings, J.A. Harvey, A. Strominger, Quantum black holes. Phys. Rev. D **46**, 638–644 (1992). arXiv:hep-th/9203042 [hep-th]
25. S.B. Giddings, W.M. Nelson, Quantum emission from two-dimensional black holes. Phys. Rev. D **46**, 2486–2496 (1992). arXiv:hep-th/9204072 [hep-th]
26. M. Alves, Dilaton gravity with a nonminimally coupled scalar field. arXiv:gr-qc/9510021 [gr-qc]
27. M. Alves, Hawking radiation in the dilaton gravity with a nonminimally coupled scalar field. Int. J. Mod. Phys. D **8**, 687–694 (1999). arXiv:hep-th/9909138 [hep-th]
28. A.R. Mikovic, V. Radovanovic, Two loop back reaction in 2-D dilaton gravity. Nucl. Phys. B **481**, 719–742 (1996). arXiv:hep-th/9606098 [hep-th]
29. K. Lochan, S. Chakraborty, T. Padmanabhan, Dynamic realization of the Unruh effect for a geodesic observer. arXiv:1603.01964 [gr-qc]
30. C. Vaz, L. Witten, Quantum naked singularities in 2-D dilaton gravity. Nucl. Phys. B **487**, 409–441 (1997). arXiv:hep-th/9604064 [hep-th]
31. A. Ashtekar, F. Pretorius, F.M. Ramazanoglu, Surprises in the evaporation of 2-Dimensional Black Holes. Phys. Rev. Lett. **106**, 161303 (2011). arXiv:1011.6442 [gr-qc]
32. S.W. Hawking, Evaporation of two-dimensional black holes. Phys. Rev. Lett. **69**, 406–409 (1992). arXiv:hep-th/9203052 [hep-th]
33. C.R. Stephens, G. 't Hooft, B.F. Whiting, Black hole evaporation without information loss. Class. Quant. Grav. **11**, 621–648 (1994). arXiv:gr-qc/9310006 [gr-qc]

34. A. Ashtekar, M. Bojowald, Black hole evaporation: a Paradigm. Class. Quant. Grav. **22**, 3349–3362 (2005). arXiv:gr-qc/0504029 [gr-qc]
35. H.M. Haggard, C. Rovelli, Quantum-gravity effects outside the horizon spark black to white hole tunneling. Phys. Rev. D **92**(10), 104020 (2015). arXiv:1407.0989 [gr-qc]

Chapter 11
Dynamic Realization of the Unruh Effect for a Geodesic Observer

11.1 Introduction

Hawking radiation from a black hole [1–10] and the Unruh radiation [3, 6–8, 11, 12] in the Rindler frame have very similar mathematical properties. In the context of an eternal black hole, the Hartle–Hawking vacuum state of a quantum field leads to a thermal density matrix for static observers in the right wedge. This arises because the modes of the quantum field in the region inaccessible to the observer are traced out. Similarly, a uniformly accelerated observer in the right wedge of the flat spacetime will describe her observations using a thermal density matrix, obtained by tracing out the modes inaccessible to her (on the left wedge) when the field is in the global, inertial, vacuum state. Both situations are time-reversal invariant; while the relevant observers see an ambient thermal radiation, they do not associate a flux of particles with this radiation.

A somewhat different situation arises in the case of a black hole formed by collapsing matter, with the quantum state being the Unruh vacuum (also known as the "In-vacuum") at very early times. In this case, at late times, observers far away from the collapsing body detect a flux of particles with a spectrum which is thermally populated. The energy carried away by the particles is ultimately obtained from the mass of the collapsing body and this leads to the concept of black hole evaporation. The mathematical description of this process takes into account; (i) the change in geometry due to the collapse process and (ii) the formation of event horizon leading to inaccessibility of a region from future asymptotic observers. With future applications in mind, let us briefly recall the key concepts.

In standard $(3 + 1)$ dimensional collapse, in which classical matter collapses to form a black hole, an apparent horizon is formed, which grows and—at a certain stage—an event horizon is formed thereby producing a black hole region. The last null ray originating from the past null infinity $\mathcal{J}^-$ at the null co-ordinate $u = d$, in the double null co-ordinate system [1, 3, 6–8], reaching future null infinity $\mathcal{J}^+$ defines the location of the event horizon. An asymptotic observer on $\mathcal{J}^+$ has no causal connection with events inside the event horizon. Further, whole of $\mathcal{J}^+$ derives its

S. Chakraborty, *Classical and Quantum Aspects of Gravity in Relation to the Emergent Paradigm*, Springer Theses, DOI 10.1007/978-3-319-63733-4_11

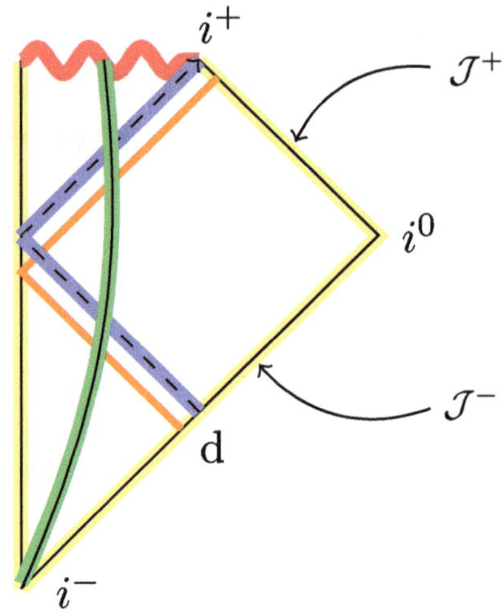

Fig. 11.1 Penrose diagram for Schwarzschild collapse

complete causal support (e.g., the thick orange line in Fig. 11.1) from only a *part of* $\mathcal{J}^-$, which lies prior to the ray forming event horizon. A wave mode which originates from $\mathcal{J}^-$ and reaches $\mathcal{J}^+$ also experiences a change in background geometry in the process. In order to obtain the Bogoliubov coefficients between the asymptotic observers on $\mathcal{J}^-$ and $\mathcal{J}^+$, we need to re-express a set of complete mode functions v_ω suited to observers on $\mathcal{J}^+$ in terms of complete set of mode functions u_ω, defined equivalently on $\mathcal{J}^-$. When we trace back the out-going modes v_ω on $\mathcal{J}^+$ through the center, (i.e., through $r = 0$ with r being the Schwarzschild radial co-ordinate) onto $\mathcal{J}^-$, we see that the mode functions suited to $\mathcal{J}^+$ have support only on the portion of mode functions u_ω on $\mathcal{J}^-$.

Therefore the Bogoliubov coefficients are obtained by taking inner products of u_ω and v_ω on a portion of $\mathcal{J}^-$, i.e., below the line $u = d$ [1, 3, 7, 8]. (This has an effect similar to that of partial tracing out of modes in the case of eternal black hole.) Furthermore, we now also have to take into account the change of geometry experienced by the mode function u_ω when we trace it back to the past null infinity. However, if we are interested only in the late time behavior of u_ω, we can use ray-optics approximation [1, 3, 7, 8] for tracing back the mode functions. *Thus, the late time behavior of the out going asymptotic future modes is essentially controlled by the tracing over of portion of* $\mathcal{J}^-$*, rather than the change in the geometry.* This is the mathematical reason why the dynamics of the collapse is irrelevant to obtain the temperature and one obtains the same result as in the case of an eternal black hole and Hartle–Hawking vacuum state. The *exact* Bogoliubov coefficients, obtained by going beyond the ray optics, will, of course, be sensitive to the geometry change as well. Further, it is the loss of time reversal invariance during the collapse which leads to a non-zero flux of energy, which is absent in the case of an eternal black hole and Hartle–Hawking vacuum state.

It turns out that a combination of these effects arise in the case CGHS black hole in (1 + 1) dimension [5, 13]. The collapse of quantum matter leads to the standard black hole evaporation scenario with observers at future right asymptotic $\mathcal{J}_R^+(t \to \infty, x \to +\infty)$ detecting a flux of thermal radiation, which is well-known in the literature. This situation is mathematically identical to what happens in the case (3 + 1) spherically symmetric collapse. But it turns out that there is another effect in the same spacetime which is very similar to Unruh effect: Observers at future left asymptotic $\mathcal{J}_L^+(t \to \infty, x \to -\infty)$ detect a thermal spectrum (at the same temperature as seen by observers at $\mathcal{J}_R^+$) but without any associated flux! This result is surprising because these are geodesic observers in a flat region of the spacetime who see no change in geometry. The effect arises due to the necessity of tracing over part of the modes for describing the physics of the observers at $\mathcal{J}_L^+$ and the mathematics closely parallels standard Unruh effect. But in this case, the necessity to trace over modes arises due to the dynamics of collapse, which takes place in a different region of the spacetime though. This effect (which, as far as we know, has been missed in the CGHS literature) constitutes a dynamic realization of Unruh effect.

11.2 A Picture Book Representation

In order to explain the concepts involved in this, rather peculiar result, we will first provide a picture-book description of how the result arises, which should demystify it.

We start with the Minkowski spacetime, for which the Penrose diagram corresponds to Fig. 11.2. The full spacetime is bounded by four null lines depicting future and past left or right null infinities, viz $\mathcal{J}_R^+$, $\mathcal{J}_R^-$, $\mathcal{J}_L^+$ and $\mathcal{J}_L^-$. In this spacetime any inertial (geodesic) observer will start from past timelike infinity i^- at $t = -\infty$ and would reach future timelike infinity i^+ at $t = \infty$. Two such geodesic observers moving leftwards and rightwards respectively, are shown by dashed curves in Fig. 11.2. In addition, there can also be some accelerated observers. An important set of such accelerated observers is the eternally accelerating Rindler observer. The trajectory of the Rindler observer starts on $\mathcal{J}_L^-$ and accelerates along a hyperbolic path to reach $\mathcal{J}_L^+$ (shown in the thick green curve in Fig. 11.2). Let us consider the causal support of the trajectories of the Rindler observer vis-á-vis the inertial observer. The inertial observer has causal access to the full spacetime, whereas the spacetime region accessed by the Rindler observer is only a part of the full Minkowski spacetime. Thus the vacuum state for the inertial observer (who can access the full spacetime) would be different from that of the Rindler observer (for whom only a part of the spacetime is allowed) leading to non zero Bogoliubov coefficients. Alternatively, the observations carried out by the Rindler observer can be described using a density matrix obtained by tracing out modes inaccessible to her. If the field is in the inertial vacuum state, the resulting density matrix will be thermal. This property will hold whenever an observer has a causal access of only a part of the spacetime and the reduced density matrix she uses will be thermal. As an illustration let us con-

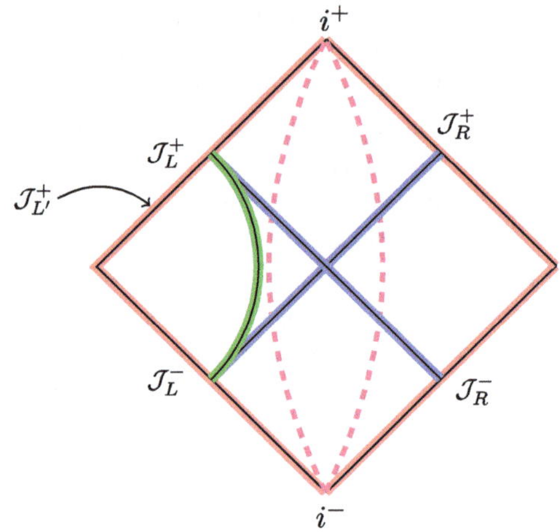

Fig. 11.2 Rindler trajectory in Minkowski spacetime

sider Fig. 11.3, where three observers are shown. The dashed observer is an inertial observer and, as in the earlier figure, has access to the full spacetime; the dotted trajectory is that of the standard, eternally accelerating Rindler observer; the third trajectory (represented by thick magenta line) represents an observer who originally started as inertial but then changed her mind and accelerates uniformly to end up on $\mathscr{J}_L^+$ just like the Rindler observer. This observer also has causal connection only to the past domain of dependence of $\mathscr{J}_{L'}^+$, which being only a portion of the full manifold, will lead to thermal density matrix. Let us now consider a hypothetical scenario in which some portion of Minkowski spacetime becomes dynamically inaccessible, as shown in Fig. 11.4 by the dashed triangular region. (Right now this is the Penrose diagram of some hypothetical spacetime; we will soon see how it actually arises in the CGHS case.) Further, we assume this truncation of spacetime is such that it requires left-moving geodesic observers to terminate on i_L^+ instead of i^+ as in Fig. 11.3. Since these observers derive their causal support only from the past of $\mathscr{J}_{L'}^+$, they are compelled to trace over a portion of field configuration on $\mathscr{J}_R^-$ and hence they will end up using a density matrix which is thermal. Whenever *any* observer, *geodesic or not*, who derives her causal support from a subset of the full spacetime, the global vacuum state will appear to be a thermal state.

The geodesic part of the above statement might appear perplexing. One may wonder how a geodesic observer can ever experience such a nontrivial effect usually associated with accelerated observers. The role of acceleration is only to make part of the spacetime region inaccessible; if we can achieve this by some other means, we will still have the same result. The situation as depicted in Fig. 11.4 appears unphysical as presented—because we have artificially removed part of the spacetime, but we will later discuss a situation in which a portion of spacetime is indeed dynamically denied to a *geodesic observer*, very much like in the spirit of the observer in Fig. 11.4 (shown in thick magenta curve).

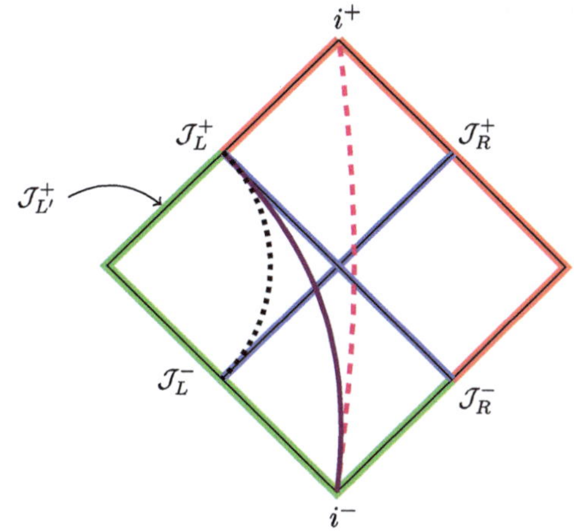

Fig. 11.3 Non-geodesic observer in Minkowski spacetime

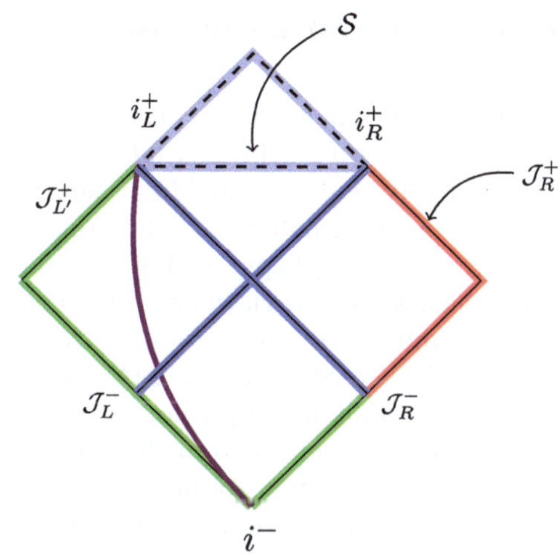

Fig. 11.4 Observers in a hypothetical spacetime

But before we do that, we will consider another example in the next section which might make all these less surprising. This will involve the collapse of a null shell forming a black hole (see Fig. 11.5). In this case, a timelike geodesic observer at $r = 0$, will remain entirely in the flat spacetime until being eaten up by the singularity and receives causal signals only from a part of $\mathcal{J}^-$ (see Fig. 11.5 below). This study will provide us useful insights towards the constructs to be used in the later sections of the chapter.

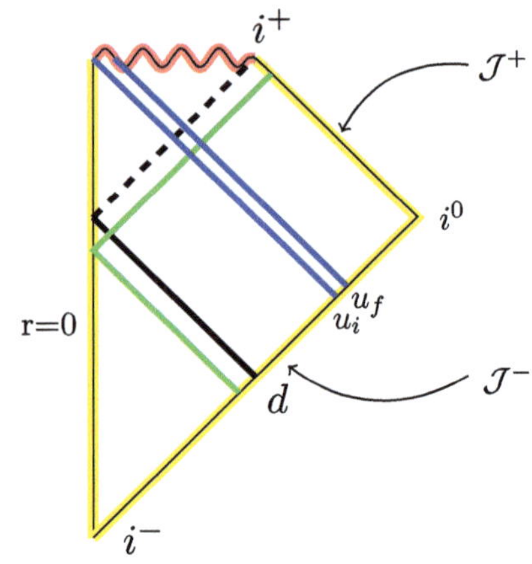

Fig. 11.5 Penrose diagram for Schwarzschild null shell collapse

11.3 A Null Shell Collapse

For a closer look at the above mentioned features, we will first consider a null shell collapsing to form a Schwarzschild hole [3, 7, 14–16] (see Fig. 11.5). In this case, the singularity gets originated from the co-ordinate $u = u_i$, the co-ordinate point of introduction of the null-shell. (This collapse model will be relevant for comparison with the null shell collapse in $(1+1)$ dimensional dilatonic gravity, forming a black hole, which is discussed later on.) The singularity starts forming from the co-ordinate $u = u_i$, while the event horizon is located at $r = 2M$, with M being the mass of the shell. The co-ordinate $r = 2M$ can be reflected back through $r = 0$ on to $\mathcal{J}^-$ at $u = d$. As discussed previously, the out-going modes v_ω on $\mathcal{J}^+$ derive their causal support completely from the portion $u < d$ on $\mathcal{J}^-$. Therefore the Bogoliubov coefficients of mode transformation, should be evaluated through the portion of mode functions u_ω in the regime $u < d$. Further, all the null rays emanating from $\mathcal{J}^-$ region $u < u_i$ also experience a change of geometry, i.e., they start moving in flat spacetime inwards, get reflected at the origin ($r = 0$) and then encounter the collapsing shell to feel the geometry changed into a Schwarzschild one.

This can be more clearly seen in an isotropic co-ordinate system (see Fig. 11.6) which uses a Cartesian x coordinate with $-\infty < x < \infty$ instead of the usual radial coordinate r with $0 < r < \infty$. The relevant Penrose diagram is shown in Fig. 11.6. In these coordinates, (see Fig. 11.6) there will be two past null infinities, namely left ($\mathcal{J}_L^-$) and right ($\mathcal{J}_R^-$), as well as two future null infinities $\mathcal{J}_L^+$ and $\mathcal{J}_R^+$. The location of event horizon in these co-ordinates is marked to (say) Δ_M, which corresponds to $r = 2M$ in Schwarzschild co-ordinates. It is usual to define the modes as left moving or right moving in these co-ordinates. A right moving mode originates from $\mathcal{J}_L^-$ and ends up on $\mathcal{J}_R^+$ and vice versa for the left-moving modes. Due to spherical symmetry, consideration of any set of past and future asymptotic observer pairs will be equivalent to any other. A time-like observer (thick green curves in Fig. 11.6), similarly starting from past time like infinity and escaping the black hole region,

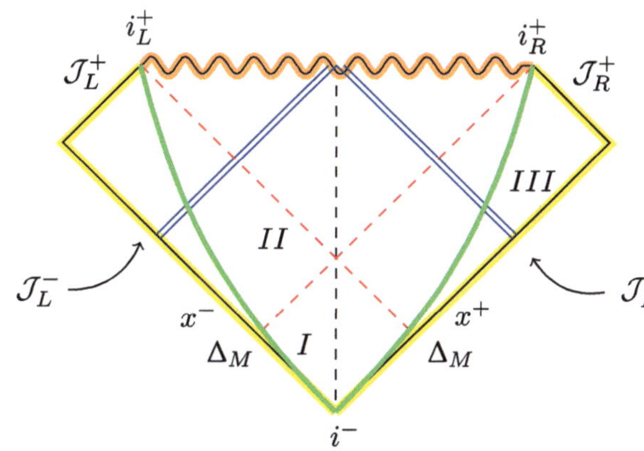

Fig. 11.6 Schwarzschild in isotropic co-ordinates

either end up on i_L^+ or on i_R^+ depending upon whether the observer moves leftwards or rightwards.

Let us consider a set of modes moving rightwards. Any null ray originating from $\mathcal{J}_L^-$ and ending up on $\mathcal{J}_R^+$ will experience a change of geometry after it crosses the collapsing null shell. For observers on $\mathcal{J}_R^+$, the isotropic co-ordinate $|x^-| = \Delta_M$ marks the location of event horizon. Therefore, the modes reaching $\mathcal{J}_R^+$ derive their causal support from the region I in Fig. 11.6. No event in the region II is connected to $\mathcal{J}_R^+$. An exactly similar picture is there for the left-moving modes.

In the standard black hole analysis, the out-going mode functions on $\mathcal{J}_R^+$ are given as

$$v_\omega = \frac{1}{\sqrt{2\omega}} e^{-i\omega u}; \qquad u \in (-\infty, \infty). \tag{11.1}$$

These mode functions provide an orthonormal basis across complete $\mathcal{J}_R^+$. Similarly the right-moving modes on $\mathcal{J}_L^-$ are spanned by

$$u_\omega = \frac{1}{\sqrt{2\omega}} e^{-i\omega v}; \qquad v \in (-\infty, \infty). \tag{11.2}$$

The Bogoliubov transformation coefficients between these modes can be evaluated by taking the covariant inner products on $\mathcal{J}_L^-$. However, in order to do that we need to express mode functions v_ω on $\mathcal{J}_L^-$ in terms of u_ω. For that purpose we need to track the out-going modes at $\mathcal{J}_R^+$ all the way down to $\mathcal{J}_L^-$. In principle, this a formidable job, since the exact form of modes in the whole spacetime is complicated at best, if obtainable in closed form. However, appealing to ray-optics approximation [1, 7] for modes very close to the horizon, we in a sense, avoid this issue. This approximation gives us the expression of modes v_ω close to $|x^-| = \Delta_M$, i.e., $u = d$ in terms of u_ω. We obtain the Bogoliubov coefficients readily as

$$\alpha_{\omega\omega'} = -2i \int_{-\infty}^{0} dx^- u_\omega \partial_- u^*_{\omega'},$$
$$\beta_{\omega\omega'} = 2i \int_{-\infty}^{0} dx^- u_\omega \partial_- u_{\omega'}. \quad (11.3)$$

However, as we discussed, this integration has to be truncated to within the region $u < d$, which gives it a Rindler kind of appearance, making it [1, 3, 7]

$$\alpha_{\Omega\omega} = \frac{1}{2\pi\kappa}\sqrt{\frac{\Omega}{\omega}} \exp\left[\frac{\pi\Omega}{2\kappa}\right] \exp\left[i(\Omega-\omega)d\right] \exp\left[\frac{i\Omega}{\kappa}\log\frac{\omega}{C}\right] \Gamma\left[-\frac{i\Omega}{\kappa}\right],$$
$$\beta_{\Omega\omega} = -\frac{1}{2\pi\kappa}\sqrt{\frac{\Omega}{\omega}} \exp\left[-\frac{\pi\Omega}{2\kappa}\right] \exp\left[i(\Omega+\omega)d\right] \exp\left[\frac{i\Omega}{\kappa}\log\frac{\omega}{C}\right] \Gamma\left[-\frac{i\Omega}{\kappa}\right]. \quad (11.4)$$

where $C = C_1C_2$ is a product of affine parameters for in-going (C_1) and out-going (C_2) null geodesics. From the rather general nature of the analysis we expect this result to give the Bogoliubov coefficients for the timelike geodesic observer stationary at $r = 0$ if we replace d by u_i. Such an observer is engulfed by a *true* singularity when the shell collapse to $r = 0$, it results in geodesic incompleteness, something we will come back to, in the last section.

In the limit $M \to 0$, the portion being traced over becomes small. Further the geometry change as well as the formation of the singularity does not occur, making $\mathcal{J}^+$ a Cauchy surface, which is not the case when $M \neq 0$. Since the effect of tracing over is intimately tied with the effect of geometry change, it is difficult to account for these effects individually in a general case in $(3+1)$ dimension. Nevertheless, one can argue that the tracing over of modes alone leads to Unruh effect with zero flux, whereas the geometry change makes the thermal radiation more 'real' with a non-zero flux [17–20]. The non vanishing of flux can be associated with moving of the geometry away from being a flat one. We will see below that these two effects can nicely be segregated in a $(1+1)$ dimensional collapse model in dilaton gravity.

11.4 Dynamical Unruh Effect in CGHS Black Hole Spacetime

The CGHS black hole [5, 13] is a $(1+1)$ dimensional gravitational collapse model of a dilatonic field ϕ interacting with gravity in the presence of cosmological constant λ and matter fields f_i. A basic introduction about the model has already been given in Sect. 10.5 of Chap. 10 and shall not be repeated here.

The CGHS model is not symmetric under left-right exchange, and hence the experiences of null ray originating from $\mathcal{J}_R^-$ or a timelike trajectory moving leftwards, will be different from what we discussed above. Such trajectories do not suffer any change in geometry in their course, hence, as we will see, the only effect a late time

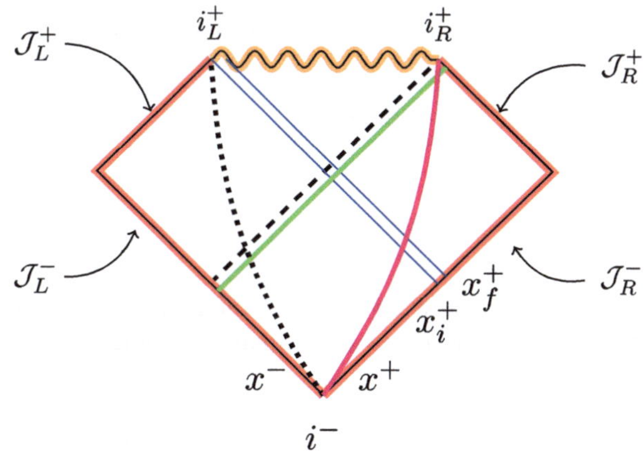

Fig. 11.7 Left moving modes in a CGHS *black* hole

observer (the dotted curve in Fig. 11.7) on $\mathscr{J}_L^+$ finds, is rooted only in the tracing over a part of Cauchy surface $\mathscr{J}_R^-$. Therefore, these observers will also obtain a thermal expectation value, but there will be no associated flux for these thermal spectrum. The spacetime metric in the region $x^+ < x_i^+$ is flat and given by Eq. (10.31) with $M, P^+ \to 0$, which on using the co-ordinates

$$x^+ = -\frac{1}{\lambda y^+}; \qquad x^- = -\frac{1}{\lambda y^-}, \tag{11.5}$$

can be written as

$$ds^2 = -\frac{dy^+ dy^-}{-\lambda^2 y^+ y^-}. \tag{11.6}$$

The singularity curve originates from the co-ordinate of the point of matter introduction, i.e., from x_i^+ which is marked through Eq. (11.5) as $y_i^+ = -1/\lambda x_i^+$. Under another set of co-ordinate transformations, the metric on $\mathscr{J}_R^-$ can also be brought into flat form. On $\mathscr{J}_R^-$ we adopt

$$\begin{aligned} e^{-\lambda\chi^+} &= -y^+, \\ e^{\lambda\chi^-} &= y^-, \end{aligned} \tag{11.7}$$

such that the metric becomes

$$ds^2 = -d\chi^+ d\chi^-, \tag{11.8}$$

with $\chi^\pm \in (-\infty, \infty)$. Therefore the complete set of left-moving mode functions corresponding to the field $f_+(\chi^+)$ can be written in these co-ordinates as

$$u_\omega^+(\chi^+) = \frac{1}{\sqrt{2\omega}} e^{-i\omega\chi^+}. \tag{11.9}$$

whereas on $\mathcal{J}_L^+$ we adopt to

$$\begin{aligned} e^{-\lambda\tilde{\chi}^+} &= -(y^+ - y_i^+), \\ e^{\lambda\tilde{\chi}^-} &= y^-, \end{aligned} \tag{11.10}$$

such that these new co-ordinates range in $\tilde{\chi}^\pm \in (-\infty, \infty)$ in the region $x^+ < x_i^+$. The metric in these new co-ordinates becomes

$$ds^2 = -\frac{d\tilde{\chi}^+ d\tilde{\chi}^-}{1 - y_i^+ e^{\lambda\tilde{\chi}^+}}. \tag{11.11}$$

The metric given by Eq. (11.11) is in a conformally flat form in these co-ordinates, hence the complete set of left-moving modes corresponding to $f_+(\tilde{\chi}^+)$ are again given as

$$v_\omega^+(\tilde{\chi}^+) = \frac{1}{\sqrt{2\omega}} e^{-i\omega\tilde{\chi}^+}, \tag{11.12}$$

since a minimally coupled scalar field is conformally invariant in $(1+1)$ dimensions. Clearly, on $\mathcal{J}_R^-$, v_ω^+ has support only in the region $x^+ < x_i^+$. The point x_i^+ is mapped to $\chi^+ \equiv \chi_i^+ = -\frac{1}{\lambda}\log(-y_i^+)$. Therefore, the Bogoliubov transformation coefficients between these two set of observers can be obtained exactly as in Eqs. (10.44), (10.45) but with the replacement $|y_i^+| \leftrightarrow P^+/\lambda$ as

$$\begin{aligned} \alpha_{\Omega\omega} &= -\frac{i}{\pi}\int_{-\infty}^{\chi_i^+} d\chi^+ v_\Omega \partial_- u_\omega^* \\ &= \frac{1}{2\pi}\sqrt{\frac{\omega}{\Omega}}\int_{-\infty}^{\chi_i^+} d\chi^+ \exp\left[\frac{i\Omega}{\lambda}\ln\left\{\left(e^{-\lambda\chi^+} - |y_i^+|\right)\right\} + i\omega\chi^+\right] \\ &= \frac{1}{2\pi\lambda}\sqrt{\frac{\omega}{\Omega}}|y_i^+|^{\frac{i(\Omega-\omega)}{\lambda}} B\left(-\frac{i\Omega}{\lambda} + \frac{i\omega}{\lambda}, 1 + \frac{i\Omega}{\lambda}\right), \end{aligned} \tag{11.13}$$

while

$$\begin{aligned} \beta_{\Omega\omega} &= \frac{i}{\pi}\int_{-\infty}^{\chi_i^+} d\chi^+ v_\Omega \partial_+ u_\omega \\ &= \frac{1}{2\pi}\sqrt{\frac{\omega}{\Omega}}\int_{-\infty}^{\chi_i^+} d\chi^+ \exp\left[\frac{i\Omega}{\lambda}\ln\left\{\left(e^{-\lambda\chi^+} - |y_i^+|\right)\right\} - i\omega\chi^+\right] \\ &= \frac{1}{2\pi\lambda}\sqrt{\frac{\omega}{\Omega}}|y_i^+|^{\frac{i(\Omega+\omega)}{\lambda}} B\left(-\frac{i\Omega}{\lambda} - \frac{i\omega}{\lambda}, 1 + \frac{i\Omega}{\lambda}\right). \end{aligned} \tag{11.14}$$

This form of Bogoliubov coefficients results from the tracing over of a portion of $\mathcal{J}_R^-$, which is causally disconnected from $\mathcal{J}_L^+$. The horizon for $\mathcal{J}_L^+$ is given as $y^+ = y_i^+$, which also marks the point of singularity. More importantly, if one is interested in the late time response, i.e., the observers reaching i_L^+, we need to take the high frequency behavior ($\omega/\lambda \gg 1$) of the Bogoliubov coefficients which are exactly the same as for the Rindler observer [6].

This effect can be more clearly understood using non-availability of Cauchy surfaces for future asymptotic observers. Let us consider a left moving timelike or null trajectory. Due to conformal flatness, a complete set of orthonormal mode functions on any surface orthogonal to them, can always be obtained as plane waves under a proper co-ordinatization (which extends as $(-\infty, \infty)$) of that surface. If we consider an orthogonal null surface for the left moving modes, before the formation of singularity (the red surface labeled '2' in Fig. 11.8), the Bogoliubov transformation coefficients between that surface and $\mathcal{J}_R^-$ will be trivial, i.e., ($\alpha_{\Omega\omega} = \delta(\Omega - \omega)$, $\beta_{\Omega\omega} = 0$). However, once the singularity forms, any null surface relating a point on $\mathcal{J}_L^-$ to the singularity (e.g., green surface labeled '1' in Fig. 11.8), fails to causally connect with the entire spacetime. Also, any timelike observer has a domain of dependence, which is not the full spacetime. A portion of $\mathcal{J}_R^-$ (right to the dashed green line) is causally denied to the past of such surfaces. Therefore, the Bogoliubov coefficients assume a non-trivial form, such as in Eqs. (10.44), (10.45) or Eqs. (11.13), (11.14), with a phase factor capturing the information about the region of the traced over modes. This is dynamically similar to a Rindler trajectory in a Minkowski spacetime, see Fig. 11.2, where an accelerated observer (thick green curve) reaches the future null infinity rather than future timelike infinity. For such observers too, the domain of dependence is only a part of the full spacetime. There is a horizon masking a portion of spacetime and hence the Bogoliubov transformation coefficients between $\mathcal{J}_R^-$ and $\mathcal{J}_{L'}^+$ (rather than the full $\mathcal{J}_L^+$) is non-trivial [3, 6, 8] and the accelerated observers obtain a thermal spectrum for a vacuum state defined on $\mathcal{J}_R^-$. However, there is no flux associated with this spectrum as the stress tensor remains identically vanishing in the flat spacetime.

Fig. 11.8 Unavailability of cauchy surface for *left*-moving modes

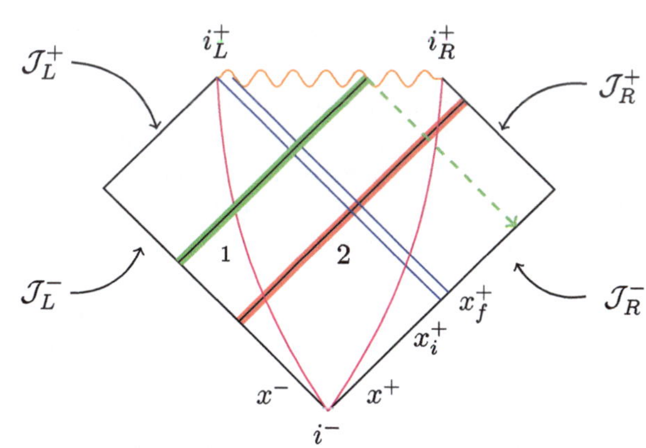

Similarly in the CGHS model, the region of spacetime $y^+ > y_i^+$ is dynamically made inaccessible to any left-moving time-like or null trajectories, which also is the case with such trajectories in the null collapse forming a Schwarzschild black hole (see for example, Fig. 11.6). However, in contrast to the Schwarzschild case, such observers in the CGHS model, do not see any change of geometry hence they do not associate any mass to the "black hole region" as seen by them. The spacetime, they move in, throughout, is flat and such observers do not receive any flux of radiation, as the vacuum expectation value of the stress energy tensor which was vanishing on $\mathcal{J}_R^-$, stays put on zero, in the region $y^+ < y_i^+$. As we discussed previously, the right moving observers tuned to right moving modes, witness a thermal radiation flux at a temperature $1/\lambda$ which is independent of the matter content of the forming black hole and depends only on the other dimensionful parameter in the theory.

Similarly, the observers at $\mathcal{J}_L^+$ coupled to left-moving modes, observe a similar kind of Bogoliubov coefficients as their right-moving counterparts Eqs. (10.44), (10.45) but with the parameter exchange $|y_i^+| \leftrightarrow P^+/\lambda$ which mark the corresponding event horizons for such observers and appear as overall phases in the transformation coefficients. Hence, we see that such form of Bogoliubov coefficients Eqs. (11.13), (11.14) are entirely due to tracing over of modes which lie in the causally denied region of spacetime, to $\mathcal{I}_L^+$ and not due to any geometry change. One can check that if the fraction of tracing over vanishes, which is marked by $|y_i^+| \to 0$, the Bogoliubov coefficients assume a trivial form, i.e., $\alpha_{\Omega\omega} \to \delta(\Omega - \omega)$ while $\beta_{\Omega\omega} \to 0$. Similarly, the Bogoliubov coefficients for the right-moving observers Eqs. (10.44), (10.45), assume a trivial form in the limit $P^+ \to 0$, marking the vanishing of event horizon, as well as, the amount of change of geometry suffered by such observers. Hence, for such observers effects of tracing over is indistinguishable from geometry change. Both these effects vanish simultaneously in that limit. Therefore, we see that the vacuum response for both left-moving or right-moving observers is indistinguishable from each other. Late time radiation for such observers on $\mathcal{J}_L^+$ or on $\mathcal{J}_R^+$ for vacuum state (of a test field) is thermal. However, unlike the left-moving observers, the right moving observers associate a flux also with the radiation and hence the black hole region shrinks as a result of the evaporation. While left-moving observers do not associate any mass to the region beyond their horizon, the location of their horizon does not shrink and the "black hole region" does not evaporate for them.

All the geodesic observers moving rightwards or remaining stationary at any finite value (with the exception of stationary observer at left infinity) end up in singularity and in this process have to undergo a geometry change (thick cyan curves in Fig. 11.9), so they witness a combined effect and hence a flux of radiation, as in the Schwarzschild scenario [15, 16, 21]. Whereas the left-moving timelike observers who end up at i_L^+, do not see any geometry change but encounter flux-free thermal radiation due to a dynamical emergence of a horizon, which traces out information of a section of field configuration, exactly in the spirit of the Rindler observer, as discussed previously.

Finally, we will comment on the back-reaction due to black hole evaporation and its implications for our result. To compute any kind of back-reaction in the spacetime

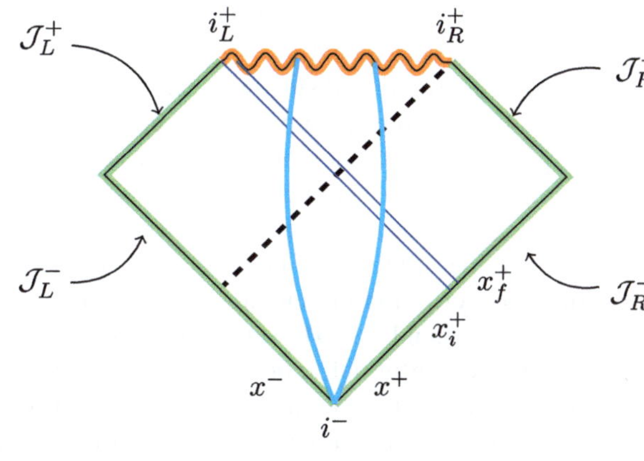

Fig. 11.9 Freely falling observers in CGHS *black* hole spacetime

in Einstein gravity, one needs to use an equation of the form $G_{ab} = \langle T_{ab} \rangle$. *Neither side of this equation can be handled without additional assumptions and, obviously, the result will depend on the additional assumptions!* The left-hand-side identically vanishes in $D = 2$ because Einstein tensor identically vanishes in two dimensions. The right hand side is: (i) divergent and needs to be regularized and (ii) depends on the chosen quantum state. The best we can do, therefore, is to give prescriptions to define both sides and the results will obviously *depend on these prescriptions*. We discuss our set of choices while describing the pertinent issues in two-dimensional gravity below.

To get a non-zero left hand side, one postulates some non-Einsteinian form of the gravity action (like the CGHS action in Eq. (10.27) and vary the metric to get the equations of motion. Obviously, this is not G_{ab} (which, of course, is zero) but can possibly act as a proxy for the same. We stress the fact that we are trying to model some gravitational features by some suitable dilatonic action and one cannot ignore the implicit ad-hocness in the procedure. For example, in such an approach, one also often takes the normal ordered, non-covariant, expectation value of the stress energy tensor as the classical source. However, in a two-dimensional spacetime, the expectation value of the stress energy tensor comes with a conformal anomaly term [5] as well. The anomaly term can then act as an extra source of stress energy and may lead to an evolution different from the classical case, if not accounted for properly. A prescription used in the literature is to add some extra terms (e.g., Polyakov action, RST action terms) to the standard CGHS action corresponding to the conformal anomaly terms and obtain these *modified* equations of motion, as a result of the variational principle. But this new action will correspond to a different physical system, though one which can also be thought of as a proxy for gravity. But in this prescription, flat spacetime will not be a solution to the vacuum state for the semiclassical equations due to the anomaly, which we consider somewhat unphysical. There is no unique way of handling this issue because, as we stressed before, $G_{ab} = 0$ in $D = 2$ and the left hand side which we work with to mimic $G_{ab} = \langle T_{ab} \rangle$ depends crucially on the model we use, with the hope that it can mimic aspects of $D = 4$ gravity. We have chosen simple, physically well-motivated choices to do this as we describe next.

To analyze the right hand side, we need a scheme for defining a $c-$number stress tensor from the quantum operator and the vacuum state has to be motivated from some specific (geometric) considerations. We have *defined* the vacuum state by a natural assumption: viz. that the geometry must remain flat when sourced by such a (vacuum) state. This is important because as we discussed, in two-dimensional spacetime, the expectation value of the stress energy tensor has a conformal anomaly term which can violate this criterion in general, if this term is not accounted for in the source terms in the right hand side. However, there exists a family of such states [22] which take care of the conformal anomaly terms and ensure that the geometry remains flat when the state is a vacuum from this family. Therefore the results discussed in this chapter, use stability of the solution under these class of states. A more detailed analysis of conformal anomaly modified evaporating black holes can be found in [5]. Our scheme to deal with such terms will be reported in details elsewhere [23]. At last, we discuss the response of an Unruh-DeWitt detector carried by such a geodesic observer.

11.5 Unruh-DeWitt Detector Response

To confirm the fact operationally, that an observer who entirely stays in the left block of spacetime and has access to only a portion of full spacetime, truly finds the vacuum state of a field defined/supported in the full spacetime as an excited state on her Hilbert space defined in the portion of spacetime, we turn to analysis of response of an Unruh-DeWitt detector carried by a left-moving inertial observer.

Such an observer describes the spacetime metric as in Eq. (11.8). We construct a co-ordinate system $\{x \equiv (T, X)\}$

$$\begin{aligned} \chi^+ &= T + X \\ \chi^- &= T - X \end{aligned} \tag{11.15}$$

to write Eq. (11.8) as

$$ds^2 = -dT^2 + dX^2 \tag{11.16}$$

such that the Wightman function, corresponding to the vacuum defined in full spacetime, in $(1 + 1)$ dimension assumes the form

$$\mathcal{W}(x, x') = -\frac{1}{2} \log |x - x'|^2 = -\frac{1}{2} \log \Delta\chi^+ \Delta\chi^-. \tag{11.17}$$

A left moving inertial observer is given by the trajectory

$$\frac{dT}{d\tau} = \gamma; \qquad \frac{dX}{d\tau} = -v\gamma,$$

where v is the inertial velocity and γ is the corresponding Lorentz factor. If the spacetime were flat throughout, i.e., no matter shell had collapsed, the inertial observer, carrying the detector, would have exhausted the full range of proper time $\tau \in (-\infty, \infty)$. The appearance of singularity due to collapse of the matter shell leads to geodesic incompleteness and hence regulate the proper time interval of the geodesic observer to $(-\infty, \tau_f)$, with τ_f marking the location i_L^+ (see the left curve in Fig. 11.8).

In $(1 + 1)$ dimension, the response of a monopole detector coupled to a (test) field $\Phi(x)$ is ambiguous [24]. Therefore, we work with derivative field detector with interaction Hamiltonian

$$\mathcal{H}_{\text{int}} = c\mu(\tau)\frac{d}{d\tau}\Phi(x(\tau)), \tag{11.18}$$

where c is a coupling constant, $\mu(\tau)$ is detector's monopole moment operator and τ is the proper time detected by observer carrying the detector. The detector response function [24], for a two level detector, with level separation ω, will be given for this case as

$$\mathcal{F}^{(1)}(\omega) = \int_{-\infty}^{\tau_f} d\tau \int_{-\infty}^{\tau_f} d\tau' e^{-i\omega(\tau-\tau')}\partial_\tau\partial_{\tau'}\mathcal{W}(\tau, \tau'). \tag{11.19}$$

In flat portion of the spacetime, on any trajectory, if the test field $\Phi(x)$ is in the vacuum state,

$$\partial_\tau\partial_{\tau'}\mathcal{W}(\tau, \tau') = \frac{1}{(\tau - \tau')^2}, \tag{11.20}$$

becomes solely a function of $\Delta\tau = \tau - \tau'$. Had the spacetime been geodesically complete, $\tau_f \to \infty$ and the detector would not have clicked. However, due to collapsing matter field and the subsequent removal of a portion of spacetime, the spacetime becomes geodesically incomplete. In such a case, the detector response will be analogous to a finite time inertial detector [25], which indeed registers clicks. However, notice that here we have not introduced any artificial switching function, but it is the inextendibility of the geodesic, which leads to the non-zero response function. In the case there is no black hole collapse, the detector will fall silent.

We can now transform to co-ordinates $\Delta\tau$ and a mean proper time $\bar{\tau} = (\tau + \tau')/2$ to re-write Eq. (11.19) as

$$\mathcal{F}^{(1)}(\omega) = \int_{-\infty}^{\tau_f} d\bar{\tau} \int_{-(2\tau_f - 2\bar{\tau})}^{(2\tau_f - 2\bar{\tau})} \frac{d\Delta\tau}{(\Delta\tau)^2} e^{-i\omega\Delta\tau}. \tag{11.21}$$

Therefore, the detection rate becomes time dependent as fit for a detector moving through an excited state [26], i.e., at different points on the trajectory the detector will click differently. Hence, we show that the detector also conforms to the notion that the state as measured by the left moving geodesic observer is non-vacuum. With a bit of exercise [25], one can evaluate the clicking rate of the detector at any general point of time as well. The detector clicks with a non-zero rate and the left-moving observer concludes that there are particles in her patch of the spacetime.

It is important to note that this detector is constructed in a manner such that it will not register any clicking, if the quantum field resides in the vacuum state of the observer carrying the detector along (e.g., Inertial observer in Minkowski spacetime in regular $(3+1)$ dimensional gravity). Therefore, the global vacuum defined on $\mathcal{J}_R^-$ fails to be a vacuum state for left-moving geodesic observers. Further, the detector does not click in a steady fashion at a thermal detection rate, but this is due to the fact that the detector response depends hugely on the trajectory unlike the Bogoliubov transformation. For instance, it is easy to demonstrate that in Fig. 11.3 the non-geodesic observer (thick magenta curve) and the Rindler observer (dotted trajectory) obtain the same Bogoliubov coefficients as they derive the casual support from the same patch, yet Unruh-DeWitt detectors carried by them behave differently. Still both such observers agree that the state they measure is non-vacuum.

References

1. S. Hawking, Particle creation by black holes. Commun. Math. Phys. **43**, 199–220 (1975)
2. S. Hawking, Black hole explosions. Nature **248**, 30–31 (1974)
3. N. Birrell, P. Davies, *Quantum Fields in Curved Space* (Cambridge University Press, Cambridge, 1982)
4. A.D. Helfer, Do black holes radiate? Rept. Prog. Phys. **66** 943–1008 (2003). arXiv:gr-qc/0304042 [gr-qc]
5. A. Fabbri, J. Navarro-Salas, *Modeling Black Hole Evaporation* (Imperial College Press, London, 2005)
6. V. Mukhanov, S. Winitzki, *Introduction to Quantum Effects in Gravity*, 1st edn. (Cambridge University Press, Cambridge, 2007)
7. L.E. Parker, D. Toms, *Quantum Field Theory in Curved Spacetime*. Cambridge Monographs on Mathematical Physics (Cambridge University Press, Cambridge, 2009), http://www.cambridge.org/de/knowledge/isbn/item2327457
8. T. Padmanabhan, *Gravitation: Foundations and Frontiers* (Cambridge University Press, Cambridge, UK, 2010)
9. M. Visser, Essential and inessential features of Hawking radiation. Int. J. Mod. Phys. D **12**, 649–661 (2003). arXiv:hep-th/0106111 [hep-th]
10. S. Takagi, Vacuum noise and stress induced by uniform accelerator: hawking-unruh effect in Rindler manifold of arbitrary dimensions. Prog. Theor. Phys. Suppl. **88**, 1–142 (1986)
11. P. Davies, Scalar particle production in Schwarzschild and Rindler metrics. J. Phys. A **8**, 609–616 (1975)
12. W. Unruh, Notes on black hole evaporation. Phys. Rev. D **14**, 870 (1976)
13. C.G. Callan, Jr., S.B. Giddings, J.A. Harvey, A. Strominger, Evanescent black holes. Phys. Rev. D **45**, 1005–1009 (1992). arXiv:hep-th/9111056 [hep-th]
14. T. Padmanabhan, Thermodynamical aspects of gravity: new insights. Rept. Prog. Phys. **73**, 046901 (2010). arXiv:0911.5004 [gr-qc]
15. S. Singh, S. Chakraborty, Black hole kinematics: the âinâ-vacuum energy density and flux for different observers. Phys. Rev. D **90**(2), 024011 (2014). arXiv:1404.0684 [gr-qc]

16. M. Smerlak, S. Singh, New perspectives on Hawking radiation. Phys. Rev. D **88**(10), 104023 (2013). arXiv:1304.2858 [gr-qc]
17. P. Davies, S. Fulling, W. Unruh, Energy momentum tensor near an evaporating black hole. Phys. Rev. D **13**, 2720–2723 (1976)
18. L. Ford, T.A. Roman, Motion of inertial observers through negative energy. Phys. Rev. D **48**, 776–782 (1993). arXiv:gr-qc/9303038 [gr-qc]
19. L. Ford, T.A. Roman, Averaged energy conditions and quantum inequalities. Phys. Rev. D **51**, 4277–4286 (1995). arXiv:gr-qc/9410043 [gr-qc]
20. L. Ford, T.A. Roman, Averaged energy conditions and evaporating black holes. Phys. Rev. D **53**, 1988–2000 (1996). arXiv:gr-qc/9506052 [gr-qc]
21. S. Chakraborty, S. Singh, T. Padmanabhan, A quantum peek inside the black hole event horizon. JHEP **1506**, 192 (2015). arXiv:1503.01774 [gr-qc]
22. K. Lochan, S. Chakraborty, T. Padmanabhan, Information retrieval from black holes. arXiv:1604.04987 [gr-qc]
23. K. Lochan, S. Charaborty, T. Padmanabhan, Quantum correlations in two dimensional black holes (2016). To appear
24. J. Louko, Unruh-DeWitt detector response across a Rindler firewall is finite. JHEP **09**, 142 (2014). arXiv:1407.6299 [hep-th]
25. L. Sriramkumar, T. Padmanabhan, Response of finite time particle detectors in noninertial frames and curved space-time. Class. Quant. Grav. **13**, 2061–2079 (1996). arXiv:gr-qc/9408037 [gr-qc]
26. K. Lochan, T. Padmanabhan, Inertial nonvacuum states viewed from the Rindler frame. Phys. Rev. D **91**(4), 044002 (2015). arXiv:1411.7019 [gr-qc]

Part V
Zero Point Length–Towards Quantum Gravity

Chapter 12
Spacetime with Zero Point Length is Two-Dimensional at the Planck Scale

12.1 Introduction and Motivation

The existence of a fundamental length scale, which provides an ultimate lower bound to measurement of spacetime intervals, is a model-independent feature of quantum gravity, since various approaches to quantum gravity lead to this result. The appearance of such a minimum length scale actually derives from basic principles of quantum mechanics, special and general relativity, and hence is often considered as a robust feature of quantum gravity. This result can have important consequences, e.g., notions of causality or distance between two events cannot be expected to have a continuous behaviour at this length scale [1–6].

It seems, therefore, natural to look for a description of spacetime at the mesoscopic scales which interpolate between Planck scale and low energy scales, by incorporating the effect of introducing a zero point length to the spacetime through modification of the spacetime metric g_{ab} to some suitable object q_{ab}, which we shall refer to as the *qmetric*. The q_{ab} must be constructed such that the geodesic distance between two points computed using q_{ab} acquires a lower bound [7]. If we can determine the qmetric in terms of g_{ab}, then we can compute all other geometrical variables (like, e.g., curvature tensor) by using the qmetric in the place of g_{ab} in the relevant expressions. Such a procedure is necessarily approximate—compared to a fully rigorous non-perturbative quantum gravitational approach—but will surely capture some of the effects at the mesoscopic scales which interpolate between Planck scale and low energy scales at which the classical metric provides an adequate description.

Unfortunately, we cannot use perturbation techniques to compute q_{ab} for a given classical geometry described by a g_{ab}. In fact, we would expect it to be non-local and singular at any given event. In the absence of explicit computability, we need a physically motivated ansatz to relate q_{ab} to g_{ab}. Such an ansatz was introduced and described in fair detail in [7]. The essential idea was to recognize that a primary effect of quantum gravity will be to endow spacetime with a zero-point length [1–6] by modifying the geodesic interval $\sigma^2(x, x')$ between two events x and x' in a Euclidean spacetime to a form like $\sigma^2 \to \sigma^2 + \ell_0^2$ (here $\ell_0 \sim L_P$, the Planck

S. Chakraborty, *Classical and Quantum Aspects of Gravity in Relation to the Emergent Paradigm*, Springer Theses, DOI 10.1007/978-3-319-63733-4_12

length). More generally, we will assume that a key effect of quantum gravity is to modify $\sigma^2 \to \mathcal{F}(\sigma^2)$ where the function $\mathcal{F}(\sigma^2)$ satisfies the constraint $\mathcal{F}(0) = \ell_0^2$. While most of our results are insensitive to the explicit functional form of $\mathcal{F}(\sigma^2)$, for illustration, we will use $\mathcal{F}(\sigma^2) = \sigma^2 + \ell_0^2$.

This approach essentially treats the $\sigma^2(x, x')$ as more fundamental than the metric $g_{ab}(x)$ itself. As we have already emphasized, there is considerable amount of evidence for the existence of such a zero-point-length of spacetime; there have also been arguments in the literature suggesting that g_{ab} may not be the most primitive variable to use while studying quantum gravitational effects (see e.g., [8]). The key claim being, while we may not know how quantum gravity modifies the classical metric a priori, we *do* have an indirect handle on it, if we assume that quantum gravity introduces a zero-point length to the spacetime in the manner described above.

One can determine [9, 10] the form of q_{ab} for a given g_{ab} by using the requirements that

1. It should lead to a spacetime interval with a zero-point-length.
2. The two-point function describing small perturbations of the metric should have a regularized non-singular coincidence limit.

It was shown in the earlier works [9, 10] that these conditions allow us to determine q_{ab} uniquely in terms of g_{ab} (and its associated geodesic interval σ^2). We find that

$$q_{ab} = Ah_{ab} + Bn_a n_b; \qquad q^{ab} = \frac{1}{A}h^{ab} + \frac{1}{B}n^a n^b \tag{12.1}$$

where $h_{ab} = g_{ab} - n_a n_b$, and

$$B = \frac{\sigma^2}{\sigma^2 + \ell_0^2}; \qquad A = \left(\frac{\Delta}{\Delta_{\mathcal{F}}}\right)^{2/D_1} \frac{\sigma^2 + \ell_0^2}{\sigma^2}; \qquad n_a = \frac{\nabla_a \sigma^2}{2\sqrt{\sigma^2}} \tag{12.2}$$

with Δ being the Van Vleck determinant related to the geodesic interval σ^2 by

$$\Delta(x, x') = \frac{1}{\sqrt{g(x)g(x')}} \det\left\{\frac{1}{2}\nabla_a^x \nabla_b^{x'} \sigma^2(x, x')\right\} \tag{12.3}$$

$\Delta_{\mathcal{F}}$ is defined by replacing σ^2 with $\mathcal{F}(\sigma^2)$ in the series expansion of Δ (see [10]). We interpret q_{ab} as the effective spacetime metric incorporating some of the non-perturbative effects of quantum gravity at Planck scales. As described in detail in the previous works [7, 11] the qmetric has the following properties:

- Unlike $g_{ab}(x)$, the qmetric $q_{ab}(x, x')$ is a bi-tensor depending on two events x, x' through σ^2. It is easy to show that this non-locality is essential if spacetime has to acquire a zero-point length. Almost by definition, any local metric will lead to a geodesic interval which vanishes in the limit of $x \to x'$.
- q_{ab} reduces to the background metric g_{ab} in the limit of $\ell_0^2 \to 0$. That is, in the classical limit of $\hbar \to 0$, qmetric reduces to the standard metric as we would expect.

- In the opposite limit of $(\sigma^2/\ell_0^2) \to 0$ the qmetric is singular at all events. This is natural when we interpret qmetric as the metric of the mesoscopic spacetime; we would not expect it to be well defined at any given event and will require some kind of smearing over Planck scales for it to be meaningful.
- If the background metric is flat, then the qmetric is also flat, i.e., there exists a coordinate mapping from $q_{ab} \to \eta_{ab}$. This is, however, rather subtle because the coordinate transformation is, in fact, singular in the coincidence limit in (regular) Cartesian coordinates, and the mapping effectively removes a geodesic region of size ℓ_0 from the spacetime around all events.
- Let $\Phi[g_{ab}(x)]$ be a scalar constructed from the background metric and possibly several of its derivatives, like, for example, the Ricci scalar $R[g_{ab}(x)]$.

We can now compute the corresponding (bi)scalar $\Phi[q_{ab}(x, x'); \ell_0^2]$ for the qmetric by replacing g_{ab} by q_{ab} in $\Phi[g_{ab}(x)]$ and evaluating all derivatives at x keeping x' ("base point") fixed. We interpret the value of this scalar by taking the limit $x \to x'$ in this expression keeping ℓ_0^2 non-zero. As noted in [7, 11], several useful scalars like R, K etc. remain finite and local in this limit even though the qmetric itself is singular when $x \to x'$ with non-zero ℓ_0^2. This arises from the algebraic fact that the following two limits do not commute:

$$\lim_{\ell_0^2\to 0}\lim_{x\to x'}\Phi[q_{ab}(x, x'); \ell_0^2] \neq \lim_{x\to x'}\lim_{\ell_0^2\to 0}\Phi[q_{ab}(x, x'); \ell_0^2] \qquad (12.4)$$

All these computations are most easily performed [12] by choosing a synchronous coordinate system for the background metric which can always be done in a local region. In this coordinate system, the equi-geodesic surface $\sigma =$ constant, which is at a constant geodesic distance from the base point, has a simple description. These facts show that we have a well-defined procedure for doing the computations in the mesoscopic spacetimes using the qmetric. More details regarding this approach will be presented elsewhere [12].

12.2 Volume and Area in Presence of Zero Point Length

In this chapter, we will concentrate on one key effect of introducing such a zero point length to the spacetime metric. This relates to the volume of the geodesic ball:

$$V_D(\ell, \ell_0) \equiv \int_{\sigma\leq\ell} \mathrm{d}\sigma \mathrm{d}\Omega_{D-1}\sqrt{q} \qquad (12.5)$$

enclosed by an equi-geodesic surface of size $\sigma = \ell$ in the D-dimensional Euclidean spacetime. Here, q is the determinant of the qmetric with

$$\sqrt{q} = \left(\frac{\Delta}{\Delta_{\mathcal{F}}}\right)\left(\frac{\sigma^2 + \ell_0^2}{\sigma^2}\right)^{D_2/2}\sqrt{g}; \quad D_2 \equiv (D-2) \tag{12.6}$$

Classically we would expect the scaling $V_D(\ell) \propto \ell^D$ for sufficiently small ℓ when we can ignore the scales involved in spacetime curvature. Let us now evaluate the same quantity for the mesoscopic spacetime. Using the synchronous coordinates, a straightforward (though lengthy) calculation, gives the following result:

$$\sqrt{q} = \sigma\left(\sigma^2 + \ell_0^2\right)^{D_2/2}\left(1 - \frac{1}{6}\mathcal{S}(x')\left(\sigma^2 + \ell_0^2\right) + \cdots\right) \tag{12.7}$$

where $\mathcal{S} \equiv R_{ab}n^a n^b$ is a scalar constructed from the background g_{ab} and is evaluated at base point x'. (Though we only need the leading order term, we have displayed next order term in the expansion which contribute in the limit of $\ell_0^2/\sigma^2 \to 0$.) Here n_a is the tangent to the geodesic and hence depends on the geodesic that connects the base point x' to the point x on the equi-geodesic surface. Thus integrating over the angular coordinates on the equi-geodesic surface amounts to averaging n_a over the solid angle. [Such an angular integral over $n^a n^b$ leads to, $(\Omega/D)g^{ab}$, where $\Omega = 2\pi^{D/2}\Gamma(D/2)^{-1}$ (see e.g., [13])]. A straightforward integration now gives the result, to the same order of accuracy, as:

$$V_D(\ell, \ell_0) = \frac{\Omega}{D}\left\{\left(\ell^2 + \ell_0^2\right)^{D/2} - \ell_0^D\right\} - \frac{\Omega}{6D(D+2)}R(x')\left\{\left(\ell^2 + \ell_0^2\right)^{(D+2)/2} - \ell_0^{D+2}\right\} \tag{12.8}$$

where R is the Ricci scalar for the background metric g_{ab}.

Let us now consider the two relevant limits of this expression. First, when $(\ell_0/\ell) \to 0$, we get

$$\lim_{\ell_0 \to 0} V_D(\ell, \ell_0) = \frac{\Omega}{D}\ell^D\left[1 - \frac{1}{6(D+2)}R(x')\ell^2\right] \tag{12.9}$$

which shows that the volume scales as ℓ^D except for curvature induced corrections captured by the second term. This correction is a standard result known in differential geometry [13]. What is more interesting is the limit of $(\ell/\ell_0) \to 0$, when we get

$$\lim_{\ell \to 0} V_D(\ell, \ell_0) = \frac{\Omega}{2}\ell_0^D\left(\frac{\ell^2}{\ell_0^2}\right)\left[1 - \frac{1}{6D}R(x')\ell_0^2\right] \to \frac{\Omega}{2}\ell_0^D\left(\frac{\ell^2}{\ell_0^2}\right) \propto \ell^2 \tag{12.10}$$

This suggests the remarkable possibility that the existence of zero-point-length makes the physical spacetime essentially 2-dimensional near Planck scale. A convenient measure of such a dimensional reduction is provided by the quantity

$$D_{\text{eff}} = D + \frac{\mathrm{d}}{\mathrm{d}\ln\ell}\left\{\ln\left(\frac{V_D(\ell,\ell_0)}{V_D(\ell,\ell_0=0)}\right)\right\} \tag{12.11}$$

which, using the above expressions, decreases from $D_{\text{eff}} = D$ for large ℓ to $D_{\text{eff}} = 2$ as $\ell \to 0$. It might seem that a simpler definition is just $D_{\text{eff}} = (\mathrm{d}\ln V_D(\ell,\ell_0)/\mathrm{d}\ln\ell)$. However, in any D-dimensional curved space with a smooth metric g_{ab}, this definition will give D only for $R\ell^2 \ll 1$; that is at scales small compared to curvature scale. The role of $V_D(\ell,\ell_0=0)$ in the definition is to remove the contribution from the background curvature to D_{eff} thereby ensuring that $D_{\text{eff}} = D$ when $\ell_0 = 0$. So any deviation of D from D_{eff} arises only due to the existence of the zero-point-length.

There have been several indications from various approaches to quantum gravity that spacetime might "look" two dimensional when probed at extremely small scales (most of these arguments refer to the so called *spectral dimension* defined via the process of random walk). Carlip [14–17] has given several independent set of arguments suggesting that such a dimensional reduction may be an inevitable feature of quantum gravity. Somewhat closer in spirit to the ideas presented here seems to be the results of [18, 19], where the authors argue that dimensional reduction to $D = 2$ might provide a mechanism by which quantum gravity "self-renormalizes" at Planck scale. A geometrically similar approach (but based on quantization of area) in the context of loop quantum gravity and spin-foam models, was discussed in [20–22], while similar result was obtained in the context of generalized uncertainty principle in [23]. The current study links such a dimensional reduction to the existence of zero-point-length in spacetime, *independent of any specific model of quantum gravity*.

To understand the *algebraic* origin of this result, one could study the case of a flat spacetime in synchronous coordinates in which σ denotes the radial geodesic distance from a chosen origin. In this case, the qmetric leads to the line element of the form (when $D = 4$)

$$\mathrm{d}s^2 = \frac{1}{A(\sigma)}\mathrm{d}\sigma^2 + A(\sigma)\sigma^2\mathrm{d}\Omega_3^2; \quad A = 1 + \frac{\ell_0^2}{\sigma^2} \tag{12.12}$$

which transforms to $ds^2 = d\bar{\sigma}^2 + \bar{\sigma}^2 d\Omega_3^2$ with $\bar{\sigma} = \sqrt{\sigma^2 + \ell_0^2}$. In regular Cartesian coordinates, this transformation is *singular* in the coincidence limit (see Appendix A of [9]) and is equivalent to removing a "hole" of radius ℓ_0 from the manifold in a specific manner. In fact, in flat spacetime, we will get the result $V_D(\ell,\ell_0) = (\Omega/D)[(\ell^2 + \ell_0^2)^{D/2} - \ell_0^D]$ which clearly has the limits $V_D(\ell,\ell_0) \propto \ell^D$ (when $(\ell_0/\ell) \to 0$) and $V_D(\ell,\ell_0) \propto \ell^2$ (when $(\ell/\ell_0) \to 0$). This clarifies the algebraic origin of the result.

12.3 Conclusions

We believe the conjecture, that the mesoscopic spacetime is described by the qmetric, is a powerful and useful one. It is well motivated by the existence of the zero-point-length in the spacetime and leads to well defined computation rules which can incorporate the effects of quantum gravity at mesoscopic scales without us leaving the comfort of a continuum differential geometry (albeit one involving a singular and non-local metric). Its power is seen once again in the current chapter, where it leads to a definite conclusion that the effective dimensionality of the spacetime at Planck scales is $D_{\text{eff}} = 2$. This opens up useful further avenues of exploration.

References

1. B. DeWitt, Gravity: a universal regulator? Phys. Rev. Lett. **13**, 114–118 (1964)
2. T. Padmanabhan, Planck length as the lower bound to all physical length scales. Gen. Relativ. Gravit. **17**, 215–221 (1985)
3. T. Padmanabhan, Physical significance of planck length. Ann. Phys. **165**, 38–58 (1985)
4. T. Padmanabhan, Duality and zero point length of space-time. Phys. Rev. Lett. **78**, 1854–1857 (1997). arXiv:hep-th/9608182 [hep-th]
5. L.J. Garay, Space-time foam as a quantum thermal bath. Phys. Rev. Lett. **80**, 2508–2511 (1998). arXiv:gr-qc/9801024 [gr-qc]
6. L.J. Garay, Quantum gravity and minimum length. Int. J. Mod. Phys. A **10**, 145–166 (1995). arXiv:gr-qc/9403008 [gr-qc]
7. D. Kothawala, T. Padmanabhan, Entropy density of spacetime as a relic from quantum gravity. Phys. Rev. **90**(12), 124060 (2014). arXiv:1405.4967 [gr-qc]
8. G. 't Hooft, Quantum gravity: a fundamental problem and some radical ideas, in *Recent Developments in Gravitation*, ed. by M. Levi, S. Deser (Plenum, New York/London, 1978)
9. D. Kothawala, Minimal length and small scale structure of spacetime. Phys. Rev. D **88**(10), 104029 (2013). arXiv:1307.5618 [gr-qc]
10. D.J. Stargen, D. Kothawala, Small scale structure of spacetime: the van Vleck determinant and equigeodesic surfaces. Phys. Rev. D **92**(2), 024046 (2015). arXiv:1503.03793 [gr-qc]
11. D. Kothawala, T. Padmanabhan, Entropy density of spacetime from the zero point length. Phys. Lett. B **748**, 67–69 (2015). arXiv:1408.3963 [gr-qc]
12. S. Chakraborty and T. Padmanabhan, "Under Preparation," *Under Preparation* (2016)
13. A. Gray, The volume of a small geodesic ball of a Riemannian manifold. Mich. Math. J **20**, 329 (1973)
14. S. Carlip, R.A. Mosna, J.P.M. Pitelli, Vacuum fluctuations and the small scale structure of spacetime. Phys. Rev. Lett. **107**, 021303 (2011). arXiv:1103.5993 [gr-qc]
15. S. Carlip, The small scale structure of spacetime," in *Proceedings, Foundations of Space and Time: Reflections on Quantum Gravity* (2009), pp. 69–84. arXiv:1009.1136 [gr-qc], https://inspirehep.net/record/867166/files/arXiv:1009.1136.pdf
16. S. Carlip, Spontaneous dimensional reduction in short-distance quantum gravity? AIP Conf. Proc. **1196**, 72 (2009). arXiv:0909.3329 [gr-qc]
17. G. Calcagni, Fractal universe and quantum gravity. Phys. Rev. Lett. **104**, 251301 (2010). arXiv:0912.3142 [hep-th]
18. J. Ambjorn, J. Jurkiewicz, R. Loll, Spectral dimension of the universe. Phys. Rev. Lett. **95**, 171301 (2005). arXiv:hep-th/0505113 [hep-th]
19. J. Ambjorn, J. Jurkiewicz, R. Loll, Reconstructing the universe. Phys. Rev. D **72**, 064014 (2005). arXiv:hep-th/0505154 [hep-th]

20. L. Modesto, Fractal structure of loop quantum gravity. Class. Quantum Gravity **26**, 242002 (2009). arXiv:0812.2214 [gr-qc]
21. G. Calcagni, D. Oriti, J. Thurigen, Laplacians on discrete and quantum geometries. Class. Quantum Gravity **30**, 125006 (2013). arXiv:1208.0354 [hep-th]
22. G. Calcagni, D. Oriti, J. ThÃŒrigen, Spectral dimension of quantum geometries. Class. Quantum Gravity **31**, 135014 (2014). arXiv:1311.3340 [hep-th]
23. V. Husain, S.S. Seahra, E.J. Webster, High energy modifications of blackbody radiation and dimensional reduction. Phys. Rev. D **88**(2), 024014 (2013). arXiv:1305.2814 [hep-th]

Part VI
Summary and Outlook

Chapter 13
Summary and Outlook

In this last chapter, we shall present a concise summary of the results we have obtained in this thesis as well as future directions of exploration.

As emphasized in the introduction, general relativity is the strongest candidate for describing gravity and has passed all the experimental tests, starting from bending of light in 1920 to detection of gravitational waves in 2015, with flying colors. Despite its outstanding successes general relativity is not a complete theory. This has to do with two facts—(i) general relativity leads to existence of singularities, where predictability, essential for being a complete theory, breaks down and (ii) being a classical theory it cannot explain the universe at sufficiently small length scales (equivalently at sufficiently high energies). It is often postulated that the two issues are interrelated, as soon as one has a "quantum theory of gravity" one would achieve a possible resolution of singularity. There are many candidates for the yet-to-be-built theory of quantum gravity, but they all lack in one perspective or another and no unique description has emerged till date.

This motivates a fresh look at the problem, which started after Bekenstein and Hawking discovered that one can associate an entropy and a temperature [1, 2] with black holes. Soon after it was shown that black holes obey the laws of thermodynamics as well [3, 4]. In 1995, Jacobson was able to show that, starting from Clausius relation one can arrive at Einstein's field equations [5], which was complemented in 2002 by Padmanabhan, who showed that Einstein's equations can be written in a thermodynamic language [6].

This brings out a paradigm shift suggesting that gravity may not be a fundamental interaction, but is an emergent phenomenon, like elasticity of a fluid body. There have been several works exploring this possibility in the past, in this thesis we have been partly engaged in extending these results further by appropriately generalizing them, as well as discovering new geometrical quantities that have nice thermodynamic interpretation. We have reviewed some of these ideas in the context of general relativity (essential for latter parts of the thesis) as well as have reviewed Lanczos-Lovelock models of gravity in Part II besides obtaining alternative geometrical variables to

S. Chakraborty, *Classical and Quantum Aspects of Gravity in Relation to the Emergent Paradigm*, Springer Theses, DOI 10.1007/978-3-319-63733-4_13

describe them. Thermodynamic interpretation of these variables as well as other geometrical constructs, e.g., Noether current, gravitational momentum have been presented in Part III. Related results arise even when one considers quantum fields in curved spacetime, which we have explored in Part IV. Using a covariant, *but* observer dependent notion of energy density and flux, we have explored the behaviour of quantum fields inside the event horizon, near the singularity using appropriate observers. We have also shown that choosing the observer and the quantum field under investigation wisely one can have a possible resolution of the information loss paradox. Further in the case of dilaton gravity it has been noticed that there are flat spacetime, inertial observers who experiences Unruh effect! In the penultimate part, Part V we have explored some interesting features that a spacetime acquires on introduction of a zero point length, in particular the spacetime becomes effectively two-dimensional near the Planck scale—a distinctive feature of quantum gravity. Having summarized a broad overview of the results presented in each of the individual parts of this thesis, we will now explore possible future directions.

In the case of general relativity the conjugate variables (f^{ab}, N^c_{ab}) lead to Einstein's equations, besides having thermodynamic significance. The variables introduced for Lanczos-Lovelock gravity have similar properties, their variations have thermodynamic significance (see Chap. 4) and also they lead to vacuum gravitational field equations. Thus it is clear from the above discussion that metric is not the only available choice for the dynamical variable, there can be other variables which can describe geometrical aspects of gravity as well as have interesting thermodynamic interpretation.

This raises an interesting question: What corresponds to a true dynamical variable of a gravity theory in higher spacetime dimensions? This question can have profound implications in our search for quantum gravity. For general relativity and in four spacetime dimensions it is well known that the quantum version of the Hamiltonian and momentum constraints become ill-posed with metric as the dynamical variable and one has to introduce suitably defined variables, e.g. tetrad and their conjugate momentum to make them well-posed. These variables, known in the literature as Ashtekar variables are the cornerstones in the loop quantum gravity formalism. However it is well known that these variables do not work in higher dimensions and also not for higher curvature theories. Thus it is not at all clear what should be the correct geometrical variables if one wants to quantize Lanczos-Lovelock gravity. Thus one needs to understand the structure of constraint equations and possible generalization of the tetrad variables to the premise of Lanczos-Lovelock gravity. Some preliminary steps in this direction have been taken recently [7–9] but much more needs to be done. The variables introduced in this thesis might play a pivotal role to simplify the constraint equations in general relativity and Lanczos-Lovelock gravity and worth pursuing in future.

In Chaps. 5, 6 and 7 we have discussed various thermodynamic properties of geometrical variables associated with both general relativity and Lanczos-Lovelock gravity. All the results derived in these chapters are mainly formal, with a few simple examples. It would be instructive to understand the results and their thermodynamic significance in more realistic situations. For example, null surfaces of Friedmann-

Lemaitre-Robertson-Walker spacetime would be a very good test bed for the results obtained. It would be nice if one can provide a natural thermodynamic meaning to the evolution of the universe. Further, all the results have been derived assuming affine connection, i.e., the torsion has been assumed to be vanishing. It would be interesting to check whether the thermodynamic properties enlisted in this thesis hold even with torsion (note that entropy of a null surface gets modified in presence of torsion), which would span a vast landscape of gravitational theories. Further, there are a few results, namely the equivalence of gravitational field equations with Navier-Stokes equation, which have been derived within the premise of general relativity alone. An important future project could be to show the equivalence of field equations in Lanczos-Lovelock gravity to that of Navier-Stokes.

In Chap. 8, using a local version of Cardy's formula we have derived the entropy area relation for a null surface, whose surface gravity and surface two-metric are independent of time and an angular coordinate. These two criteria are very important, since without them the Virasoro algebra is not closed. Hence when these two criteria are not satisfied one does not obtain the entropy area relation. It is not clear whether this is a fundamental restriction or merely a limitation of this approach.

We also want to comment on the remarkable robustness of Cardy's formula in this context. The relation between the Virasoro algebra and the entropy (through Cardy formula) was originally derived in the context of flat spacetime physics [10, 11]. It is not obvious that the result should generalize to curved spacetime and black hole horizons, which was a bit of a surprise in Carlip's approach [12]. (This is possibly related to local nature of the result and the fact that, in the local inertial frame, we can use the special relativistic physics.) In this thesis we have extended the applicability of Cardy's formula even further. As we have shown, working with densitized objects, one can reproduce the entropy area relation correctly. It is important to understand why this approach works and whether it can be derived from a more local procedure. This might, in turn, throw more light on the applicability of Cardy's formula in the context of curved geometry.

Chapter 9 shows a peculiar feature—the energy density of a quantum field near the singularity diverges much faster compared to the matter energy density that forms the black hole. This signals that effect of backreaction due to the quantum fields would be significant and can potentially alter the classical picture. This effect can be understood in more detail if one considers the effect of back reaction, i.e., the Einstein's equations with the right hand side being replaced by vacuum expectation value of the stress energy tensor of the quantum field. Further in this thesis we have considered a simple collapse model, it would be interesting to see if such features persist in more general inhomogeneous collapse models, e.g., Lemaitre-Tolman-Bondi collapse model as well.

There are many interesting future implications of the ideas we have presented in Chap. 10. Study of such non-vacuum distortions could be undertaken for more general set ups, which will involve computation of Bogoliubov coefficients for a more realistic scenario involving angular momentum, charge etc. Even at the vacuum level, there are different sources of non-thermal vacuum response as suggested in [13]. A realistic situation encompassing all such distortions will involve Bogoliubov coefficients modified in a precise manner. Therefore, even at the classical level, there should be a more realistic assessment of the allowed symmetry of the initial data viz-á-viz the information encryption in the distorted spectrum. It is also worthwhile to study the field configurations corresponding to states which encode maximal information and their field theoretic interpretations.

Apart from dealing with the more generalized situations and the Bogoliubov coefficients therein, a rigorous analysis has to be done for different kinds of initial data. For instance, in Chap. 10 our discussion was limited to pure states which are eigenstates of the number operator. Analysis regarding the most general set of initial data—such as one which is not an eigenstate of the number operator—will be required to exhaust the full Hilbert space. A classical initial data, after all, might correspond to a coherent state-like description. Similarly, the analysis could be translated to the language of wave packets for a realistic physical response in the out-configuration. Further, analogous field theoretic analysis has to be done for a mixed state description, e.g., a thermal initial data. In these cases, the analysis of higher order correlation will certainly become more important. In any case, we need to study the resulting non thermal spectra from the point of view of various aspects in unitary evolution of the black hole such as strength of the correction, first bit release time etc. and its potential to make the evolution unitary in the spirit of [14]. Lastly, a truthful implementation of back-reaction in realistic collapse scenarios in higher dimensions remains due. Recently,'t Hooft has proposed a model [15], which potentially captures the information about back-reaction through a shift in a null geodesic due to discontinuity across a matter field geodesic. Supplementing the effect of non-vacuum distortion with the shift arising from the back-reaction is expected to reveal more aspects of issues related to information accessible by future asymptotic observers. Further the results derived in Chaps. 10 and 11 can be applied in a broader context as well. In both these cases we have been dealing with a situation where the initial data on a full Cauchy surface can be derived from the knowledge of the data on a smaller portion of a later Cauchy surface. This can have interesting consequences in algebraic quantum field theory.

Finally, the results regarding qmetric as one introduces a zero point length in the spacetime immediately suggest some interesting future avenues to explore. For example, using the qmetric if one derives the curvature tensor components and take the coincidence limit, it will lead to curvature tensors for the background metric with $\mathcal{O}(\ell_0)$ corrections. These corrections will modify the background geometry at the Planck scale, e.g., this will lead to a modification of the horizon structure for black holes. Further introduction of zero point length in principle should eliminate all the singularities from the background spacetime. However the Hamilton-Jacobi and other differential/integral relations becomes inapplicable near the singularity, which might modify the near singularity structure of the qmetric. Further exploration of black hole entropy in the context of qmetric and possible Planck length correction to it would be of much interest.

It should be evident from the above discussions that we are, in all likelihood merely exploring the tip of an iceberg. The actual theory of quantum gravity is buried deep within. In this thesis we have used a three step ladder to quantum gravity. First, from the classical viewpoint we have made quiet a bit of progress by identifying natural dynamical variables in general relativity and also to its higher curvature cousin, the Lanczos-Lovelock gravity. On the other hand, the reinterpretation of classical gravity as an emergent phenomenon has been put on a firm ground by showing the thermodynamic structure in the most general scenario possible, which might imply that quantization of gravity by standard means is futile, since gravity is like elasticity. Going one step further we have introduced quantum fields but have kept the gravitational interaction to be classical. There we have demonstrated the importance of back reaction, tried to provide a resolution to the information loss paradox, a long standing puzzle in semi-classical gravity and have explored some puzzling features of dilaton gravity. As the final frontier we have explored possible consequences of a zero point length, a generic quantum gravity effect. The lessons learnt and results obtained in this thesis if pursued carefully might shed some light to the ongoing search for quantum gravity. Thus the end is not really an end, but the beginning of new ideas, new physics and possibly a theory of quantum gravity.

References

1. J.D. Bekenstein, Black holes and entropy. Phys. Rev. D **7**, 2333–2346 (1973)
2. S. Hawking, Particle Creation by Black Holes. Commun. Math. Phys. **43**, 199–220 (1975)
3. S. Hawking, Black holes and thermodynamics. Phys. Rev. D **13**, 191–197 (1976)
4. J.M. Bardeen, B. Carter, S. Hawking, The four laws of black hole mechanics. Commun. Math. Phys. **31**, 161–170 (1973)
5. T. Jacobson, Thermodynamics of space-time: the Einstein equation of state. Phys. Rev. Lett. **75**, 1260–1263 (1995). arXiv:gr-qc/9504004 [gr-qc]
6. T. Padmanabhan, Classical and quantum thermodynamics of horizons in spherically symmetric space-times. Class. Quantum Gravity **19**, 5387–5408 (2002). arXiv:gr-qc/0204019 [gr-qc]
7. N. Bodendorfer, T. Thiemann, A. Thurn, New variables for classical and quantum gravity in all dimensions I. Hamiltonian analysis. Class. Quantum Gravity **30**, 045001 (2013). arXiv:1105.3703 [gr-qc]
8. N. Bodendorfer, T. Thiemann, A. Thurn, New variables for classical and quantum gravity in all dimensions II. Lagrangian analysis. Class. Quantum Gravity **30**, 045002 (2013). arXiv:1105.3704 [gr-qc]
9. N. Bodendorfer, T. Thiemann, A. Thurn, New variables for classical and quantum gravity in all dimensions III. quantum theory. Class. Quantum Gravity **30**, 045003 (2013). arXiv:1105.3705 [gr-qc]
10. J.L. Cardy, Operator content of two-dimensional conformally invariant theories. Nucl. Phys. B **270**, 186–204 (1986)
11. H.W.J. Bloete, J.L. Cardy, M.P. Nightingale, Conformal invariance, the central charge, and universal finite size amplitudes at criticality. Phys. Rev. Lett. **56**, 742–745 (1986)
12. S. Carlip, Entropy from conformal field theory at killing horizons. Class. Quantum Gravity **16**, 3327–3348 (1999). arXiv:gr-qc/9906126 [gr-qc]
13. M. Visser, Thermality of the Hawking flux. JHEP **07**, 009 (2015). arXiv:1409.7754 [gr-qc]
14. G. Dvali, Non-Thermal Corrections to Hawking Radiation Versus the Information Paradox. arXiv:1509.04645 [hep-th]
15. G.t. Hooft, Diagonalizing the Black Hole Information Retrieval Process. arXiv:1509.01695 [gr-qc]

MIX
Papier aus verantwortungsvollen Quellen
Paper from responsible sources
FSC® C105338

If you have any concerns about our products,
you can contact us on
ProductSafety@springernature.com

In case Publisher is established outside the EU,
the EU authorized representative is:
Springer Nature Customer Service Center GmbH
Europaplatz 3, 69115 Heidelberg, Germany

Printed by Libri Plureos GmbH
in Hamburg, Germany